MICROCOMPUTERS IN
NUMERICAL ANALYSIS

MATHEMATICS AND ITS APPLICATIONS
Series Editor: G. M. BELL, Professor of Mathematics,
King's College London (KQC), University of London

NUMERICAL ANALYSIS, STATISTICS AND OPERATIONAL RESEARCH
Editor: B. W. CONOLLY, Emeritus Professor of Mathematics (Operational Research),
Queen Mary College, University of London

Mathematics and its applications are now awe-inspiring in their scope, variety and depth. Not only is there rapid growth in pure mathematics and its applications to the traditional fields of the physical sciences, engineering and statistics, but new fields of application are emerging in biology, ecology and social organization. The user of mathematics must assimilate subtle new techniques and also learn to handle the great power of the computer efficiently and economically.

The need for clear, concise and authoritative texts is thus greater than ever and our series will endeavour to supply this need. It aims to be comprehensive and yet flexible. Works surveying recent research will introduce new areas and up-to-date mathematical methods. Undergraduate texts on established topics will stimulate student interest by including applications relevant at the present day. The series will also include selected volumes of lecture notes which will enable certain important topics to be presented earlier than would otherwise be possible.

In all these ways it is hoped to render a valuable service to those who learn, teach, develop and use mathematics.

Mathematics and its Applications
Series Editor: G. M. BELL, Professor of Mathematics, King's College London
(KQC), University of London

Series continued at back of book

MICROCOMPUTERS IN NUMERICAL ANALYSIS

G. R. LINDFIELD
Department of Computer Science and Applied Mathematics
Aston University, Birmingham

and

J. E. T. PENNY
Department of Mechanical and Production Engineering
Aston University, Birmingham

ELLIS HORWOOD LIMITED
Publishers · Chichester

Halsted Press: a division of
JOHN WILEY & SONS
New York · Chichester · Brisbane · Toronto

First published in 1989 by
ELLIS HORWOOD LIMITED
Market Cross House, Cooper Street,
Chichester, West Sussex, PO19 1EB, England
The publisher's colophon is reproduced from James Gillison's drawing of the ancient Market Cross, Chichester.

Distributors:

Australia and New Zealand:
JACARANDA WILEY LIMITED
GPO Box 859, Brisbane, Queensland 4001, Australia

Canada:
JOHN WILEY & SONS CANADA LIMITED
22 Worcester Road, Rexdale, Ontario, Canada

Europe and Africa:
JOHN WILEY & SONS LIMITED
Baffins Lane, Chichester, West Sussex, England

North and South America and the rest of the world:
Halsted Press: a division of
JOHN WILEY & SONS
605 Third Avenue, New York, NY 10158, USA

South-East Asia
JOHN WILEY & SONS (SEA) PTE LIMITED
37 Jalan Pemimpin # 05–04
Block B, Union Industrial Building, Singapore 2057

Indian Subcontinent
WILEY EASTERN LIMITED
4835/24 Ansari Road
Daryaganj, New Delhi 110002, India

© **1989 G. Lindfield and J. E. T. Penny/Ellis Horwood Limited**

British Library Cataloguing in Publication Data
Microcomputers in numerical analysis.
1. Numerical analysis. Applications of microcomputer systems
I. Title   II. Penny, John E. T.
III. Series
591.4′028′5416

Library of Congress CIP available

ISBN 0–85312–781–6 (Ellis Horwood Limited)
ISBN 0–470–21415–5 (Halsted Press)

Printed in Great Britain by Hartnolls, Bodmin

Table of contents

5 Numerical differentiation

6 The solution of differential equations

Preface

This book has been written to assist non-specialists in numerical analysis and seeks to show how relatively advanced and sophisticated numerical analysis can be carried out using low-cost microcomputers. By non-specialist we mean those who use numerical methods but do not necessarily require a detailed theoretical knowledge of the subject or need to solve very large or complex problems. However, even the non-specialist must have a clear understanding of numerical methods and their range of application if procedures and the programs that implement them are to be used successfully. The non-specialist will include many undergraduate and postgraduate students, practising engineers and scientists and even sixth formers undertaking project work. They will have easy access to microcomputers and in this they are fortunate compared with their predecessors who, before the advent of the microcomputer, were forced to gain experience of numerical procedures by undertaking lengthy computation by hand.

Learning numerical methods with a microcomputer gives the user the ability to try many problems and to explore the range of applicability of algorithms. This book includes more than sixty-five BASIC programs suitable for use on microcomputers and the clarity of the description of the numerical procedures is enhanced by illustrating them with the numerical output from these programs. In addition the programs allow the user to learn from experience by experimentation and to this end a selection of problems is included in each chapter.

The overall structure of the book is as follows. Chapter 1 provides an elementary introduction to relevant aspects of computers and computing and to some of the more widely applied concepts of numerical analysis. Chapters 2 to 10 are concerned with the development of procedures and programs to solve a wide range of numerical problems. We rarely consider procedures that were developed for hand calculation, nor those which can only be implemented realistically on a main-frame computer. Emphasis is placed on the practical use of procedures and when they are not developed with full mathematical rigour, references are given. Finally in Chapter 11 we compare the performance, in terms of speed and accuracy, of several popular microcomputers and we use them to solve a selection of problems taken from Chapters 2 to 10. The mathematical prerequisites for a full understanding of the text are limited to a knowledge of simple algebra and trigonometry, basic differential and integral calculus including Taylor's series and the mean-value theorem, an understanding of the notation and rules for the manipulation of vectors and matrices and the algebra of complex numbers.

Choosing a programming language for this text was not easy. Pascal and BASIC were the most obvious candidates, FORTRAN was rejected because it is not widely used on microcomputers. Pascal has the clear advantage of being a well structured language. In contrast, BASIC is not standardised and some versions lack important features such as procedures and the *if-then-else* construct, leading to an over use of *goto* statements. However, a feature of many numerical algorithms that should not be overlooked is that their relative simplicity allows them to be programmed without loss of structural clarity using a restricted set of programming constructs. The advantages of BASIC are that it is still the most widely available programming language on microcomputers and is almost certainly the most widely known, used and understood. Consequently, after careful consideration, we have chosen to present programs in BASIC.

The programs given are written so that they will work satisfactorily, with very little modification, in most versions of BASIC and on most microcomputers. The programs could be developed by refining their structure and also their output presentation but these improvements would be at the expense of portability between machines. The robustness of the programs could be enhanced and data validation included but this would make them longer and more intricate and such refinements may mask the essential features of the numerical procedure. It should be relatively easy for readers to employ the facilities of their microcomputer to modify the programs to meet their particular requirements. It should be noted that whilst every effort has been made to check the correctness of programs we cannot guarantee that all the programs are error free. However, many of the programs have been used by the authors and their students for several years and are well and truly tested. It is also important to emphasise that even error free programs must be used with care and the reader must decide whether the application of a program to a particular problem is appropriate. All the programs have been run on the Apple Macintosh and the BBC microcomputers and the vast majority have also been run on the IBM PC-AT. In addition, many programs have been tested on other machines.

Finally the authors wish to acknowledge the help of their colleagues Gordon Betteley, Dr Barry Martin and Dr David Wilson, and also the help and patience of the staff of Ellis Horwood.

George Lindfield
John Penny
Aston University
September 1988.

1

Using microcomputers to solve numerical problems

1.1 THE IMPACT OF THE MICROCOMPUTER

In almost every field of computation, data processing and information technology the microcomputer, as a consequence of its flexibility, user-friendliness, low cost and wide availability is becoming increasingly competitive with the large, multi-user, mainframe computer. Whilst the microcomputer cannot equal the performance of a modern mainframe computer in computing power, the modern microcomputer far exceeds the performance of many of the early large mainframe computers. An interesting illustration of this was described by Sinnott (1984). He reported that a series of lengthy and complex astronomical calculations (to determine the ephemerides of the five outer planets of the Solar system for a period of 400 years) was carried out in 1948 using the large IBM SSEC and took 120 hours. In 1984 an amateur astronomer repeated the calculations using his TRS80 microcomputer and the process took 10 hours 25 minutes. It was also estimated that it would take an unaided human calculator, working 40 hours a week, approximately 80 years to perform the same calculations. Such anecdotes are interesting in giving a feel for the rapid improvement in computer technology. The real point of this illustration is that in 1948 (or, indeed, in 1975) only the professional scientist and engineer had access to significant computing power. Today any school, college or university, any student, amateur scientist or computer hobbyist can have access to significant computing power for the expenditure of a few hundred pounds.

The microcomputer plays an increasingly important role in education and industry and much of the work that was once done on mainframe computers has now been transferred to the microcomputer. One area where both microcomputers and mainframe computers have been used for many years is for the numerical solution and analysis of a wide range of mathematical problems. The large mainframe computer remains an essential tool for solving extremely large and complex problems despite the astonishing developments in microcomputing. However, in general users will prefer to apply a microcomputer to a numerical problem if the microcomputer memory capacity allows and if accuracy and speed of calculation are not critical factors. The microcomputer user is more in control of the computing environment than when competing with many other users for the resources of a mainframe computer. Furthermore the fast speed of operation of the mainframe computer may not be utilised during periods of high demand because the user may have to wait a considerable time before being able to access the required part of the system. Despite these comments, care must be taken to avoid the indiscriminate and uncritical use of the microcomputer and a judgement must be made based on the requirements of the particular problem. The estimation of memory requirement is normally fairly simple but the determination of the accuracy required from the system is more difficult. For some classes of numerical problem the accuracy with which the solution may be obtained is not related in any simple way to the accuracy with which the computer stores and operates on numerical values. This problem receives special consideration in later sections of this book.

The performance of a specific microcomputer is related to the hardware design and the software used by the machine. The speed of execution of programs depends on a number of factors, the most important of which are the maximum speed rating of the microprocessor chip and the form of the implementation of the computer language being used. The accuracy that can be obtained in performing computations is primarily controlled by the software implementation and in some implementations it is possible to trade off computational speed and effective memory size for increased accuracy.

The effectiveness of the microcomputer has been considerably enhanced by the dramatic increase in the memory that can be economically be incorporated into their design. At the beginning of the 1980s most microcomputers had memory capacities between 16k and 48k. By the mid-1980s memories of 512k were common and by 1987 memory capacities of 1 megabyte (1024k) were available. However, the memory accessible to programs and data may be more limited than at first appears because some memory may be used by the operating system and other software or because of poor software design. A simple one line program can be written to find the memory available and such a program has been used to obtain the results for a range of microcomputers as shown in Table 1.1.1.

The hardware, the software and the design philosophy of microcomputers continues to develop rapidly. One exciting design development which is having, and will continue to have, an impact on the implementation of numerical algorithms is the introduction of parallel processors, that is computers that can execute several elementary arithmetic operations simultaneously. By redesigning numerical

Table 1.1.1. Maximum number of elements that can be used in an integer (%), real or double precision (#) arrays in the computer indicated. The memory accessed is often dependent on the version of BASIC used.

Microcomputer	Maximum number of elements n		
	DIM A%(n)	DIM A(n)	DIM A#(n)
BBC (32k)	6329	5062	—
Amstrad CP464	21753	8700	—
Mac (512k)	159000	82350	37850
Apricot (256k)	31117	15558	7778
CBM 64	19442	7776	—
Apple	18161	7264	—
Amstrad (PC1512)	30986	15492	7745

algorithms so that they can make use of parallel processing machines, major improvements in the speed of execution of the numerical process can be obtained.

1.2 COMPUTER LANGUAGES ON THE MICROCOMPUTER

A wide range of computer languages have now been implemented on microcomputers and the number of implementations continues to grow. The purpose of these languages is to allow the user to interact as easily as possible with the computer and languages have been developed to serve the particular needs of different users. The language which is most widely used on microcomputers is BASIC. Other languages which are also available on microcomputers are FORTRAN, COBOL, Pascal, C, PILOT, LOGO and specialised languages such as APL, LISP and Prolog. These languages each have their special features which make them suitable for particular tasks. FORTRAN, BASIC, Pascal, C and APL may be used for solving scientific problems but full implementations of FORTRAN, Pascal, C and APL are available only on the more powerful microcomputers. FORTRAN was developed specifically to solve scientific problems, APL was designed to solve problems which involved extensive manipulation of matrices and Pascal and C are more general-purpose languages. COBOL is a business-orientated language which is suitable for handling large file structures containing numeric and non-numeric data. PILOT and LOGO are specialised languages, PILOT allows the microcomputer to be used efficiently for the preparation of educational material and LOGO is a graphics language. LISP and Prolog are used for research into artificial intelligence and symbolic manipulation.

All these languages are called high-level languages. High-level languages allow the user to solve a problem by writing computer executable statements which are close to natural language statements. Clearly this simplifies the task of translating the problem-solving algorithm into a form that can be executed by the computer. Within the computer the instructions and data are stored and a procedure is required to translate from the high-level language to a machine-executable form, called machine code. This translation is achieved by using a special program known as a compiler or an interpreter. These compilers or interpreters are produced by software companies for a wide range of microcomputers. They may be supplied either on floppy discs or in ROM permanently resident within the computer. The main difference between the compiler and the interpreter is that while the compiler provides a complete and, after correction, executable machine code program the interpreter translates each statement of the high-level program as it is encountered. Consequently, since the compiled version of the program can be executed directly without need for any further translation, it should run significantly faster than an interpreted program where each statement is translated each time it is accessed.

The language used for the programs in this book is BASIC, the letters of which are an acronym for Beginner's All-purpose Symbolic Instruction Code. This high-level language is the most widely available on microcomputers and, because of its relatively simple structure, provides an easy introduction to programming for the microcomputer user. The language does have significant disadvantages because it is not well structured. This means that, because of its lack of some important programming features, programs written in BASIC may not have a clear logical structure and consequently may be difficult to understand and correct. This presents major problems for large programs, but programs to solve problems in numerical methods are often surprisingly short and consequently BASIC is usually a very adequate language for implementing these techniques.

The statement that BASIC is not well structured must be qualified because recently many new dialects of BASIC have become available that have features which allow structured programming. More recent versions of Microsoft BASIC, Hewlett Packard BASIC and BBC BASIC are examples of dialects that allow structured programming to some extent. The multiplicity of different dialects of BASIC leads to problems if a program is required to work on a wide range of machines, i.e. if it is to be portable. If we require portability, then the program must be written using only those features that are available in virtually all versions of BASIC so that it can be easily adapted to work on any microcomputer. The programs in this book are written, wherever possible, to satisfy this requirement.

1.3 NUMERICAL PROCESSES AND ERROR ANALYSIS

In general terms numerical analysis is the application of the fundamental arithmetic operations to numerical data to provide numerical solutions to problems posed in mathematical form. The microcomputer provides an excellent tool for carrying out numerical work. Even those microcomputers with small memory can be used to solve a wide range of numerical problems since, as will be seen from later chapters of this book, the majority of programs require only a small amount of memory. However, although memory size may not be a problem in terms of program storage, it does present problems in the way numerical data is stored, since this will often lead to numerical errors.

Errors arise in computations performed on any computer because the finite storage capacity will place a restriction on the storage space available for each number. Numbers are automatically rounded to the number of significant places specified for the particular machine. The type of errors which arise from this process are known as rounding errors. For example the value given for 1/3 on many microcomputers is .333333333, clearly this number has been rounded to nine significant digits.

Although these rounding errors in themselves may seem insignificant, when expressions which involve many calculations are performed using the standard arithmetic operators with rounded numbers the errors will tend to accumulate.

As an example of how errors are magnified, consider the problem of solving the quadratic equation

$$ax^2 + bx + c = 0 \qquad (1.3.1)$$

The solution of this equation is given by the well known formula

$$x = [-b \pm \sqrt{(b^2 - 4ac)}]/(2a) \qquad (1.3.2)$$

Consider the calculation of the root

$$x = [-b + \sqrt{(b^2 - 4ac)}]/(2a) \qquad (1.3.3)$$

This formula may give inaccurate results when $4ac$ is small relative to b^2. The inaccuracy is caused by the cancellation of many significant digits due to the subtraction of nearly equal values. To avoid this problem we can multiply the numerator and denominator of (1.3.3) by $b + \sqrt{(b^2 - 4ac)}$ to give

$$x = -2c/[b + \sqrt{(b^2 - 4ac)}] \qquad (1.3.4)$$

This equation has removed the cancellation errors introduced by the subtraction of nearly equal quantities. Table 1.3.1 shows a comparison between the use of (1.3.3) and (1.3.4) to determine one root of a quadratic equation.

Having established that small errors can lead to significantly greater errors during calculations, we must now examine how computers store numbers and how arithmetic operations may magnify errors.

Table 1.3.1. Solution of quadratics using (1.3.3) and (1.3.4). The right-hand pair of columns check the accuracy by substituting the root back into the original quadratic. The asterisk denotes that if further decimal places were displayed these roots would differ.

Coeff's of quadratic			Root using formula :		Quadratic value using	
a	b	c	(1.3.3)	(1.3.4)	(1.3.3)	(1.3.4)
0.5	1000	0.1	-1.231E-4	-0.0001	-0.02207	7.451E-9
0.5	3000	0.1	0	-3.333E-5	1	0
0.5	4	0.1	-2.507E-2*	-2.507E-2*	3.278E-7	-7.451E-9

1.4 HOW COMPUTERS STORE NUMBERS

Unless numbers are specified to be integers, they are stored in the computers in what is known as floating-point form. Floating-point form is similar to the scientific notation used as a compact form for writing very small or very large numbers. For example the number 0.0000123 may be written in scientific notation as

$$0.123 \times 10^{-4}$$

We can write the general form of a floating-point number as

$$a \times r^{E} \quad \text{where } (1/r) \leq a \leq 1 \text{ or } -(1/r) \geq a > -1 \tag{1.4.1}$$

where

$$a = \sum_{k=1}^{t} d_k r^{-k} \tag{1.4.2}$$

Here a is called the mantissa, E the exponent and r the base, d_k is the value of the kth digit and t is the maximum number of digits allowed in the number.

The base for ordinary arithmetic is, of course, 10. However, bases other than 10 may be used. The most common bases used for work on digital computer are 2 and 16; numbers using these bases are known respectively as binary and hexadecimal numbers.

All computers represent decimal numbers by a finite number of binary digits, called bits. Thus any decimal number in the continuous spectrum of numerical values can only be approximated by the closest base 2 number and their distribution

Table 1.4.1. Errors caused by representing decimals number by 4 binary digits.

Decimal	Nearest binary	Nearest decimal	Error
0.1	0.0010	0.1250	0.0250
0.2	0.0011	0.1875	-0.0125
0.3	0.0101	0.3125	0.0125
0.4	0.0110	0.3750	-0.0250
0.5	0.1000	0.5000	0
0.6	0.1010	0.6250	0.0250
0.7	0.1011	0.6875	-0.0125
0.8	0.1101	0.8125	0.0125
0.9	0.1110	0.8750	-0.0250

will lead to some decimal numbers being less accurately represented than others. Table 1.4.1 illustrates this problem. To analyse these errors we introduce the concepts of relative and absolute error. Let the actual value of number be x and the value of its machine representation be m. Then the absolute error is simply the difference between the two, $x - m$, and its magnitude is given by $|x - m|$. The relative error is given by

$$R = (x - m)/x$$

When working to base 2 and assuming the number is rounded, the maximum error in the representation of the mantissa will be 2^{-t-1} which is the maximum value of the $(t + 1)$th term in (1.4.2). Consequently the absolute error in the representation of a number defined by (1.4.1) is 2^{E-t-1}. This error includes the effect of the exponent. Thus the relative error R is given by

$$R = 2^{E-t-1}/(a2^E) = 2^{-t-1}/a$$

since the minimum magnitude of a is half the upper bound for the value of the relative error is 2^{-t}.

The machine representation m of the number x can be calculated in terms of the relative error as

$$m = x(1 - R)$$

This is called 'floating the number' and an upper bound for the error obtained from this representation is xR.

1.5 COMPUTATION USING FLOATING-POINT NUMBERS

The definitions and estimates of error considered in section 1.4 can be used to calculate the errors produced by applying arithmetic operations to the computer representation of numbers. The formulas used to calculate these errors are derived below.

Consider the addition of two numbers which are represented by the machine as m_1 and m_2. The result of this may be written as m_0 where

$$m_0 = m_1 + m_2 \tag{1.5.1}$$

To find the relative error R_0 of the result m_0 we note that it will be the result of floating the true value, x_0. This gives

$$m_0 = x_0(1 - R_0) \tag{1.5.2}$$

Since m_1 and m_2 will also be the result of floating true values x_1 and x_2, we have similarly

$$m_1 = x_1(1 - R_1) \tag{1.5.3}$$

$$m_2 = x_2(1 - R_2) \tag{1.5.4}$$

Now the result of the addition will itself be floated to the machine precision. Denoting the relative error of this operation by R we have from (1.5.1), (1.5.2), (1.5.3) and (1.5.4)

$$m_0 = x_0(1 - R_0) = [x_1(1 - R_1) + x_2(1 - R_2)](1 - R)$$

$$= [x_1 + x_2 - x_1R_1 - x_2R_2](1 - R)$$

Noting that $x_0 = x_1 + x_2$ we may rewrite this as

$$m_0 = x_0[1 - (x_1R_1 + x_2R_2)/(x_1 + x_2)](1 - R) \tag{1.5.5}$$

On multiplying out the right-hand side of (1.5.5) and neglecting products of relative error which will be small we have

$$m_0 = x_0[1 - (x_1R_1 + x_2R_2)/(x_1 + x_2) - R] \tag{1.5.6}$$

Comparing this with (1.5.1) gives an expression for the relative error for the addition in terms of the relative errors of the input values

$$R_0 = [(x_1 R_1 + x_2 R_2) + R(x_1 + x_2)]/(x_1 + x_2)$$

We can similarly obtain results for the other arithmetic operations:

for subtraction

$$R_0 = [(x_1 R_1 - x_2 R_2) + R(x_1 - x_2)]/(x_1 - x_2)$$

for multiplication

$$R_0 = R_1 + R_2 + R$$

for division

$$R_0 = R_1 - R_2 + R$$

R is called the machine rounding error.

Repeated application of these formulae allow us to find the resulting errors from repeated use of the arithmetic operations. These results will not be derived here but they show there may be considerable accumulation of error after many operations since to obtain the resulting errors of repeated multiplications and divisions we obtain an upper bound by adding the magnitudes of the relative errors.

Error analysis may be extended to general functions and formulae for this are now derived. These formulae are usually sufficiently simple to allow the hand computation of an upper bound estimate of the overall error in a computation given upper bounds on the errors. Consider a general function of n variables $f(x_1, x_2, \ldots, x_n)$. Assuming errors of ε_i are made in each of the variables x_i for $i = 1, 2, \ldots, n$ then an approximation to the resulting error in the function is given by

$$\Delta f = \varepsilon_1 \partial f / \partial x_1 + \varepsilon_2 \partial f / \partial x_2 + \ldots + \varepsilon_n \partial f / \partial x_n$$

Here we have used Taylor's series for an n variable function to expand the function f in terms of the errors and neglected all terms of degree higher than one in these errors. An upper bound for the error in the function is given by

$$|\Delta f| \leq |\partial f / \partial x_1| |\varepsilon_1| + |\partial f / \partial x_2| |\varepsilon_2| + \ldots + |\partial f / \partial x_n| |\varepsilon_n|$$

To illustrate the use of this formula we give four examples. Consider the functions:

(1) $f(x_1,x_2, \dots ,x_n) = x_1 + x_2 + \dots + x_n$

(2) $f(x_1,x_2, \dots ,x_n) = x_1 x_2 \dots x_n$

(3) $f(x_1,x_2,x_3) = x_1 x_2 / x_3$

(4) $f(x_1,x_2) = x_1 \cos x_2$

Using the notation used for errors described above we may obtain upper bounds on the resulting error in the functions as follows:

(1) $|\Delta f| \leq |\varepsilon_1| + |\varepsilon_2| + \dots + |\varepsilon_n|$

(2) $|\Delta f|/|f| \leq |\varepsilon_1|/|x_1| + |\varepsilon_2|/|x_2| + \dots + |\varepsilon_n|/|x_n|$

(3) $|\Delta f|/|f| \leq |\varepsilon_1|/|x_1| + |\varepsilon_2|/|x_2| + |\varepsilon_3|/|x_3|$

(4) $|\Delta f| \leq |\varepsilon_1||\cos x_2| + |\varepsilon_2||x_1 \sin x_2|$

In cases (2) and (3) above we give the relative error for simplicity. Notice the results given for (1), (2) and (3) are consistent with our original analysis of page 9.

From these formula numerical values for the error are easily calculated in specific cases. In particular, we note that for a base 10 number rounded to n decimal places, an upper bound on the magnitude of the rounding error is 0.5×10^{-n}.

The theoretical description of error given here could be more detailed and comprehensive but the above description is sufficient to indicate its significance and we shall illustrate its practical consequences in later chapters.

Having discussed the effects of errors in numerical operations using the standard arithmetic operators we must now consider other sources of error which arise from the method used to solve the numerical problem. This will be illustrated by considering the simple Taylor's method for the evaluation of functions.

1.6 FUNCTION EVALUATION ON THE COMPUTER

The values of the functions $\cos x$, $\sin x$, e^x, $\log_e x$ and a number of others are frequently required in a wide range of applications of the computer. Microcomputers employ programming languages which make functions such as these directly available to the user. We now consider ways in which the calculation of the values of such functions can be performed on the computer. A simple approach to the evaluation of functions is to use Taylor's series expansion to provide a polynomial approximation to the required function by taking a finite number of terms of the expansion. The Taylor's series expansion for a function $f(x)$ may be written in the form

$$f(x + h) = f(h) + xf'(h) + \{x^2/2!\}f''(h) + \ldots$$
$$+ \{x^{n-1}/(n-1)!\}f^{(n-1)}(h) + R_n \qquad (1.6.1)$$

Here R_n is called the remainder term and this may be evaluated by using the formula given by (1.6.2)

$$R_n = \{x^n/n!\}f^{(n)}(h + tx) \quad \text{where} \quad 0 < t < 1 \qquad (1.6.2)$$

By setting $h = 0$ in (1.6.1) and applying this series expansion to the function arcsinh x we obtain

$$\text{arcsinh } x = x - x^3/6 + 3x^5/40 - 15x^7/336 + \ldots \quad \text{for } |x| < 1 \qquad (1.6.3)$$

Now, although this expansion reduces the evaluation of the function to the application of the standard arithmetic operations available on any computer, various problems arise which we will now examine in some detail.

   The expansions obtained by applying Taylor's series usually contain an infinity of terms. To make calculations possible only a finite number of terms are used. This process of using a smaller number of terms to represent a function than are given by the Taylor's series expansion is called truncation of the series. To make the calculations efficient the smallest number of terms should be used that provides the required accuracy. The simple expansions provided by the application of

Table 1.6.1. Values of the approximation (1.6.3) for given values of x.
The largest neglected term becomes significant for x close to 0.9
and consequently the approximation is less accurate.

x	Approximate arcsinh(x)	First neglected term	Exact value
0	0	0	0
0.1	9.983407E-02	6.076389E-11	9.9834079E-02
0.2	0.1986892	3.111112E-08	0.198690110
0.3	0.2956562	1.196016E-06	0.295673048
0.4	0.3899063	1.592889E-05	0.399003532
0.5	0.4805803	1.186795E-04	0.481211825
0.6	0.5664995	6.123602E-04	0.568824899
0.7	0.6456345	2.452044E-03	0.652666566
0.8	0.7142766	8.155597E-03	0.732668256
0.9	0.7658467	0.0235412	0.808866936

Taylor's series may converge slowly and for some values of x may not converge at all. The expansion given in (1.6.3), for example, is not suitable for computational use because, even if all the four given terms are used for values of x close to 1, the next term in the series is significant and therefore ensures that the approximation is inaccurate for these values. Table 1.6.1 illustrates the errors that may arise when using this type of expansion.

In Chapter 7 we consider the problem of obtaining efficient approximations to some mathematical functions which are suitable for use on the digital computer. One measure of efficiency of function approximations is the number of simple arithmetic operations required to achieve a specific accuracy, the smaller this number the more efficient the approximation. To illustrate this point in the next section we examine how an efficient procedure can be developed for evaluating a polynomial.

1.7 HORNER'S METHOD FOR EVALUATING POLYNOMIALS

A common feature of approximations to standard functions is that they require the evaluation of polynomials. The efficient evaluation of a polynomial requires a careful layout of the calculations involved. An excellent illustration of this is Horner's method for the evaluation of a polynomial. This is a simple procedure which can be illustrated by considering the evaluation of the following polynomial.

$$2x^4 + 3x^3 - x^2 + 17x + 4$$

One approach to the evaluation of this polynomial would be to evaluate each power of x by repeated multiplication then multiply each power by the appropriate constant and add or subtract the terms to obtain the required result. This procedure would require 14 simple arithmetic operations. If we write this polynomial in the equivalent form

$$(((2x + 3)x - 1)x + 17)x + 4$$

which is known as nesting, the number of simple arithmetic operations required is 8. This gives a considerable saving in computational effort and in addition will sometimes lead to significant reduction in the accumulated rounding error. In general, for small x, Horner's method is more accurate. For large x the direct evaluation method may be more accurate. Horner's algorithm for the evaluation of the polynomial

$$a_0 x^n + a_{n-1} x^{n-1} + \ldots + a_n$$

begins by setting $p_0 = a_0$ and then using the following pair of equations recursively

$$p_k = xp_{k-1} + a_k \qquad \text{for } k = 1, 2, \dots, n \qquad (1.7.1)$$

In general Horner's method for the evaluation of an nth degree polynomial requires only n multiplications and n additions. The alternative approach described earlier requires $n(n + 1)/2$ multiplications and n additions. Table 1.7.1 gives the results of evaluating a polynomial using both techniques for a given set of values of the independent variable and using single and double precision. These results indicate there is little difference in the accuracy obtained by the two methods.

Table 1.7.1. Comparison of simple polynomial evaluation with Horner's method. The performance of the two methods is similar. Double precision working gives results which are much closer to the exact values, -10^{-5}, -10^{-10} and -10^{-15}.

	Simple polynomial evaluation		Horner's method	
X	Single pre.	Double precision	Single pre.	Double precision
1.1	-1.0729E-5	-1.0000000001D-5	-1.0133E-5	-1.0000000000051D-5
1.01	1.1921E-5	-9.9997787828D-11	5.9605E-8	-9.99997862294D-16
1.001	-5.961E-7	-4.4408920985D-16	4.1723E-7	-6.61338147751D-16

```
100   REM -------------------------------------------------------------
110   REM EVALUATION OF POLYNOMIAL:HORNER'S METHOD OR NESTING GL/JP (C)
120   REM -------------------------------------------------------------
130   PRINT "HORNER'S METHOD FOR EVALUATING POLYNOMIALS"
140   PRINT "ENTER DEGREE OF POLYNOMIAL";
150   INPUT N
160   DIM A(N)
170   FOR I=0 TO N
180       PRINT"ENTER COEFFICIENT OF X^";I;
190       INPUT A(I)
200   NEXT I
210   PRINT " X";TAB(15);"POLYNOMIAL VALUE"
220   FOR C=1 TO 3
230       P=A(0)
240       READ X
250       FOR K=1 TO N
260           P=P*X+A(K)
270       NEXT K
280       PRINT X;TAB(15);P
290   NEXT C
295   REM SAMPLE VALUES FOR X
300   DATA 1.1,1.01,1.001
```

Program 1.7.1

Program 1.7.1 was used to produce the results given in Table 1.7.1 . It should be noted that this program uses only one array to store the polynomial coefficients. There is no need to store the coefficients p_k of (1.7.1) since we require only the final result obtained from using (1.7.1). This is the value of the polynomial and is stored in the variable P in the program.

1.8 RICHARDSON'S EXTRAPOLATION

Most numerical techniques involve error and sometimes this error is dependent on a parameter which can be adjusted. For example, in certain numerical methods this parameter is a step size which we may call h. Richardson's Extrapolation method provides a method of error reduction based on a systematic reduction in the step size parameter h. Consider the following equation:

$$V = V_1 + d_p h^p + \sum_{i=p+1}^{\infty} d_i h^i \tag{1.8.1}$$

Here V_1 is the approximation to the true value V. The dominant error term involves h^p and the remaining error terms are also included. If we reduce the value h to ch, where c is less than one, we can obtain a new approximation V_2 such that

$$V = V_2 + d_p c^p h^p + \sum_{i=p+1}^{\infty} d_i c^i h^i \tag{1.8.2}$$

Now if we multiply (1.8.1) by $c^p h^p$ and (1.8.2) by h^p and subtract the results we eliminate the term in h^p and obtain

$$V = (V_2 - c^p V_1)/(1 - c^p) + \sum_{i=p+1}^{\infty} d_i^* h^i \tag{1.8.3}$$

where d_i^* are modified constant coefficients. The dominant term is now of order h^{p+1} so the order of the error has been reduced and consequently

$$V \approx (V_2 - c^p V_1)/(1 - c^p) \tag{1.8.4}$$

provides a more accurate approximation for V. In practice c is often chosen to be 0.5 and, for example, p may be 2 so that (1.8.4) becomes

$$V \approx (4V_2 - V_1)/3 \tag{1.8.5}$$

This method has important applications in evaluating derivatives and integrals and solving differential equations, see Chapters 4, 5 and 6.

1.9 COMPUTATIONAL COMPLEXITY

In this section we give a brief introduction to the important concept of the complexity of an algorithm. The complexity of a particular algorithm is related to the number of elementary operations, i.e. additions, subtractions, multiplications and divisions, required to solve a specific type of problem. For example, in Chapter 3 it is shown that it takes an order of n^3 elementary operations to solve a system of n linear equations in n variables using the Gauss elimination algorithm. Clearly the time taken by the computer is directly related to the number of elementary operations required. When an algorithm is such that it can solve a problem in a time which is of order n^k, where n gives the size of the problem, then it is said to be a polynomial-time algorithm. Most numerical algorithms are members of this class. Polynomial-time algorithms can be distinguished from exponential-time algorithms. For exponential-time algorithms the execution time increases as an exponential function of the operations n for example as 3^n, e^n or n^n notice that n appears as an exponent. The polynomial-time algorithm will always be faster then the exponential-time algorithms for sufficiently large n.

There are two main classes of problems:

(i) A problem is a member of the class called P if it can be solved by a deterministic algorithm, by which we mean a well defined specific algorithm, in a time which is of order n^k, (i.e. polynomial-time).

(ii) A class of problems called NP which includes the class P as a subset but also contains other problems which cannot, currently, be solved by deterministic polynomial-time algorithms. This type of problem have algorithms of the exponential type available for them and therefore, theoretically can be solved but the solution methods are not efficient. This does not exclude the possibilty that deterministic polynomial-time algorithms may be found for them. NP stands for nondeterministic polynomial, a nondeterministic algorithm is one which, for example, may contain an element of chance and NP problems can be solved using this type of algorithm.

A further class of problem, which is a subset of NP, is the 'NP complete' set of problems called the NPC class. These are NP problems of such difficulty that *if any one of them can be shown to belong to class* P *then all members of* NP *will belong to* P and consequently NP = P. To show that NP = P is still one of the great unsolved mathematical problems. The Linear Programming problem illustrates some aspects of computational complexity. Linear Programming is concerned with maximising a linear function subject to linear constraints and has wide and important commercial and scientific applications. The Simplex algorithm and its variants were, from 1945 on, the only means of solving this problem; the solution time could be exponential in terms of the size of the problem and consequently some problems could be intractable because of their large size. In 1979 the Russian mathematician Khachiyan produced a deterministic algorithm, see Khachiyan (1979), which he showed to be a polynomial-time algorithm. This algorithm was not successful practically but its discovery established that the problem was a member of class P rather than of class NP. This discovery led to renewed interest in the search for polynomial-time algorithms for Linear Programming problems. In 1984 Karmarkar, working for AT&T Bell Laboratories, produced an improved polynomial time algorithm, Karmarkar (1984), which has been practically successful and has provided a startling improvement over the Simplex algorithm. For an interesting report on the effectiveness of the Karmarkar algorithm see Gill *et al.* (1986). It is clear that these concepts, which appear to be of only theoretical interest, can be a crucial stimulus in the development of new and more efficient algorithms. An interesting and illuminating introduction to this difficult area is given by Lewis and Papadimitriou, (1977).

1.10 SOME NOTES ON THE HISTORY OF NUMERICAL ANALYSIS

What is understood by numerical analysis, as distinct from numerical computation, can be considered to have evolved since at least the 16th century, although, as Hollingdale (1972) observes, the term numerical analysis only came into general use as recently as 1947. The impetus to develop numerical analysis arose from the need to solve practical problems in such diverse areas as astronomy and navigation, surveying, civil and military engineering, and accountancy. Here we describe some early examples of numerical analysis.

Since the earliest times astronomy has posed many difficult mathematical problems. The need to solve these problems arose not only from philosophical studies but also from practical requirements such as navigation. One such problem, which was solved iteratively, *circa* 1400, was the problem of calculating sin 1°, given sin 3°. Now sin $3\theta = 3 \sin \theta - 4 \sin \theta^3$. Thus by making $\theta = 1°$ and letting $x = \sin 1°$ and $B = \sin 3°$ it is necessary to solve $B = 3x - 4x^3$ where B is a known constant. Another important problem was the solution of Kepler's equation. The equation has the form $E - e \sin E = M$ and the problem is to determine E knowing e and M. Kepler produced an iterative method for its solution. The equation is used to predict planetary positions and in the 17th century was closely involved in the solution of the central astronomical problem of the time which was to explain the motion of the planet Mars. The solution of this problem was ultimately to place the Sun at the centre of the known Universe and to change the philosophical views of humanity. Kepler's method of solution was by a laborious process of trial and error. Another problem studied by Kepler and Newton arose from the geometry of circles and required the solution of $\sqrt{[\{\sqrt{(x^2 + a^2)} + d\}^2 - b^2]} + c = x$. Kepler used the values $a = 476.5, b = 90668, c = 13971$ and $d = 65656.5$ and gave the solution as $x = 25772$. Newton provided a much more efficient algorithm for solving non-linear equations which bears his name today and despite its early discovery is extremely effective for solving many equations.

An important type of non-linear equation is the polynomial. Gauss had shown in 1799 that every polynomial of degree n can be resolved into a product of n root-factors. However, there does not exist specific algebraic formula for solving any polynomial and this was demonstrated theoretically by Abel in 1824 and Galois in 1832. They showed that fifth degree polynomials and above could not be solved by algebraic methods.

The need to find the areas of irregular plane surfaces attracted the attention of the mathematicians. Determining areas is equivalent to the mathematical process of integration and as early as 1639 Cavalieri determined a geometric form of the rule given algebraically by Simpson in 1743. Gauss's method of integration dates from 1814. Both of these methods are still widely used to provide good approximations to many integrals. Gauss applied his method to the following problem

$$\int_{100000}^{200000} \frac{dx}{\log x} = 8406.2431211$$

Another area of considerable importance is interpolation using polynomial approximation. The theoretical basis for this is Weierstrass's theorem, presented in 1886, which means broadly that any function can be approximated by a sufficiently high degree polynomial. The early algorithms for implementing this were based on the method of finite differences developed by Newton in the 17th century. This method is not particularly suitable for modern computer implementation and consequently it has been largely superseeded by other techniques. Newton's interest in interpolation arose much earlier than the work of Weierstrass and stemmed mainly from the consideration of a practical problem. This was to compile a table of values of $n^{1/p}$ for $p = 2, 3, 4$ and $n = 1, 2, 3, \ldots , 10000$. Newton's solution was to compute directly an initial 100 values and then interpolate the remaining ones. A further application of Newton's finite difference methods was in the field of astronomy in plotting the paths of comets from a limited number of initial observations.

An important area for the application of numerical techniques is the solution of differential equations. The crucial early work in this area was by Euler, who produced a simple numerical algorithm for the solution of the equation $y' = f(x,y)$ Cauchy, in lectures published in Paris in 1840 formalised this procedure and stated the conditions under which it converges. Rudolph Lipschitz in 1877 showed that Cauchy's conditions could be relaxed and replaced by an inequality on the boundaries of the function called the Lipschitz bound.

Developments in celestial mechanics, ballistics and other areas gave further impetus for the discovery of more powerful methods for solving differential equations. Heun in 1900, Runge in 1905 and Kutta in 1901 made major contributions to solving these problems. Interestingly it was Heun's method that was used by the early computer ENIAC to solve differential equations. An alternative technique was developed in 1883 by the astronomer and mathematician Adams and further improvements were made by F. R. Moulton in 1925. Real progress in solving partial differential equations was not achieved until as recently as 1929 with the work of Courant,

Friedrichs and Lewy.

An important procedure used in the analysis of a large number of observations which are prone to error is the method of least squares, first published by Legendre in 1805 in a memoir on comets but used by Gauss for at least 10 years before this. The application of the method of least squares leads to systems of linear equations that must be solved. Consequently, because of his interest in the problem, produced both an elimination and iterative method for solving systems of linear equations. In 1845 Jacobi developed an iterative technique for solving both systems of linear equations and eigenvalue problems.

Readers wishing to learn more of the history of numerical analysis should refer to the very readable text by Goldstine (1977).

1.11 THE RELIABILITY OF COMPUTER PROGRAMS

This chapter ends with an important reminder to the reader concerning all programs, including the ones in this book. Even with careful program development and testing, minor typographical errors in program code may lead to significant errors in the results of the execution of the program. Initial testing may fail to discover such an error because a section of a program may only be explored for a particular type of rare or unpredicted problem. Even for well tested programs more subtle types of failure can occur for the following reasons.

(1) A program is used to solve a problem for which it is not designed. Because of the limitations of the algorithm it implements, this may result in the program failing completely or giving inadequate results. An example of this is using a program which can only reliably find the distinct roots to solve an equation when the equation only has multiple roots.

(2) A program is used to solve a problem which is intrinsically difficult. For example a problem in which rounding errors in the data leads to very much larger errors in the solution.

(3) A convergence criterion is used in a program to terminate an iteration which does not take into account the precision of the particular computer being used.

Despite the above comments programs can be designed to give satisfactory solutions for the majority of problems.

In this chapter we have been concerned with introducing some generally important concepts which will be used directly or indirectly in later chapters. We consider the problem of error generation on a range of specific microcomputers in Chapter 11.

2

Non-linear algebraic equations

2.1 INTRODUCTION

A major problem which arises in science and engineering is to find the value of a variable x which makes a specific function of that variable equal zero. Stated in general terms we wish to find x such that

$$f(x) = 0$$

This value is sometimes called a root or a zero of the equation. In many cases there may be more than one root for a given equation. Some examples of different types of equation are

(1) $4x^4 - 2x^3 + x^2 - 3x + 1.7 = 0$

(2) $xe^{-x} - 1 = 0$

(3) $\tanh x \tan x - x = 0$

Type (1) is known as a polynomial equation and these equations constitute an important subset of non-linear algebraic equations in that specialised methods can be used for their solution. The remaining equations are known as transcendental equations.

Fig. 2.1.1 illustrates how we can locate, approximately, the position of the roots of a given equation by graphical means. The diagram shows that the function $f(x)$

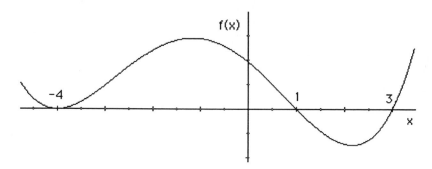

Fig. 2.1.1 Graph of a function with two coincident roots at $x = -4$ and
two seperate roots at $x = 1$ and $x = 3$.

has several roots and these occur where the graph of $f(x)$ crosses or touches the x axis.
In the case where the graph touches or is coincident with the x-axis this reveals that
the equation has a multiple root at this point. Fig. 2.1.2 and Fig. 2.1.3 illustrate this
case. Fig. 2.1.2 illustrates the behaviour of the function

$$f(x) = (x-2)^2$$

In this case the root is said to be of multiplicity two. In Fig. 2.1.3 the function
is

$$f(x) = (x-2)^3$$

In general a function with a root of multiplicity s may be written in the form

$$f(x) = (x-d)^s p(x)$$

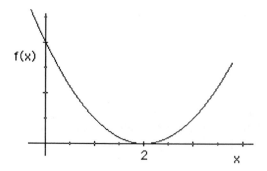

Fig. 2.1.2. Graph of the function with two coincident roots at $x = 2$.

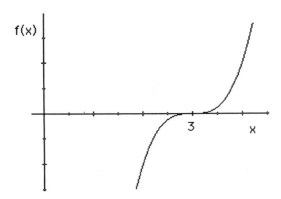

Fig. 2.1.3. Graph of the function with three coincident roots at $x = 3$.

where $p(x)$ is not divisible by $(x-d)$. The root is said to be of even multiplicity if s is even and odd multiplicity if s is odd. It should be noted at this point that finding multiple roots of equations presents significant difficulties for many numerical algorithms. This and other problems will be discussed in a later section.

An alternative graphical method of finding the root or roots of an equation can be used when it is possible to split the equation into two separate parts which can easily be graphed. To illustrate this consider the function

$$x^3 - 4x - 1 = 0$$

We can easily draw the graphs $y = x^3$ and $y = 4x + 1$ and the x coordinates of the points of intersection will give us an approximation to the value of the roots. The accuracy of these approximations will depend on the accuracy of the graph. This graphical method of solution does not provide us with a practical means of solving equations since it requires extensive computations to give approximations of low accuracy. However it does give a means of producing rough initial approximations for some of the algorithms we shall discuss in this chapter.

We can see from the Fig. 2.1.1 that, except in the case of multiple roots, the value of the function $f(x)$ changes sign after passing through the root. This observation leads to a simple but reliable algorithm, which is described below, for finding roots of non-linear equations.

2.2 METHOD OF BISECTION

This method assumes that we have a specified function $f(x)$ and that we know values of the independent variable x enclosing the root we wish to find. If we let the values

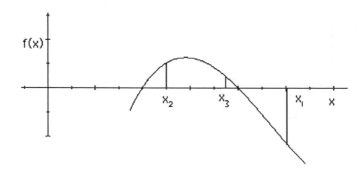

Fig. 2.2.1 Illustration of the bisection method.

enclosing the root be x_1 and x_2 then we can illustrate the situation in Fig. 2.2.1. We note, as before, that $f(x_1)$ and $f(x_2)$ have different signs. If we now find the point midway between x_1 and x_2 this will be given by

$$x_3 = (x_1 + x_2)/2$$

From Fig. 2.2.1 we can see that the root now lies between x_1 and x_3 in an interval half the size of the original interval. We continue in this way by taking a point midway between x_1 and x_3 and thus obtain an even smaller interval in which the root lies. This procedure can be repeated, reducing the size of the interval each time, until we locate the root to the accuracy we require. Fig. 2.2.1 helps to establish where the root will lie after finding a new mid-point. However we must avoid the considerable effort of drawing a graph for each equation that we wish to solve. Consequently we must find some algebraic alternative.

We have noted already that the value of the function changes sign after passing a root. To find the interval in which a root lies we form the product of the function at the mid-point with the function calculated at one of the end points of the current interval. If this product is negative then a root lies between the mid-point and the end-point where the function has been calculated, otherwise it lies between the mid-point and the other end-point. This description may be expressed in algebraic terms as follows. If

$$f(x_1)f(x_3) < 0$$

then the root lies in the interval (x_3, x_1), otherwise it lies in the interval (x_3, x_2). This technique is implemented in Program 2.2.1. From Fig. 2.1.2 it can be seen that this method cannot be used to find roots of even multiplicity since there will be no change in sign of the function as the independent variable values pass through the root.

```
100   REM -----------------------------------
110   REM   BISECTION METHOD PROGRAM GL/JP (C)
120   REM -----------------------------------
130   PRINT "BISECTION METHOD"
135   REM DEFINE FUNCTION AT STATEMENT 140
140   DEF FNA(X)=X^3-10*X^2+29*X-20
150   PRINT "ENTER TWO APPROXIMATIONS ENCLOSING ROOT";
160   INPUT X0,X1
165   REM ENSURE ROOT LIES IN INTERVAL
170   FA= FNA(X0):FB= FNA(X1)
180   FPA=FA*FB
190   IF FPA>0 THEN PRINT "INTERVAL MAY NOT CONTAIN ROOT." : GOTO 150
200   PRINT "ENTER ACCURACY YOU REQUIRE ";
210   INPUT EPS
220   PRINT TAB(5);"APPROXIMATION";TAB(25);"FUNCTION
VALUE";TAB(45);"ITERATION"
230   E2=EPS*EPS
240   IF ABS(FPA)>E2 THEN 270
250   IF ABS(FA)<EPS THEN  PRINT "ROOT IS ";X0 : GOTO 380
260   PRINT "ROOT IS ";X1 : GOTO 380
265   REM FIND IMPROVED APPROXIMATION BY BISECTION
270   IT=IT+1
280   X2=(X0+X1)/2
290   FC= FNA(X2)
300   PRINT TAB(5);X2;TAB(25);FC;TAB(45);IT
310   FS=FC*FA
320   IF ABS(FS)<E2 THEN FS=0
330   N=SGN(FS)
340   ON N+2 GOTO 350,360,370
350   X1=X2 : GOTO 270
360   PRINT "ROOT = ";X2;"NUMBER OF ITERATIONS=";IT : GOTO 380
370   X0=X2: FA=FC: GOTO 270
380   END
```

Program 2.2.1 .

Table 2.2.1 illustrates the performance of the bisection method for the equation

$$f(x) = x^3 - 10x^2 + 29x - 20 = 0$$

It can be seen that convergence is slow as the change in x at each iteration is halved and the reduction of the interval does not accelerate. Although the method is reliable in many cases its slow convergence means that we should seek alternative methods having a faster rate of convergence.

Table 2.2.1. The convergence of the bisection method to the root at
$x = 1$ of the equation given in the text using program 2.2.1.

```
BISECTION METHOD
ENTER TWO APPROXIMATIONS ENCLOSING ROOT? 0,2.5
ENTER ACCURACY YOU REQUIRE ? .005
        APPROXIMATION           FUNCTION VALUE         ITER'N
            1.25                   2.578125                1
            .625                  -5.537109                2
            .9375                 -.7775879                3
            1.09375                1.064301                4
            1.015625               .1857948                5
            .9765625              -.2854137                6
            .9960938              -4.698181E-02            7
            1.005859               7.007217E-02            8
            1.000977               1.171112E-02            9
            .9985352              -1.759338E-02           10
            .9997559              -2.929688E-03           11
            1.000366               4.394531E-03           12
ROOT =  1.000366  NUMBER OF ITERATIONS=  12
```

2.3 SIMPLE ITERATIVE OR FIXED POINT METHODS

Consider the equation

$$f(x) = 0$$

This may be rewritten in the form

$$x = g(x)$$

and it can be used as an iterative procedure by writing it as

$$x_{r+1} = g(x_r) \quad r = 0, 1, 2, ...$$

It is assumed that an initial approximation x_0 is available. As an example consider
the equation

$$x^3 - 10x^2 + 29x - 20 = 0$$

We can rewrite this as

$$x = (20 - x^3 + 10x^2)/29$$

or in iterative form

$$x_{r+1} = (20 - x_r^3 + 10x_r^2)/29 \quad r = 0, 1, 2, \dots \qquad (2.3.1)$$

If we have some initial approximation to one of the roots we may use (2.3.1) in an attempt to find an improved approximation. Looking again at our original equation we can see that a number of alternative iterative formulae could be derived and some of these are listed below:

$$x_{r+1} = 10 - 29/x_r + 20/x_r^2 \qquad r = 0, 1, 2, \dots \qquad (2.3.2)$$

$$x_{r+1} = 0.1x_r^2 - 2/x_r + 2.9 \qquad r = 0, 1, 2, \dots \qquad (2.3.3)$$

$$x_{r+1} = \sqrt{[(x_r^3 + 29x_r - 20)/10]} \quad r = 0, 1, 2, \dots \qquad (2.3.4)$$

An important question arises; which, if any, of the above iterative formula will converge to the root of the equation we wish to find? Table 2.3.1 illustrates the behaviour of these formulae using an initial approximation $x = 0.6$ which is close to the root of the original equation, $x = 1$. Formula (2.3.4) has been excluded since it leads to the square root of a negative number at the first iteration. Table 2.3.1 shows that (2.3.1) converges to the root we require. Formula (2.3.2) converges to another root and (2.3.3) diverges.

Table 2.3.1. Performance of the simple iterative method in attempting to find the root of the equation given in the text.

```
SIMPLE ITERATIVE METHOD FOR SOLVING EQUATIONS
ACCURACY = 0.005
```

FORMULA	(2.3.1)	(2.3.2)	(2.3.3)
ITER			
1	.8063449	17.22222	-30.39733
2	.8957807	8.383558	95.95774
3	.9415667	6.825408	923.4803
4	.9665771	6.180481	85284.47
5	.9806781	5.831392	7.273441E+08
6	.9887637	5.615063	5.290294E+16
7	.9934437	5.469658	2.798721E+32

2.4 CONVERGENCE OF SIMPLE ITERATIVE METHODS

In view of the above results we clearly need to establish some general rule for testing in advance if a particular iterative method will converge or not. If we let the exact root of

$$f(x) = 0$$

be s, then we can write successive iterates as $x_{r+1} = s + \varepsilon_{r+1}$ and $x_r = s + \varepsilon_r$ where ε_{r+1} and ε_r are the errors in the $r + 1$th and r th iterates respectively. Hence the general iterative formula becomes:

$$s + \varepsilon_{r+1} = g(s + \varepsilon_r) \tag{2.4.1}$$

Now, by the mean-value theorem, there exists t such that in the interval $a < x < b$

$$g(b) - g(a) = (b - a)g'(t) \tag{2.4.2}$$

Consequently putting $a = s$ and $b = s + \varepsilon_r$ we have

$$g(s + \varepsilon_r) = g(s) + \varepsilon_r g'(t_r) \tag{2.4.3}$$

where t_r lies between s and $s + \varepsilon_r$. By virtue of (2.4.1) this gives

$$s + \varepsilon_{r+1} = g(s) + \varepsilon_r g'(t_r) \tag{2.4.4}$$

However $f(s) = 0$ and therefore $s - g(s) = 0$. Note that s is the fixed point which maps onto itself, hence the name *fixed point method*. From (2.4.4) we have

$$\varepsilon_{r+1} = \varepsilon_r g'(t_r)$$

Assuming the upper bound for $|g'(t_r)|$ is M then we have:

$$|\varepsilon_{r+1}| < M |\varepsilon_r| \quad r = 0, 1, 2, \ldots$$

After repeated application of this formula we obtain

$$|\varepsilon_{r+1}| < M^{r+1} |\varepsilon_0| \tag{2.4.5}$$

Clearly if $M < 1$ then M tends to zero as r increases and so by virtue of (2.4.5) the error also tends to zero. Consequently we conclude that if $M < 1$ then the method converges. Hence our iterative procedure will converge if

$$|g'(t_r)| < 1 \qquad\qquad (2.4.6)$$

Applying the convergence requirement to the iterative formulae (2.3.1), (2.3.2), (2.3.3) we find:

for (2.3.1) $|g'(0.6)| = 0.3765517 < 1$ and therefore the method converges,

for (2.3.2) $|g'(0.6)| > 1$ and therefore the method diverges,

for (2.3.3) $|g'(0.6)| > 1$ and therefore the method diverges.

The graphs in Fig. 2.4.1 illustrate how the condition $|g'(0.6)| < 1$ governs convergence. When $|g'(0.6)| < 1$, since $x_{r+1} = g(x_r)$, the horizontal projections of the points P and Q on the line $y = x$ give successively improved approximations to x and this process may be repeated until convergence is achieved.

In general the methods we have considered converge slowly. The aim of the next section is to establish conditions which enable us to classify methods according to their rate of convergence.

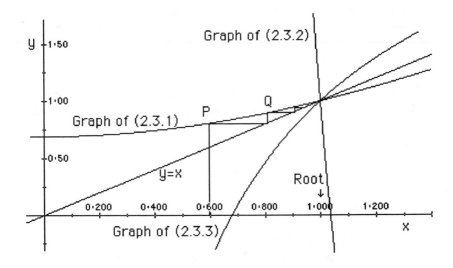

Fig 2.4.1. Only iterative formula (2.3.1) converges since its slope is less than one. The other two formula do not converge since their slopes are clearly greater than one.

2.5 CONVERGENCE AND ORDER OF AN ITERATIVE METHOD

Consider the general iterative form

$$x_{r+1} = g(x_r) = g(s + (x_r - s))$$

Expanding by Taylor's theorem we have

$$x_{r+1} = g(s) + (x_r - s) g'(s) + (x_r - s)^2 g''(t_r)/2! \qquad (2.5.1)$$

where t_r lies in the interval (x_r, x_{r+1}) and we have for, the moment, ignored higher order terms. This expansion assumes that $g(x)$ may be differentiated twice. Now $x_{r+1} = s + \varepsilon_{r+1}$ and $s = g(s)$ so that inserting in (2.5.1) gives

$$\varepsilon_{r+1} = (x_r - s) g'(s) + (x_r - s)^2 g''(t_r)/2! \qquad (2.5.2)$$

$$\varepsilon_{r+1} = \varepsilon_r g'(s) + \varepsilon_r^2 g''(t_r)/2! \qquad (2.5.3)$$

If we neglect ε_r^2 and higher powers of ε_r then we have

$$\varepsilon_{r+1} = \varepsilon_r g'(s) \qquad (2.5.4)$$

This indicates that the current error is directly proportional to the value of the previous error. This is called a first order method. If g is such that $g'(s) = 0$ this gives

$$\varepsilon_{r+1} = \varepsilon_r^2 g''(t_r)/2! \qquad (2.5.5)$$

which indicates that the current error is proportional to the square of the previous error. This implies that if $g'(s) = 0$ then the convergence of the method is much faster. Such a method is known as a second order method.

   In general a method is said to be of order p if

$$g'(s) = g''(s) = \dots = g^{(p-1)}(s) = 0$$

and

$$g^{(p)}(s) \neq 0$$

where the superscript (p) denotes the pth derivative of the function g. By using Taylor's series to expand $g(x_{r+1})$ up to the $(p + 1)$th term these conditions give

$$\varepsilon_{r+1} = g^{(p)}(t_r)\varepsilon_r^p / p! \qquad (2.5.6)$$

This shows that the higher the order of the method the faster the convergence. As an example of a higher order method we will now consider Newton's method.

2.6 NEWTON'S METHOD

We wish to solve the equation

$$f(x) = 0.$$

Let x_r be the rth approximation to the root then

$$f(x) = f(x_r + (x - x_r))$$

By using Taylor's series expansion and ignoring higher order terms

$$f(x) = f(x_r) + (x - x_r)f'(x_r) + \ldots \tag{2.6.1}$$

Now if x_r is replaced by an improved approximation to the root x_{r+1} such that

$$f(x_{r+1}) \approx 0$$

we have

$$f(x_{r+1}) \approx 0 \approx f(x_r) + (x_{r+1} - x_r)f'(x_r)$$

and hence we obtain the iterative formula

$$x_{r+1} = x_r - f(x_r)/f'(x_r) \tag{2.6.2}$$

It can be shown that Newton's method is a second order method. The method is illustrated graphically in Fig. 2.6.1.

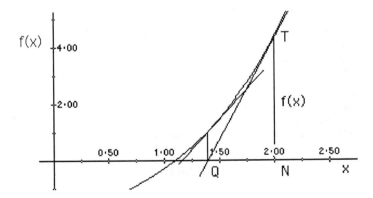

Fig 2.6.1. The tangent at the point N strikes the x-axis at a point
Q which is closer to the root.

Table 2.6.1. Results of Newton's method for finding a root of the equation given in the text. Note the rapid convergence.

```
NEWTON'S METHOD FOR SOLVING EQUATIONS
ENTER INITIAL APPROXIMATION TO ROOT? 0
ENTER THE ACCURACY YOU REQUIRE? .005
        APPROXIMATION           FUNCTION VALUE          ITERATION
         .6896552                 -4.428226                1
         .9558743                 -.5432243                2
         .9989067                 -1.312828E-02            3
ROOT = .9999993  ITERATIONS= 4
```

Table 2.6.1 gives the results of Newton's method for the equation $x^3 - 10x^2 + 29x - 20 = 0$. By comparing this table with Table 2.3.1 it can be seen that the convergence of Newton's method is significantly faster than the simple iterative method. Program 2.6.1 implements Newton's method; it requires the user to supply both the equation to be solved and its derivative.

Starting with a complex initial approximation convergence to a complex root can be achieved if computation is performed in complex arithmetic. This would require considerable modification to program 2.6.1.

```
100   REM ------------------------------------
110   REM   NEWTON'S METHOD PROGRAM GL/JP (C)
120   REM ------------------------------------
130   PRINT "NEWTON'S METHOD FOR SOLVING EQUATIONS"
135   REM DEFINE FUNCTION AND DERIVATIVE AT 140 AND 150, FNB(X) IS THE
DERIVATIVE OF FNA(X)
140   DEF FNA(X)=X^3-10*X*X+29*X-20
150   DEF FNB(X)=3*X^2-20*X+29
160   PRINT "ENTER INITIAL APPROXIMATION TO ROOT";
170   INPUT X0
180   PRINT "ENTER THE ACCURACY YOU REQUIRE";
190   INPUT EPS
200   PRINT TAB(5);"APPROXIMATION";TAB(25);"FUNCTION VALUE"; TAB(45);
 "ITERATION"
205   REM MAIN NEWTON ITERATION
210   D=FNA(X0)/FNB(X0)
220   IF ABS(D)<EPS THEN PRINT "ROOT = ";X0-D;" ITERATIONS=";IT+1:GOTO 270
230   X1=X0-D
240   IT=IT+1:X0=X1
250   PRINT TAB(5);X1;TAB(25);FNA(X1);TAB(45);IT
260   GOTO 210
270   END
```

Program 2.6.1

2.7 THE SECANT METHOD

To avoid the calculation of $f'(x)$ required by Newton's method a simple numerical approximation can be made to $f'(x)$. If x_r and x_{r-1} are successive approximations to the root then $f(x_r)$ may be approximated by

$$f(x) \approx (f(x_r) - f(x_{r-1}))/(x_r - x_{r-1}) \tag{2.7.1}$$

substituting this approximation into (2.6.2) gives

$$x_{r+1} \approx x_r - f(x_r)(x_r - x_{r-1})/(f(x_r) - f(x_{r-1}))$$

This gives the iterative form

$$x_{r+1} = [x_{r-1}f(x_r) - x_r f(x_{r-1})]/[f(x_r) - f(x_{r-1})] \quad r = 0, 1, 2, \ldots \tag{2.7.2}$$

This method, which is known as the secant method, is clearly not the same as the general iterative form $x_{r+1} = g(x_r)$. The relation between successive errors can be found by substituting for x_{r-1}, x_r, x_{r+1} in terms of the errors and after approximating and rearranging an expression of the form $\varepsilon_{r+1} = K\varepsilon_r\varepsilon_{r-1}$ can be obtained. This gives an equivalent order for the method which is not an integer and lies between one and two. The method requires two initial approximations to the solution to start the iterative process and the procedure is illustrated graphically in Fig. 2.7.1. The line PQ is, of course, a secant which strikes the x-axis at a point closer to the root than either of the two starting points. The process can be repeated by drawing the secant through Q and R and then repeated until the required accuracy is obtained. The major advantage of this method is that it does not require the calculation of the derivative of $f(x)$.

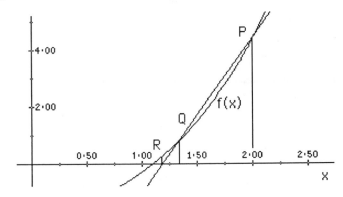

Fig. 2.7.1. An illustration of how the secant method provides improved
approximations to the root of an equation $f(x) = 0$.

An important feature of the secant method is the way in which the values of the variables x_{r+1}, x_r, x_{r-1} should be interchanged as we proceed from one iteration to the next. In the secant method the secant is always derived from the two most recent points. Thus once x_{r+1} has been generated we set $x_r = x_{r+1}$ and $x_{r-1} = x_r$ for the next iteration. An alternative to the secant method is to chose, at each iteration, values which enclose the root. This is known as the method of *regula falsi*.

Table 2.7.1 shows that the performance of the secant method is comparable to Newton's method. The calculations are again performed for the solution of the equation

$$f(x) = x^3 - 10x^2 + 29x - 20 = 0$$

Program 2.7.1 implements the secant method; note the user need only supply the equation to be solved and two initial starting values.

Table 2.7.1. Use of program 2.7.1 to solve the problem given in the text.

```
SECANT METHOD
ENTER TWO INITIAL APPROXIMATIONS TO THE ROOT? 0,2.5
ENTER ACCURACY REQUIRED? .005
        APPROXIMATION          FUNCTION VALUE          ITERATIONS
         1.95122                5.941582                   1
        12.25069              673.0521                     2
         1.859488               5.77774                    3
         1.769513               5.544769                   4
         -.3719019            -32.2197                     5
         1.4551                 4.105644                   6
         1.248604               2.56599                    7
          .9044586             -1.211266                   8
         1.014817                .1762695                  9
         1.000797               9.563446E-03              10
          .999993              -8.392334E-05              11
ROOT = .999993  ITERATIONS= 11
```

```
100   REM -------------------------------
110   REM    SECANT METHOD PROGRAM GL/JP (C)
120   REM -------------------------------
130   PRINT "SECANT METHOD"
135   REM DEFINE FUNCTION AT STATEMENT 140
140   DEF FNA(X) = X^3-10*X^2+29*X-20
150   PRINT "ENTER TWO INITIAL APPROXIMATIONS TO THE ROOT";
160   INPUT X0,X1
170   PRINT "ENTER ACCURACY REQUIRED";
180   INPUT EPS
190   PRINT TAB(5);"APPROXIMATION";TAB(25);"FUNCTION VALUE";
TAB(45);"ITERATIONS"
195   REM CALCULATE INITIAL FUNCTION VALUES
200   FA= FNA(X0):FB= FNA(X1)
210   IT=IT+1
215   REM CALCULATE IMPROVED APPROXIMATION
220   X2=X1-(X1-X0)*FB/(FB-FA)
230   PRINT TAB(5);X2;TAB(25); FNA(X2);TAB(45);IT
240   D= ABS(X2-X1)
250   IF D>EPS THEN X0=X1: X1=X2: FA=FB : FB= FNA(X1): GOTO 210
260   PRINT "ROOT = ";X2; "ITERATIONS=";IT
270 END
```

Program 2.7.1.

2.8 AITKEN'S METHOD FOR ACCELERATING CONVERGENCE

In section 2.4 we obtained a relationship between successive errors for the simple iterative method which had the form

$$\varepsilon_{r+1} = K\varepsilon_r \quad r = 0, 1, 2, \dots \tag{2.8.1}$$

where

$$x_r = s + \varepsilon_r \quad \text{for } r = 0, 1, 2, \dots \tag{2.8.2}$$

Consider the expression

$$(x_r x_{r+2} - x_{r+1}^2)/(x_{r+2} - 2x_{r+1} + x_r) \tag{2.8.3}$$

By using (2.8.2) we may obtain expressions for x_{r+2}, x_{r+1}, x_r in terms of $\varepsilon_{r+2}, \varepsilon_{r+1}, \varepsilon_r$. Substituting for these in (2.8.3) we find that expression (2.8.3) is approximately equal to s. In fact, if K of (2.8.1) is constant for all r, the expression (2.8.3) gives the exact value of s. In general however K is not constant for all r but the formula (2.8.3) may be used to give an improved approximation to the root s. Hence, given any three approximations x_{r+2}, x_{r+1}, x_r to s, we can use the formula

$$s = (x_r x_{r+2} - x_{r+1}^2)/(x_{r+2} - 2x_{r+1} + x_r) \qquad (2.8.4)$$

for $r = 0, 1, 2, ...$ to give an improved approximation for s.

2.9 DEALING WITH THE PROBLEM OF MULTIPLE ROOTS

The methods discussed so far cannot be guaranteed to converge efficiently for all problems. In particular, when a given function has a multiple root which we require, the methods we have described will either not converge at all or converge more slowly. For example, Newton's method will not have order two convergence for a multiple root. However in some cases simple modifications can be made to the methods to maintain the rate of convergence. Two such methods are described below.

If we wish to determine a root of known multiplicity m for the equation $f(x) = 0$ then Schroder's method may be used. It has the form

$$x_{r+1} = x_r - mf(x_r)/f'(x_r) \text{ for } r = 0, 1, 2, ... \qquad (2.9.1)$$

It is assumed we have an initial approximation x_0. The similarity to Newton's method is obvious and like Newton's method it has order two convergence. The major disadvantage of this method is that the multiplicity of the root must be known in advance.

An alternative approach to this problem that does not require any knowledge of the multiplicity of the root is to replace the function $f(x)$ in the equation by $p(x)$ where

$$p(x) = f(x)/f'(x) \qquad (2.9.2)$$

If

$$f(x) = (x - s)^m h(x)$$

then

$$f'(x) = m(x - s)^{m-1}h(x) + (x - s)^m h'(x)$$

On division we find that $p(x)$ has the root s to multiplicity one. For example, with this modification Newton's method becomes

$$x_{r+1} = x_r - p(x_r)/p'(x_r) \text{ for } r = 0, 1, 2, ... \qquad (2.9.3)$$

where

$$p(x_r) = f(x_r)/f'(x_r)$$

A similar modification can be made to the secant method. The disadvantage of this method is that we must calculate a further higher derivative.

Table 2.9.1 compares the performance of Schroder's method, Newton's method and Newton's method modified in the manner described above for finding a multiple root. Programs 2.9.1 and 2.9.2 provide implementations of Newton's method modified for multiple roots and Schroder's method respectively.

Table 2.9.1. Comparison in terms of the number of iterations required to solve the problem with the starting value is 0.5. When the difference between successive approximations is 0.00005 the estimate is accepted.

Function	Method :		
	Newton	Mod. Newt'n	Schroder
$(x-3)^{10}$	81	2	2
$(x^2-3)^{10}$	70	6	6
$(e^x-x)^2$	10	3	3

```
100   REM ---------------------------------------------
110   REM NEWTON'S METHOD FOR MULTIPLE ROOTS GL/JP (C)
120   REM ---------------------------------------------
130   PRINT "NEWTON'S METHOD FOR SOLVING AN EQUATION WITH MULTIPLE ROOTS"
134   REM DEFINE FUNCTION AND FIRST AND SECOND DERIVATIVES AT 140,150,160
135   REM FNB(X) AND FNC(X) ARE THE FIRST AND SECOND DERIVATIVES OF FNA(X)
140   DEF FNA(X)=(X^2-3)^10
150   DEF FNB(X)=20*(X^2-3)^9*X
160   DEF FNC(X)=20*(X^2-3)^8*(19*X*X-3)
170   PRINT "ENTER INITIAL APPROXIMATION TO ROOT";
180   INPUT X0
190   PRINT "ENTER ACCURACY YOU REQUIRE ";
200   INPUT EPS
205   REM IMPROVE APPROXIMATION
210   IT=IT+1
220   IF FNA(X0)=0 THEN PRINT "ROOTS = ";X0; "ITERATIONS=";IT:GOTO 300
230   D= FNA(X0)/FNB(X0)
240   D1=1-D* FNC (X0) / FNB (X0)
250   D=D/D1
260   IF ABS (D)<EPS THEN PRINT "ROOTS = ";X0; "ITERATIONS=";IT:GOTO 300
270   X1=X0-D
280   X0=X1
290   GOTO 210
300   END
```

Program 2.9.1

```
100   REM --------------------------------------------------
110   REM SCHRODER'S METHOD FOR MULTIPLE ROOTS  GL/JP (C)
120   REM --------------------------------------------------
130   PRINT "SCHRODER'S METHOD FOR MULTIPLE ROOTS"
135   REM DEFINE FUNCTION AND DERIVATIVE AT 140 AND 150, FNB(X) IS THE
DERIVATIVE OF FNA(X)
140   DEF FNA(X)=(EXP(-X)-X)^2
150   DEF FNB(X)=2*(EXP(-X)-X)*(-EXP(-X)-1)
160   PRINT "ENTER INITIAL APPROXIMATION TO ROOT";
170   INPUT X0
180   PRINT "ENTER ACCURACY REQUIRED";
190   INPUT EPS
200   PRINT "ENTER THE MULTIPLICITY OF THE ROOT";
210   INPUT M
215   REM IMPROVE APPROXIMATION
220   IT=IT+1
230   IF FNB(X0)=0 THEN PRINT "ROOT=" ;X1;"ITERATIONS= ";IT: GOTO 290
240   D=M* FNA(X0)/ FNB(X0)
250   IF ABS(D)<EPS THEN PRINT "ROOT=" ;X1;"ITERATIONS= ";IT: GOTO 290
260   X1=X0-D
270   X0=X1
280   GOTO 220
290   END
```

Program 2.9.2.

2.10 NUMERICAL PROBLEMS

The following major problems can occur in the solution of non-linear equations.

> (1) Finding good initial approximations.
> (2) Ill-conditioned functions.
> (3) Deciding on the most suitable convergence criteria.
> (4) Discontinuities in the equation to be solved.
> (5) Finding all the solutions of a given equation.

We now consider each of these in detail.

(1)   Finding a good initial approximation can be a major difficulty if the function is complicated and little is known of its behaviour. Simple techniques that may be tried are to take sample values of the independent variable until a change in the sign occurs in the calculated values of the function; this gives us an interval in which the root lies. For simpler functions the traditional sketch graph can be useful in approximately locating the root. The micro-computer can be a considerable help in this area since many systems allow the use of plotting packages to display specified functions over given ranges.

(2) Ill-conditioning in a non-linear equation means that if the coefficients which
     define an equation vary slightly then the location of the roots of the equation
     change by a far greater amount. In the case of an ill-conditioned equation
     we must ensure the coefficients are accurate to many more significant figures
     than we require for the accuracy of the solution. The classic example of an
     ill-conditioned equation is given by Wilkinson (1963) as the polynomial
     equation

$$f(x) = (x + 1)(x + 2)(x + 3) \dots (x + 20) = 0$$

The roots of this polynomial are obviously the integers -1 to -20. If however
we add a small increment to the coefficient of x^{19} such that

$$f(x) + 2^{-23}x^{19} = 0$$

this changes the coefficient of x^{19} from 210 to 210.0000001 approximately.
This minute change in this coefficient alters the values of the roots dramatically. For example, working to nine decimal places, the roots -18 and -19
in the original equation become the complex pair

$$-19.502439400 + 1.940330347i$$
and $\quad -19.502439400 - 1.940330347i.$

The solution of ill-conditioned equations can present major difficulties for
those microcomputers which provide relatively low accuracy for a computation.

(3) Deciding on a convergence criteria for the solution of a non-linear equation
     appears deceptively simple. If we decide to rely on checking that successive
     iterates are converging, this leads to consideration of the quantity

$$|x_{r+1} - x_r| \text{ for } r = 0, 1, 2, \dots$$

In this case our stopping criteria could be to finish when

$$|x_{r+1} - x_r| < \varepsilon$$

where ε is some small pre-set value. In certain cases this criteria may be
misleading, particularly in the case of functions which vary rapidly for small
changes in the independent variable. In this case we should monitor both the
closeness of successive iterates and the function values.

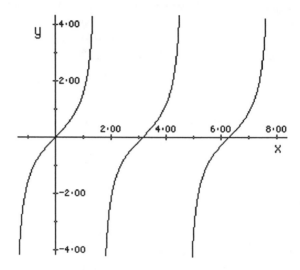

Fig. 2.10.1. Graph of y = tan x showing discontinuities.

(4) Discontinuities. The graph of tan x, given in Fig. 2.10.1, shows that the function tan x has many discontinuities. A typical discontinuity lies close to $3\pi/2$ and small variations in x around this value result in violent changes in the value of tan x as is illustrated in Table 2.10.1. Finding the solution of equations which involve discontinuities presents problems for most root finding procedures, particularly if the root lies close to a discontinuity. Examples of equations which involve discontinuities are $x = \tan x$ and $\tanh x = \tan x$.

Table 2.10.1. Values of tan x close to a discontinuity.

x value	tanx
4.68	30.86374
4.685	36.50183
4.69	44.65748
4.695	57.50245
4.7	80.71463
4.705	135.3416
4.7100	418.6781
4.7150	-382.899
4.7200	-131.3731
4.7250	-79.28613
4.7300	-56.77362

(5)   Finding all the roots of a given non-linear equation is an extremely difficult
      problem and in the case where there are an infinity of roots it is clearly
      impossible. In the case of transcendental functions, methods have been
      suggested but they are not guaranteed to converge to all roots. However, for
      polynomial equations a number of efficient techniques are available to
      find all roots and we will discuss some of these methods in later sections.

In the next section we shall describe a powerful method designed to overcome
some of the difficulties listed above.

2.11 THE METHOD OF BRENT

Some problems may present particular difficulties for algorithms which in general
work well. For example, algorithms which have fast ultimate convergence may
initially diverge. One way to improve the reliability of an algorithm is to ensure
that at each stage the root is confined to a known interval and the method of bisection
may be used to provide an interval in which the root lies. Thus a method which
combines bisection with a rapidly convergent procedure may be able to provide both
rapid *and* reliable convergence.

The method of Brent combines inverse quadratic interpolation with bisection
to provide a powerful method that has been found to be successful on a wide range
of difficult problems. The method is easily implemented on the microcomputer and
the user has only to supply the equation to be solved together with an initial interval
in which the root lies. The detailed description of the algorithm may be found in
Brent, (1971). It can be seen from the program that the algorithm is not particularly
complex. Similar algorithms of comparable efficiency have been developed by
Dekker (1969). Program 2.11.1 implements Brent's method.

```
100   REM ------------------------------------------------
110   REM BRENT'S METHOD FOR SOLVING EQUATIONS GL/JP (C)
120   REM ------------------------------------------------
130   PRINT "SOLUTION OF EQUATION BY BRENT'S METHOD"
135   REM DEFINE FUNCTION AT STATEMENT 140
140   DEF FNA(X)=2*COS(X)-EXP(X)
150   PRINT "INPUT TWO INITIAL APPROXIMATIONS ENCLOSING ROOT";
160   INPUT X0,X1
170   PRINT "INPUT ACCURACY";
180   INPUT EPS
190   F0= FNA(X0) : F1 = FNA(X1)
200   IF F0*F1>0 THEN PRINT " NO ROOT IN INTERVAL  " : GOTO 150
210   X2=X0:F2=F0:D=X1-X0:E=D
220   IF ABS(F2)<ABS(F1)  THEN X0=X1:X1=X2 :X2=X0:F0=F1:F1=F2:F2=F0
```

Program 2.11.1 (continues).

```
230   IT=IT+1
235   REM BISECTION STAGE
240   M=(X2-X1)/2
250   IF ABS(M) < EPS OR F1 =0 THEN 440
260   IF ABS(E) <EPS OR ABS(F0) <= ABS(F1) THEN D=M:E=M : GOTO 370
270   S=F1/F0
280   IF X0=X2 THEN N1=(X2-X1)*S:D1=1-S:  GOTO 320
285   REM QUADRATIC INTERPOLATION STAGE
290   D1=F0/F2:R=F1/F2
300   N1=S*(2*M*D1*(D1-R)-(X1-X0)*(R-1))
310   D1=(D1-1)*(R-1)*(S-1)
320   IF N1>0 THEN D1=-D1: GOTO 340
330   N1=-N1
340   S=E:E=D
350   IF 2*N1<3*M*D1-ABS(EPS*D1) AND N1<ABS(S*D1/2) THEN D=N1/D1:GOTO 370
360   D=M:E=M
370   X0=X1:F0=F1
380   IF ABS (D) > EPS THEN X1=X1+D : GOTO 400
390   X1=X1+((M<0)-(M>0))*EPS
400   F1 = FNA(X1)
410   IF F1>0  AND F2>0 THEN 210
420   IF F1<= 0 AND F2 <=0 THEN 210
430   GOTO 220
440   PRINT "SOLUTION AFTER";IT;" ITERATIONS=";X1
450   END
```

Program 2.11.1.

2.12 COMPARATIVE STUDIES

Table 2.12.1 provides a comparison of the methods of Newton, bisection, secant and Brent. The methods are tested on the following problems:

$$(1) \quad x^2 - (x-1)^5 = 0$$
$$(2) \quad x - e^{-x} = 0$$
$$(3) \quad \cos((x^2 + 5)/(x^4 + 1)) = 0$$
$$(4) \quad x^3 - 10x^2 + 29x - 20 = 0$$
$$(5) \quad x - \tan x = 0$$
$$(6) \quad (x-1)^5 = 0$$
$$(7) \quad 2\cos x - e^x = 0$$

The entries in the table are the number of function evaluations required to obtain a specified accuracy which was the same for each problem. It should be noted that the Newton method alone requires the user to provide both the function and its derivative. The results supply some evidence to indicate that the best method is Brent's method. The secant and bisection methods perform quite well but require more computation to find the root.

The methods we have considered enable us to find a particular solution of the

Table 2.12.1. Comparisons of the number of function evaluations required to find the root. The number of iterations is shown in brackets when it differs from the number of function evaluations. The starting point(s) are given. Convergence is achieved when the function value is reduced to less than 0.00005. Fail1 indicates failure to converge, Fail2 indicates convergence to a root outside the interval.

Meth Func.	Newton	Bisection	Secant	Brent	Start. Point	End point
1	28(14)	19	FAIL2	12(11)	1	5
2	8(4)	17	6	6(5)	0	5
3	10(5)	15	4	7(6)	-0.5	-1
4	12(6)	19	FAIL2	10(9)	-1.5	3
5	Fail1	19	FAIL2	9(8)	3	4.7
6	98(49)	Low accur. 7	63	45(44)	-10	10
7	8(4)	14	5	6(5)	0.4	1

general non-linear equation. In the next sections we will consider methods specifically designed to find all the roots of polynomial equations.

2.13 BAIRSTOW'S METHOD

Assuming an initial approximation to the root of an equation is available, the methods we have considered above enable us to find a more accurate value of the root. However, there are equations which have many roots and we may wish to find not one but all of them. An obvious example of an equation with many roots is

$$a_0 x^n + a_1 x^{n-1} + a_2 x^{n-2} + \ldots + a_n = 0 \tag{2.13.1}$$

This is a polynomial equation of degree n and it has n roots. A common approach for locating the roots of a polynomial is to find all its quadratic factors. These will have the form

$$x^2 + ux + v$$

where u and v are the constants we wish to determine. Once the quadratic factors are found it is easy to solve the quadratics to find the roots we require.

To develop the equations to find u and v we consider the following equality

$$(x^2 + ux + v)(b_0 x^{n-2} + b_1 x^{n-3} + b_2 x^{n-4} + \dots + b_{n-2}) + R(x)$$
$$= x^n + a_1 x^{n-1} + a_2 x^{n-2} + \dots + a_n \tag{2.13.2}$$

where a_0 has been taken as one and $R(x)$, the remainder after the division of polynomial (2.13.1) by the quadratic factor, has the form

$$R(x) = (x + u)b_{n-1} + b_n \tag{2.13.3}$$

The notation b_{n-1} and b_n for the constants in the above equation has been chosen for convenience in the development of the algorithm. On equating coefficients of both sides of (2.13.2) we obtain the following system of equations

$$
\left.
\begin{aligned}
b_0 &= 1 \\
b_1 &= a_1 - u \\
b_2 &= a_2 - b_1 u - v \\
&\dots\dots\dots\dots\dots\dots \\
&\dots\dots\dots\dots\dots\dots \\
b_k &= a_k - b_{k-1}u - b_{k-2}v \\
&\dots\dots\dots\dots\dots\dots \\
&\dots\dots\dots\dots\dots\dots \\
b_n &= a_n - b_{n-1}u - b_{n-2}v
\end{aligned}
\right\} \tag{2.13.4}
$$

Clearly, if we require $x^2 + ux + v$ to be an exact quadratic factor then the remainder $R(x)$ should be zero. Therefore we require that

$$b_{n-1} = b_n = 0$$

Now from (2.13.4) we know that b_{n-1} and b_n are functions of u and v. Consequently we wish to find u and v such that

$$b_{n-1}(u,v) = 0$$
$$b_n(u,v) = 0 \tag{2.13.5}$$

Assuming some initial approximation to u and v we require improved values $u + du$ and $v + dv$ which drive (2.13.5) closer to zero. By using Taylor's series we have

$$b_{n-1}(u+du,v+dv) = 0 = b_{n-1}(u,v)+\{\partial b_{n-1}/\partial u\}du +\{\partial b_{n-1}/\partial v\}dv + \dots$$

$$b_n(u+du,v+dv) = 0 = b_n(u,v)+\{\partial b_n/\partial u\}du +\{\partial b_n/\partial v\}dv + \dots$$

$$(2.13.6)$$

If we assume that we can neglect terms containing higher powers of du and dv then these equations are linear in du and dv and can be written in the form

$$\begin{aligned} b_{n-1} &= c_{n-2}du + c_{n-3}dv \\ b_n &= c_{n-1}du + c_{n-2}dv \end{aligned}$$

$$(2.13.7)$$

Having found du and dv we can use the improved values of u and v in (2.13.6) to calculate new coefficients $c_{n-1}, c_{n-2}, c_{n-3}, b_{n-1}$ and b_n and the process is repeated until the values of b_{n-1} and b_n become sufficiently small. However, before we can do this a major problem still needs to be resolved. How do we find the coefficients $c_{n-1}, c_{n-2}, c_{n-3}, b_{n-1}$ and b_n in terms of u and v?

This is achieved as follows. Differentiating the general equation of (2.13.4) with respect to u we have

$$\partial b_k/\partial u = -\partial(b_{k-1}u)/\partial u - \partial(b_{k-2}v)/\partial u$$

$$= -\{u\partial b_{k-1}/\partial u + b_{k-1}\}- v\partial b_{k-2}/\partial u$$

$$= -b_{k-1}- u\partial b_{k-1}/\partial u - v\partial b_{k-2}/\partial u$$

Denoting $\partial b_k/\partial u$ by c_k for all k we have

$$c_k = -b_{k-1}- uc_{k-1}- vc_{k-2} \text{ where } c_{-2} = c_{-1} = 0$$

Similarly, differentiating with respect to v and denoting $\partial b_{k+1}/\partial v$ by d_k we have

$$d_{k-1} = -b_{k-1}- ud_{k-2}- vd_{k-3}$$

These are essentially the same recurrence formulae and so d_k and c_k are equal and this leads to the following recurrence formula for c_k

$$c_k = -b_k - c_{k-1}u - c_{k-2}v \qquad k = 0, 1, 2, \dots, n-1$$

where $c_{-2} = c_{-1} = 0$. We obtain the required derivatives from

$$\partial b_{n-1}/\partial u = c_{n-2}; \ \partial b_n/\partial u = c_{n-1}; \ \partial b_{n-1}/\partial v = c_{n-3}; \ \partial b_n/\partial v = c_{n-2}$$

Having found a quadratic factor by applying this process iteratively until the required accuracy is obtained we can repeat this process with the residual polynomial determined by the b_k coefficients until all the quadratic factors have been found.

Tables 2.13.1 and 2.13.2 give the results of Bairstow's method applied to solve two polynomials. Program 2.13.1 implements this method. In this program the initial values of u and v are arbitrarily set to one.

Table 2.13.1. Solution of $x^4-6x^3+11x^2+2x-28 = 0$ using Bairstow's method.

```
BAIRSTOW'S METHOD SOLVING FOR POLYNOMIALS

ENTER THE DEGREE OF THE POLYNOMIAL?  4
ENTER COEFFICIENTS OF  POLYNOMIAL
IT IS ASSUMED THE HIGHEST DEGREE COEFFICIENT IS 1
ENTER COEFFICIENT OF X^ 3  ? -6
ENTER COEFFICIENT OF X^ 2  ?  11
ENTER COEFFICIENT OF X^ 1  ?  2
ENTER COEFFICIENT OF X^ 0  ? -28
ENTER REQUIRED ACCURACY?  .00005

   ROOT        REAL PART           IMAGINARY PART
    4          3.236068             0
    3         -1.236068             0
    2          2                    1.732051
    1          2                   -1.732051
```

Table 2.13.2. Solution $x^5-3x^4-10x^3+10x^2+44x+48 = 0$ using Bairstow's method.

```
BAIRSTOW'S METHOD SOLVING FOR POLYNOMIALS

ENTER THE DEGREE OF THE POLYNOMIAL?  5
ENTER COEFFICIENTS OF  POLYNOMIAL
IT IS ASSUMED THE HIGHEST DEGREE COEFFICIENT IS 1
ENTER COEFFICIENT OF X^ 4  ? -3
ENTER COEFFICIENT OF X^ 3  ? -10
ENTER COEFFICIENT OF X^ 2  ?  10
ENTER COEFFICIENT OF X^ 1  ?  44
ENTER COEFFICIENT OF X^ 0  ?  48
ENTER REQUIRED ACCURACY?  .00005

   ROOT        REAL PART           IMAGINARY PART
    5          3                    0
    4         -2                    0
    3         -1                    1
    2         -1                   -1
    1          4                    0
```

```
100    REM --------------------------------------------
110    REM BAIRSTOWS METHOD FOR POLYNOMIALS GL/JP (C)
120    REM --------------------------------------------
130    DIM A(50),B(50),C(50)
140    PRINT "BAIRSTOW'S METHOD SOLVING FOR POLYNOMIALS":PRINT
150    PRINT "ENTER THE DEGREE OF THE POLYNOMIAL";
160    INPUT N
170    PRINT "ENTER COEFFICIENTS OF  POLYNOMIAL "
180    PRINT "IT IS ASSUMED THE HIGHEST DEGREE COEFFICIENT IS 1"
190    FOR I=1 TO N
200        PRINT "ENTER COEFFICIENT OF X^"; N-I;
210        INPUT A(I)
220    NEXT I
230    PRINT "ENTER REQUIRED ACCURACY";
240    INPUT EPS:PRINT
250    PRINT TAB(3);"ROOT";TAB(12);"REAL PART";TAB(30);"IMAGINARY PART"
260    TX=0
265    REM SOLVE QUADRATIC
270    IF N <=2 THEN GOSUB 630 : GOTO 890
275    REM FIND QUADRATIC FACTOR
280    U0=1:V0=1
290    U=U0:V=V0
300    IT=1
310    B(1)=A(1)-U
320    B(2)=A(2)-B(1)*U-V
330    FOR K=3 TO N
340        B(K)=A(K)-B(K-1)*U-B(K-2)*V
350    NEXT K
360    C(1)=B(1)-U
370    C(2)=B(2)-C(1)*U-V
380    FOR K=3 TO N-1
390        C(K)=B(K)-C(K-1)*U-C(K-2)*V
400    NEXT K
405    REM CALCULATE CHANGE IN U AND V
410    C1=C(N-1):B1=B(N)
420    CB=C(N-1)*B(N-1)
430    C2=C(N-2)*C(N-2)
440    BC=B(N-1)*C(N-2)
450    IF N>3 THEN C1=C1*C(N-3):B1=B1*C(N-3)
460    DN=C1-C2
470    IF DN=0 THEN PRINT " DIVISION BY ZERO" : STOP
480    DU=(B1-BC)/DN
490    DV=(CB-C(N-2)*B(N))/DN
500    U=U+DU : V=V+DV
510    ST= SQR(DU*DU+DV*DV)
520    IF ST < EPS THEN 550
530    IF IT>50 THEN PRINT "CONVERGENCE IS SLOW"
540    IT=IT+1 : GOTO 310
550    TX=1
560    GOSUB 630
570    FOR I=1 TO N
580        A(I)=B(I)
590    NEXT I
```

Program 2.13.1. (continues)

```
600   IF N<3 THEN GOSUB 630
610   IF N>2 THEN 280
620   END
625   REM SUBROUTINE FOR QUADRATICS
630   IF N<= 0 THEN  890
640   IF N> 2 AND TX=0 THEN 880
650   IF TX=1 AND N>2 THEN 720
660   ON N+1 GOTO 890,670,700
670   R=-A(1) :IM=0:IT=1
680   GOSUB 900
690   RETURN
700   U=A(1):V=A(2)
710   IT=1
720   D=U*U-4*V
730   S=SGN(D)
740   ON S+2 GOTO 750,860,810
750   D=-D
760   IM=SQR(D)/2 :R=-U/2
770   GOSUB 900
780   IM=-IM
790   GOSUB 900
800   RETURN
810   R=(-U+SQR(D))/2 : IM=0
820   GOSUB 900
830   R=(-U-SQR(D))/2 : IM=0
840   GOSUB 900
850   RETURN
860   R=-U:IM=0
870   GOSUB 900
880   RETURN
890   END
895   REM OUTPUT SUBROUTINE
900   PRINT TAB(5);N; TAB(12);R; TAB(30); IM
910   N=N-1
920   RETURN
```

Program 2.13.1

2.14 THE QUOTIENT-DIFFERENCE ALGORITHM

The importance of the quotient-difference (QD) algorithm lies in its ability to provide approximations to all the roots of a given polynomial simultaneously. The major draw back is that the convergence is slow and for this reason the method is often used to find good starting approximations for techniques such as Bairstow's method. The QD algorithm itself does not require any starting approximations.

Given the polynomial

$$a_0 x^n + a_1 x^{n-1} + a_2 x^{n-2} + ... + a_n = 0 \qquad (2.14.1)$$

The QD method is based on working with the difference equation related to the polynomial (2.14.1). This has the form

$$a_0 x_s + a_1 x_{s-1} + a_2 x_{s-2} + \dots + a_{s-n} = 0$$

From this equation a sequence of values x_s can be generated using

$$x_s = -(a_1 x_{s-1} + a_2 x_{s-2} + \dots + a_{s-n})/a_0$$

The quotients $q_s = x_{s+1}/x_s$ can then be calculated. It can be shown that the sequence q_s tends to a root of the equation. The initial values are $x_0 = 1$ and $x_s = 0$ for $s = -1, -2, -3, \dots, -n$.

The method can be illustrated for the polynomial equation

$$x^2 - 6x + 5 = 0$$

The corresponding difference equation is

$$x_s = 6x_{s-1} - 5x_{s-2} \text{ where } x_0 = 1 \text{ and } x_{-1} = 0.$$

Taking the values $s = 1, 2, \dots$ we generate the sequence $x_1 = 6$, $x_2 = 31$, $x_3 = 156$, $x_4 = 781$, $x_5 = 3906$, etc. Consequently the sequence of values for q_s is: 6, 5.167, 5.0323, 5.0064, etc. This sequence is clearly converging to the dominant root of the equation at $x = 5$. The QD method is a generalisation of this technique to find all the roots of the equation.

The steps required to implement the algorithm are given but the theoretical justification of the method is complex and is not provided. The QD scheme of calculations can be generated either by row or by column. The method of generation by row has the important advantage that the scheme is stable, that is the errors do not build up without bound. The calculations for the QD algorithm are performed by using the following simple recurrence relations

$$q^{(k)}_{j+1} = d^{(k)}_j - d^{(k-1)}_{j+1} + q^{(k)}_j$$

$$d^{(k)}_{j+1} = [q^{(k+1)}_j / q^{(k)}_{j+1}] d^{(k)}_j$$

(2.14.2)

for $k = 1, 2, \dots, n$ and $j = 1, 2, 3$, etc. Where $q^{(k)}_{j+1}$ gives the $j + 1$th approximation to the kth root. The value of j is increased until convergence is achieved to the required accuracy. The initial two rows of the table are defined by the relations

$$q^{(1)}_0 = -a_1/a_0 \text{ and } q^{(k)}_{1-k} = 0 \text{ for } k = 2, 3, \dots, n$$

$$d^{(k)}_{1-k} = a_{k+1}/a_k \text{ for } k = 1, 2, \dots, n-1$$

Table 2.14.1.

	$-a_1/a_0$	0	0	0
0	a_2/a_1	a_3/a_2	a_4/a_3	
	$q^{(1)}{}_1$	$q^{(2)}{}_0$	$q^{(3)}{}_{-1}$	$q^{(4)}{}_{-2}$
0	$d^{(1)}{}_1$	$d^{(2)}{}_0$	$d^{(3)}{}_{-1}$	
	$q^{(1)}{}_2$	$q^{(2)}{}_1$	$q^{(3)}{}_0$	$q^{(4)}{}_{-1}$
0	$d^{(1)}{}_2$	$d^{(2)}{}_1$	$d^{(3)}{}_0$	

In addition

$$d^{(0)}{}_j = d^{(n)}{}_j = 0 \quad \text{for all } j \text{ values.}$$

Table 2.14.1 gives the layout of calculations usually adopted for hand computation. It should be noted that this algorithm requires that the coefficients of the original polynomial should be non zero otherwise a division by zero will occur. This difficulty can be avoided by making a suitable substitution in the polynomial of the form $y = x - c$ where c is some constant. It can be shown that for real roots each of the $d^{(k)}$ values will tend to zero while the $q^{(k)}$ values will tend to the roots of the equation. This is illustrated in Table 2.14.2. If the values of the $d^{(k)}$ do not tend to zero this indicates that some of the roots are complex. The quadratic factors corresponding to these pairs of complex conjugate roots may be found by using the relations

$$a^{(k)}{}_j = q^{(k+1)}{}_{j+1} + q^{(k+2)}{}_j$$

$$b^{(k)}{}_j = q^{(k+1)}{}_j \, q^{(k+2)}{}_j$$

The quadratic factors are then given by

$$x^2 + a^{(k)}{}_p x + b^{(k)}{}_p$$

where p was the value of j when sufficient accuracy was reached. Program 2.14.1 implements this method and Table 2.14.2 shows the output from the program with a specified polynomial. Table 2.14.3 illustrates the QD method applied to a polynomial with complex roots.

```
100   REM --------------------------------------------------------
110   REM QUOTIENT DIFFERENCE METHOD FOR POLYNOMIALS GL/JP (C)
120   REM --------------------------------------------------------
130   PRINT "QUOTIENT DIFFERENCE ALGORITHM FOR SOLVING POLYNOMIALS "
135   REM N.B. SUCCESSIVE COEFFICIENTS SHOULD BE DIFFERENT
136   REM CONVERGENCE MAY BE SLOW
140   PRINT "ENTER THE DEGREE OF THE POLYNOMIAL";
150   INPUT N
160   DIM A(N),E(N),Q(N),E1(N),Q1(N+1),X(N),SU(N),PR(N),R(N)
170   PRINT "ENTER COEFFICIENTS OF POLYNOMIAL"
180   FOR I = 0 TO N
190       PRINT "ENTER COEFFICIENT OF X^";N-I;
200       INPUT A(I)
210   NEXT I
220   PRINT "ENTER THE ACCURACY YOU REQUIRE";
230   INPUT EPS
240   PRINT
250   FOR I=1 TO N
260       PRINT "  Q(";I;")";TAB(12*I)
270   NEXT I
280   PRINT
285   REM COMPUTE Q AND E VALUES
290   Q(1)=-A(1)/A(0)
300   FOR J=2 TO N
310       Q(J)=0
320   NEXT J
330   FOR J=1 TO N-1
340       E(J)=A(J+1)/A(J)
350   NEXT J
360   E(0)=0: K=0
370   FOR I=1 TO N
380       Q1(I)=E(I)-E(I-1)+Q(I)
390       PRINT Q1(I); TAB(12*I);
400   NEXT I
410   PRINT
420   FOR I=1 TO N
430       E(I)=Q1(I+1)/Q1(I)*E(I)
440   NEXT I
450   S=0
460   FOR I=1 TO N
470       S=S+(Q(I)-Q1(I))^2
480       IF ABS(Q1(I)-Q(I))<EPS THEN R(I)=1
490   NEXT I
495   REM TEST FOR CONVERGENCE
500   IF  SQR(S)<EPS OR K>40 THEN 550
510   FOR I=1 TO N
520       Q(I)=Q1(I)
530   NEXT I
540   K=K+1: GOTO 370
550   S1=0
560   FOR I=1 TO N
570       S1=S1+E(I)*E(I)
```

Program 2.14.1. (continues)

```
580   NEXT I
585   REM OUTPUT REAL ROOTS
590   FOR I=1 TO N
600       IF R(I)=1 THEN PRINT "X(";I;")=";Q1(I)
610   NEXT I
615   REM CHECK FOR QUADRATIC FACTORS
620   IF SQR(S1)<.01 THEN 730
630   PRINT : PRINT "QUADRATIC FACTORS ARE: "
640   FOR I=1 TO N
650       A$="":B$=""
660       IF ABS(E(I))<.01 THEN 720
670       SU(I)=Q(I)+Q(I+1)
680       PR(I)=Q(I)*Q1(I+1)
690       IF -SU(I)>0 THEN A$="+"
700       IF PR(I)>0 THEN B$="+"
710       PRINT "X*X";A$;-SU(I);"*X ";B$;PR(I)
720   NEXT I
730   END
```

Program 2.14.1

Table 2.14.2. Solution of $x^3 - 2x^2 - 5x + 6 = 0$ using the QD method.
This polynomial has real roots.

```
QUOTIENT DIFFERENCE ALGORITHM FOR SOLVING POLYNOMIALS
ENTER THE DEGREE OF THE POLYNOMIAL? 3
ENTER COEFFICIENTS OF POLYNOMIAL
ENTER COEFFICIENT OF X^ 3 ? 1
ENTER COEFFICIENT OF X^ 2 ? -2
ENTER COEFFICIENT OF X^ 1 ? -5
ENTER COEFFICIENT OF X^ 0 ? 6
ENTER THE ACCURACY YOU REQUIRE? .00005
```

Q(1)	Q(2)	Q(3)
4.5	-3.7	1.2
2.444444	-1.255255	.8108108
3.5	-2.562201	1.062201
2.707~~~		
2.999554	-1.999554	.9999995
3.000297	-2.000297	1
2.999802	-1.999802	.9999999
3.000132	-2.000132	1
2.999912	-1.999912	1
3.000058	-2.000059	1
2.99996	-1.999961	1
3.000026	-2.000026	1
2.999982	-1.999982	1
3.000011	-2.000011	1

```
X( 1 )= 3.000011
X( 2 )=-2.000011
X( 3 )= 1
```

Table 2.14.3. Solution of $x^3 + 3x^2 + 4x + 2 = 0$ using the QD method. The quadratic factor gives complex roots and the real root is -1.

```
QUOTIENT DIFFERENCE ALGORITHM FOR SOLVING POLYNOMIALS
ENTER THE DEGREE OF THE POLYNOMIAL? 3
ENTER COEFFICIENTS OF POLYNOMIAL
ENTER COEFFICIENT OF X^ 3 ? 1
ENTER COEFFICIENT OF X^ 2 ? 3
ENTER COEFFICIENT OF X^ 1 ? 4
ENTER COEFFICIENT OF X^ 0 ? 2
ENTER THE ACCURACY YOU REQUIRE? .00005


   Q( 1 )         Q( 2 )          Q( 3 )
 -1.666667      -.8333334       -.5
 -.9999999      -1.2            -.8
 -.1999996      -1.800001       .........

 .........      -1.045621       -.999992
 .1060777       -2.106061       -.9999956
 -20.94791      18.94793        -.9999973
 -1.904032      -9.594917E-02
                                -.9999972
 -.9443604      -1.055621       -.9999962
 .1283757       -2.128358       -.9999954
 -17.65666      15.65668        -.9999949
 -1.886147      -.1138353       -.9999949
 -.934344       -1.065639       -.9999952
 X( 3 )=-.9999952

 QUADRATIC FACTORS ARE:
 X*X+ 1.999983 *X + 2.009951
```

2.15 LAGUERRE'S METHOD

Laguerre's method provides a rapidly convergent procedure for locating the roots of a polynomial. The method is most successful when applied to finding the roots of a polynomial where they are real and distinct. The convergence of the method is significantly slower when attempting to locate a multiple root.

The method takes the following form when applied to the polynomial

$$p(x) = x^n + a_1 x^{n-1} + a_2 x^{n-2} + \dots + a_n$$

Starting with an initial approximation x_1 we apply the following iterative formula to the polynomial $p(x)$.

$$x_{i+1} = x_i - np(x_i)/[p'(x_i) \pm \sqrt{\{h(x_i)\}}] \tag{2.15.1}$$

where

$$h(x_i) = (n-1)[(n-1)\{p'(x_i)\}^2 - np(x_i)\,p''(x_i)],$$

n is the degree of the polynomial and $i = 1, 2, 3, ...$ The sign taken in (2.15.1) is determined so that it is the same as the sign of $p'(x_i)$.

Given an initial approximation the method will converge to a root of the polynomial which we can denote by r. To obtain the other roots of the polynomial we divide the polynomial $p(x)$ by the factor $(x - r)$ which provides another polynomial of degree $n - 1$ and we can now apply iteration (2.15.1) to this polynomial and repeat the whole procedure again. The process is repeated until all roots are found to the required accuracy. The process of dividing by the factor $(x - r)$ is known as deflation and can be performed in a simple and efficient way which is described below.

Since we have a known factor $(x - r)$ then

$$a_0 x^n + a_1 x^{n-1} + a_2 x^{n-2} + ... + a_n$$

$$= (x - r)(b_0 x^{n-1} + b_1 x^{n-2} + b_2 x^{n-3} + ... + b_{n-1}) \qquad (2.15.3)$$

On equating coefficients of the powers of x on both sides we have

$$b_0 = a_0$$

$$b_i = a_i + rb_{i-1} \qquad \text{for } i = 1, 2, ... , n - 1 \qquad (2.15.4)$$

This process is known as synthetic division. Care must be taken here, particularly if the root is found to low accuracy since ill-conditioning can magnify the effect of small errors in the coefficients of the deflated polynomial. Again the low accuracy of some microcomputers may present problems.

This completes the description of the method but a few important points should be noted. Assuming sufficient accuracy can be maintained in calculations the method of Laguerre will converge for any value of the initial approximation. Convergence to complex roots and multiple roots can be achieved but at a slower rate. In the case of a complex root the value of the function $h(x_i)$ becomes negative and consequently the algorithm must be adjusted to deal with this case. A key feature that should be considered is that the derivatives of the polynomial can be found efficiently by repeated synthetic division. The program reflects this fact since the user is not required to provide any derivatives only the coefficients of the polynomial to be solved. Table 2.15.1 gives the results of the method applied to a specific polynomial. Program 2.15.1 implements this method for *real roots only*.

Table 2.15.1. Results of solving $x^3 - 2x^2 - 5x + 6 = 0$ using Laguerre's method.

```
LAGUERRE'S METHOD FOR SOLVING POLYNOMIALS
FINDS REAL ROOTS ONLY

ENTER DEGREE OF POLYNOMIAL (>2)?  3
ENTER COEFFICIENTS OF POLYNOMIAL
ENTER COEFFICIENT OF X^ 3 ?  1
ENTER COEFFICIENT OF X^ 2 ?  -2
ENTER COEFFICIENT OF X^ 1 ?  -5
ENTER COEFFICIENT OF X^ 0 ?  6
ENTER THE ACCURACY YOU REQUIRE?  .00005
ENTER THE INITIAL APPROXIMATION?  10
ROOTS OF EQUATION ARE :
  3    1   -2
NUMBER OF ITERATIONS =   4
```

```
100   REM --------------------------------------------
110   REM    LAGUERRE'S METHOD  POLYNOMIALS GL/JP (C)
120   REM --------------------------------------------
130   PRINT "LAGUERRE'S METHOD FOR SOLVING POLYNOMIALS"
140   PRINT "FINDS REAL ROOTS ONLY" : PRINT
150   PRINT "ENTER DEGREE OF POLYNOMIAL (>2)";
160   INPUT N
170   IF N<=2 THEN 150
180   DIM A(N),P(3),B(N),S(N)
190   PRINT "ENTER COEFFICIENTS OF POLYNOMIAL"
200   FOR I =0 TO N
210      PRINT "ENTER COEFFICIENT OF X^"; N-I;
220      INPUT A(I)
230   NEXT I
240   PRINT "ENTER THE ACCURACY YOU REQUIRE";
250   INPUT EPS
260   PRINT "ENTER THE INITIAL APPROXIMATION";
270   INPUT X
280   NR=N-1: IT=0: S=1
290   FOR R=1 TO NR
300      FOR I=1 TO N
310         B(I)=A(I)
320      NEXT I
330      N1=N-1:N2=N
340      FOR J=0 TO 2
350         FOR I=1 TO N2
360            A(I)=A(I)+X*A(I-1)
370         NEXT I
380         P(J)=A(N2)
390         N2=N2-1
400      NEXT J
410      P(2)=2*P(2)
```

Program 2.15.1. (continues).

```
420        FOR I=1 TO N
430            A(I)=B(I)
440        NEXT I
450        IF ABS(P(0))<1E-10 THEN S(R)=X1 :X=X1: GOTO 540
460        IF P(1)<0 THEN S=-1
465        REM NOW USING LAGUERRE'S TO FORMULA FIND ROOT
470        IT=IT+1
480        H=N1*(N1*P(1)*P(1)-N*P(0)*P(2))
490        IF ABS(H)<.0000005 THEN H=0
500        IF H<0 THEN 660
510        X1=X-N*P(0)/(P(1)+S*SQR(H))
520        IF ABS(X1-X)>EPS THEN X=X1 : GOTO 330
530        S(R)=X1
535        REM NOW DEFLATE AND RETURN TO FIND NEXT ROOT
540        FOR I=1 TO N1
550            A(I)=A(I)+X1*A(I-1)
560        NEXT I
570        N=N-1
580    NEXT R
590    S(NR+1)=-A(1)/A(0)
600    PRINT "ROOTS OF EQUATION ARE : "
610    FOR I=1 TO NR+1
620        PRINT S(I);"  ";
630    NEXT I
640    PRINT : PRINT "NUMBER OF ITERATIONS = ";IT
650    GOTO 670
660    PRINT "ROOT IS COMPLEX"
670    END
```

Program 2.15.1

2.16 MOORE'S METHOD

This method may be applied to polynomials having real or complex coefficients. The basis of the method is to express the polynomial in terms of its real and complex parts as follows

$$f(z) = u + iv \qquad (2.16.1)$$

where u is the real part and v the imaginary part of the polynomial. The polynomial is defined as

$$f(z) = (a_0 + ib_0) + (a_1 + ib_1)z + ... + (a_n + ib_n)z^n$$

where z is given by

$$z = x + iy \qquad (2.16.2)$$

It can be shown that the for a specified polynomial the values of u and v can be calculated by using functions known as Siljak functions. Each of the powers of z can be defined in terms of these functions X_r, Y_r as follows

$$z^r = X_r + iY_r \quad \text{for } r = 0, 1, 2, \dots, n \qquad (2.16.3)$$

where

$$X_r = X_r(x, y) \text{ and } Y_r = Y_r(x, y)$$

These expressions allow the polynomial $f(z)$ to be separated easily into its real and imaginary parts u and v. The Siljak functions can in turn be calculated from the following recurrence relations

$$X_{r+2} - 2xX_{r+1} + (x^2 + y^2)X_r = 0$$
$$Y_{r+2} - 2xY_{r+1} + (x^2 + y^2)Y_r = 0$$

$$(2.16.4)$$

for $r = 0, 1, 2, \dots, n$. The initial values are taken as

$$X_0 = 1; \ X_1 = x; \ Y_0 = 0; \ Y_1 = y$$

Solving the polynomial can now be viewed as finding the x and y values that make u and v simultaneously zero. Initial values of x and y may be taken 0 and 1 or some other values. This can be achieved by minimising the function $g(x, y)$ defined by

$$g(x, y) = u^2 + v^2 \qquad (2.16.5)$$

The minimum of g will clearly be obtained at zero when both u and v are zero. Moore uses the method of steepest descent to minimise the function g and a description of this method and the details of the algorithm can be found in Moore (1967). Once a root is found the polynomial is deflated and the process repeated. The results of the method applied to a given polynomial are shown in Table 2.16.1 It should be noted the method can be applied to polynomials with real or complex coefficients. Program 2.16.1 implements this method.

Table 2.16.1. Moore's method used to solve $x^4 - 6x^3 + 11x^2 + 2x - 28 = 0$.

```
MOORE'S METHOD FOR SOLVING POLYNOMIALS
ENTER THE DEGREE OF THE POLYNOMIAL?  4
ENTER THE COEFFICIENTS OF THE POLYNOMIAL
ENTER REAL AND IMAG COEFFICIENT OF X^ 4 ?  1,0
ENTER REAL AND IMAG COEFFICIENT OF X^ 3 ?  -6,0
ENTER REAL AND IMAG COEFFICIENT OF X^ 2 ?  11,0
ENTER REAL AND IMAG COEFFICIENT OF X^ 1 ?  2,0
ENTER REAL AND IMAG COEFFICIENT OF X^ 0 ?  -28,0
ENTER ACCURACY?  .00005

SOLUTION
ROOT  1 =-1.236068  -  1.192093E-07 I
ROOT  2 = 2  +  1.732051I
ROOT  3 = 2  -  1.732051I
ROOT  4 = 3.236068  +  1.018634E-09 I
NUMBER OF ITERATIONS =  29
```

```
100    REM ------------------------------------------------
110    REM J.B.MOORE'S METHOD FOR POLYNOMIALS GL/JP (C)
120    REM ------------------------------------------------
130    DIM A(20),B(20),R(20),I(20),XR(20),YR(20),ZR(20),ZI(20)
140    PRINT "MOORE'S METHOD FOR SOLVING POLYNOMIALS"
150    PRINT "ENTER THE DEGREE OF THE POLYNOMIAL";
160    INPUT N
170    PRINT "ENTER THE COEFFICIENTS OF THE POLYNOMIAL"
180    FOR I =0 TO N
190        PRINT "ENTER REAL AND IMAG COEFFICIENT OF X^";N-I;
200        INPUT A(N-I),B(N-I)
210    NEXT I
220    PRINT "ENTER ACCURACY";
230    INPUT EPS
240    N2=N
250    IT=0
260    X=.5:Y=1
270    F=1E+30:Q=0
280    TE=0
290    R(0)=1:I(0)=0
300    R(1)=X:I(1)=Y
305    REM MAIN LOOP
310    IT=IT+1
320    S=2*R(1)
330    T=R(1)*R(1)+I(1)*I(1)
340    N1=N-1
350    IF Q>=50 THEN PRINT "METHOD NOT CONVERGING": STOP
355    REM GENERATE SILJAK'S FUNCTIONS
360    FOR K=1 TO N1
370        R(K+1)=S*R(K)-T*R(K-1)
380        I(K+1)=S*I(K)-T*I(K-1)
390    NEXT K                              Program 2.16.1 (continues).
```

```
395    REM CALCULATE U,V,DU,DV
400    FOR K=0 TO N
410        ZR(K)=A(K)*R(K)-B(K)*I(K)
420        ZI(K)=A(K)*I(K)+B(K)*R(K)
430    NEXT K
440    U=0:V=0
450    DU=0:DV=0
460    FOR K=0 TO N
470        U=U+ZR(K)
480        DU=DU+(K+1)*(A(K+1)*R(K)-B(K+1)*I(K))
490        V=V+ZI(K)
500        DV=DV+(K+1)*(A(K+1)*I(K)+B(K+1)*R(K))
510    NEXT K
520    NF=U*U+V*V
530    IF ABS(NF)< EPS *EPS THEN XR(N)=R(1):YR(N)=I(1): GOTO 620
540    IF NF>F THEN DX=DX/4 :DY=DY/4:Q=Q+1:R(1)=X+DX:I(1)=Y+DY: GOTO 310
545    REM CACULATE INCREMENT IN X AND Y
550    DF=DU*DU+DV*DV
560    DX=(-U*DU-V*DV)/DF
570    DY=(U*DV-V*DU)/DF
580    X=R(1):Y=I(1)
590    R(1)=X+DX:I(1)=Y+DY
600    F=NF
610    Q=0: GOTO 310
615    REM DO SYNTHETIC DIVISION
620    X=R(1):Y=I(1)
630    S1=A(N1):S2=B(N1)
640    A(N1)=A(N):B(N1)=B(N)
650    A(N)=0:B(N)=0
660    FOR K=N-2 TO 0 STEP -1
670        T1=A(K):T2=B(K)
680        A(K)=S1+X*A(K+1)-Y*B(K+1)
690        B(K)=S2+X*B(K+1)+Y*A(K+1)
700        S1=T1:S2=T2
710    NEXT K
720    N=N-1
730    IF N=1 THEN 780
740    FOR I=1 TO N
750        R(I)=0 : I(I)=0
760    NEXT I
770    GOTO 260
780    D=A(1)*A(1)+B(1)*B(1)
790    XR(1)=-(A(0)*A(1)+B(0)*B(1))/D
800    YR(1)=(A(0)*B(1)-B(0)*A(1))/D
810    PRINT: PRINT "SOLUTION"
820    FOR I=1 TO N2
830        S$=" + "
840        IF SGN(YR(I))<0 THEN S$=" - "
850        PRINT "ROOT ";I;"=";XR(I);S$;ABS(YR(I));" I"
860    NEXT I
870    PRINT "NUMBER OF ITERATIONS = ";IT
880    END
```

Program 2.16.1

2.17 PROBLEMS ENCOUNTERED WHEN SOLVING POLYNOMIALS

The type of problems which present difficulties for the methods described in this chapter are those which have equal roots or roots closely clustered in a small interval. In the case of multiple roots convergence is generally slower but, if the accuracy of the given microcomputer permits, the roots can in general be found. In the case of closely clustered roots it may be impossible to find the roots with sufficient accuracy to separate them. The major problem in finding roots to high accuracy for methods that use deflation is that the deflated polynomial coefficients may not be found with adequate accuracy. There will be some error in the calculation of the deflated polynomial since it uses the best approximation to the most recently found root to perform the deflation. These errors may be greatly magnified in subsequent roots if the deflated polynomial is ill-conditioned.

One way of dealing with the problem of decreasing accuracy in roots found using deflation is to use a hybrid approach. We may use a method that does not use deflation, for example the QD method, to find initial approximations and supply these to a more rapidly convergent method such as Bairstow's or Newton's to find the roots we require thus reducing the number of iterations required. It can be shown that to increase accuracy during the deflation process it is important to obtain the roots in order of magnitude starting with the smallest root.

We now consider the following problem which presents difficulties because it has close roots.

$$x^3 - 3.001x^2 + 3.002x - 1.001 = 0 \quad \text{(with roots = 1, 1 and 1.001)}$$

In attempting to solve this equation we find that the Bairstow and Moore methods succeed but unless high precision arthrimetic is used the method of Laguerre fails. An examination of this problem using these methods is left as an exercise for the reader.

2.18 SOLVING SYSTEMS OF NON-LINEAR EQUATIONS

The methods considered so far have been concerned with finding one or all the roots of a non-linear algebraic equation with one independent variable. We now consider methods for solving systems of non-linear algebraic equations in which each equation is a function of a specified number of variables. We can write such a system in the form

$$f_i(x_1, x_2, \dots, x_n) = 0 \text{ for } i = 1, 2, 3, \dots, n \tag{2.18.1}$$

A simple approach for solving this system, based on Newton's method for the single variable equation is described below.

To illustrate this procedure we first consider a system of two equations in two variables.

$$f_1(x_1, x_2) = 0$$

$$f_2(x_1, x_2) = 0$$

(2.18.2)

Given initial approximations to x_1 and x_2 as x_1^0 and x_2^0 we wish to find improved approximations

$$x_1^1 = x_1^0 + dx_1^0$$

$$x_2^1 = x_2^0 + dx_2^0$$

These approximations should be such that they drive the values of the functions closer to zero, so that

$$f_1(x_1^1, x_2^1) \approx 0$$

$$f_2(x_1^1, x_2^1) \approx 0$$

(2.18.3)

or

$$f_1(x_1^0 + dx_1^0, x_2^0 + dx_2^0) \approx 0$$

$$f_2(x_1^0 + dx_1^0, x_2^0 + dx_2^0) \approx 0$$

(2.18.4)

Applying a two dimensional Taylor's series expansion to (2.18.4) gives

$$f_1(x_1^0, x_2^0) + \{\partial f_1/\partial x_1\}dx_1^0 + \{\partial f_1/\partial x_2\}dx_2^0 + \dots \approx 0$$

$$f_2(x_1^0, x_2^0) + \{\partial f_2/\partial x_1\}dx_1^0 + \{\partial f_2/\partial x_2\}dx_2^0 + \dots \approx 0$$

(2.18.5)

If we neglect terms involving powers of dx_1^0 and dx_2^0 higher than one, then system (2.18.5) represents a system of two linear equations in two unknowns. The zero superscript denotes the function is to be calculated at the initial approximation and dx_1^0 and dx_2^0 are the unknowns we wish to find. Having solved the two simultaneous linear equations of system (2.18.5) we can obtain our new improved approximations and then repeat the process until we have obtained the accuracy we require. A common convergence criteria is to continue iterations until

$$\sqrt{\{(dx_1^r)^2 + (dx_2^r)^2\}} < \varepsilon$$

where r denotes the iteration number and ε is a small positive quantity preset by the user. It is a simple step to generalise this procedure for any number of variables and equations. We may write the general system of equations as

$$\mathbf{f}(\mathbf{x}) = \mathbf{0}$$

where $\mathbf{f}$ denotes the column vector of n components $(f_1, f_2, \ldots f_n)^T$ and $\mathbf{x}$ is a column vector of n components $(x_1, x_2, \ldots x_n)^T$ where T denotes the transpose. Let $\mathbf{x}^{r+1}$ denote the value of $\mathbf{x}$ at the $r + 1$th iteration then

$$\mathbf{x}^{r+1} = \mathbf{x}^r + d\mathbf{x}^r \quad \text{for } r = 0, 1, 2, \ldots$$

If $\mathbf{x}^{r+1}$ is an improved approximation to $\mathbf{x}$ then

$$\mathbf{f}(\mathbf{x}^{r+1}) = \mathbf{0}$$

or

$$\mathbf{f}(\mathbf{x}^r + d\mathbf{x}^r) = \mathbf{0} \tag{2.18.6}$$

Expanding (2.18.6) by using an n-dimensional Taylor's series expansion gives

$$\mathbf{f}(\mathbf{x}^r + d\mathbf{x}^r) = \mathbf{f}(\mathbf{x}^r) + d\mathbf{x}^r \, \nabla \mathbf{f}(\mathbf{x}^r) + \ldots \tag{2.18.7}$$

where ∇ is a vector operator, $\nabla \mathbf{f}$ is called the gradient of $\mathbf{f}$ and is an array whose components are the first order partial derivatives of $\mathbf{f}$ with respect to each of the n components of $\mathbf{x}$. If we neglect higher order terms in $(d\mathbf{x}^r)^2$ this gives, by virtue of (2.18.6)

$$\mathbf{f}(\mathbf{x}^r) + d\mathbf{x}^r \, \mathbf{J}_r = \mathbf{0} \tag{2.18.8}$$

Where $\mathbf{J}_r = \nabla \mathbf{f}(\mathbf{x}^r)$. $\mathbf{J}_r$ is called the Jacobian matrix. The subscript r denotes that the matrix is evaluated at the point $\mathbf{x}^r$ and it can be written in component form as

$$\mathbf{J}_r = [\partial f_i(\mathbf{x}^r)/\partial x_j]$$

and thus

$$d\mathbf{x}^r = -\mathbf{f}(\mathbf{x}^r) \, \mathbf{J}_r^{-1} \quad \text{for } r = 0, 1, 2, \ldots$$

Hence

$$\mathbf{x}^{r+1} = \mathbf{x}^r - \mathbf{f}(\mathbf{x}^r) \, \mathbf{J}_r^{-1} \quad \text{for } r = 0, 1, 2, \ldots$$

The matrix $\mathbf{J}_r$ may be singular and in this situation the inverse, $\mathbf{J}_r^{-1}$, cannot be calculated.

This is the general form of Newton's method. However, there are two major disadvantages with this method:

(1)   The method may not converge unless the initial approximation is a good one.

(2)   The method requires the user to provide the derivatives of each function with respect to each variable. The user must therefore provide n^2 derivatives and any computer implementation must evaluate the n functions and the n^2 derivatives at each iteration.

Table 2.18.1 illustrates the application of Newton's method to solve the system of two equations in two variables:

$$x^2 + y^2 = 4$$
$$xy = 1$$

Program 2.18.1 implements this method.

```
100   REM ---------------------------------------------
110   REM NEWTON'S METHOD FOR TWO VARIABLES GL/JP (C)
120   REM ---------------------------------------------
130   PRINT "NEWTON'S METHOD FOR TWO VARIABLES" :PRINT
140   PRINT "ENTER APPROXIMATIONS FOR X AND Y RESPECTIVELY";
150   INPUT X ,Y
160   PRINT "ENTER THE ACCURACY YOU REQUIRE";
170   INPUT EPS
175   REM GOTO SUBROUTINE FOR THE CALCULATION OF FUNCTIONS AND DERIVATIVES
180   GOSUB 290
190   DE=D1*D4-D2*D3
200   DX=(D2*F2-D4*F1)/DE
210   DY=(D3*F1-D1*F2)/DE
220   IF SQR(DX*DX+DY*DY)>EPS THEN X=X+DX:Y=Y+DY :IT=IT+1: GOTO 180
230   PRINT: PRINT "SOLUTIONS OF THE EQUATIONS ARE X=";X
240   PRINT "                              Y=";Y
250   PRINT "VALUES OF THE FUNCTIONS ARE F1=";F1
260   PRINT "                         F2=";F2
270   PRINT "NUMBER OF ITERATIONS=";IT
280   END
284   REM CALCULATE DERIVITIES AND FUNCTION VALUES
285   REM INDEPENDENT VARIABLES MUST BE NAMED X AND Y
286   REM DEFINE D1 AS PARTIAL DERIVATIVE OF F1 WRT X
287   REM DEFINE D2 AS PARTIAL DERIVATIVE OF F1 WRT Y
288   REM DEFINE D3 AS PARTIAL DERIVATIVE OF F2 WRT X
289   REM DEFINE D4 AS PARTIAL DERIVATIVE OF F2 WRT Y
290   F1=X*X+Y*Y-4
300   F2=X*Y-1
310   D1=2*X : D2=2*Y
320   D3=Y    : D4=X
330   RETURN
```

Program 2.18.1

Table 2.18.1. Use of Newton's method to solve the pair of
simultaneous equations given in the text.

```
NEWTON'S METHOD FOR TWO VARIABLES

ENTER APPROXIMATIONS FOR X AND Y RESPECTIVELY?  3,-1.5
ENTER THE ACCURACY YOU REQUIRE?  .00005

SOLUTIONS OF THE EQUATIONS ARE X=  1.931852
                               Y= .5176381
VALUES OF THE FUNCTIONS ARE F1=-2.384186E-07
                            F2= 0
NUMBER OF ITERATIONS= 5
```

2.19 BROYDEN'S METHOD FOR SOLVING A SYSTEM OF NON-LINEAR EQUATIONS

The method of Newton described in section 2.18 does not provide a practical procedure for solving any but the smallest systems of non-linear equations. As we have seen, the method requires the user to provide not only the function definitions but also the definitions of the n^2 partial derivatives of the functions. Thus, for a system of ten equations in ten unknowns, the user must provide one hundred and ten function definitions!

To deal with this problem a number of techniques have been proposed but the group of methods which appears most successful is the class of methods known as the quasi-Newton methods. The quasi-Newton methods avoid the calculation of the partial derivatives by obtaining approximations to them involving only the function values. The set of derivatives of the functions may be written in the form of the Jacobian matrix

$$\mathbf{J} = (\partial f_i / \partial x_j) \quad \text{where } i,j = 1, 2, \dots ,n \qquad (2.19.1)$$

The quasi-Newton methods provide an updating formula which gives successive approximations to the Jacobian for each iteration. Broyden and others have shown that under specified circumstances these updating formula provide satisfactory approximations to the inverse Jacobian. The structure of the algorithm suggested by Broyden is:

(1)   Input an initial approximation to the solution. Set the counter i ; $i = 0$

(2)   Calculate an initial approximation to the inverse Jacobian denoted by $\mathbf{B}^i$.

(3)   Calculate $\mathbf{p}^i = -\mathbf{B}^i \, \mathbf{f}^i$

(4)   Determine the scaler parameter t such that

$$\|\mathbf{f}(\mathbf{x}^i + t_i \, \mathbf{p}^i)\| < \|\mathbf{f}^i\|$$

the symbols $\| \ \|$ denote the norm of the vector is to be taken. If the Euclidian norm is used this means the square root of the sum of the squares of the components of the vector placed between $\|$ and $\|$ is calculated.

(5)   Calculate $\mathbf{x}^{i+1} = \mathbf{x}^i + t_i \, \mathbf{p}^i$

(6)   Calculate $\mathbf{f}^{i+1} = \mathbf{f}(\mathbf{x}^{i+1})$. If $\|\mathbf{f}^{i+1}\| < \varepsilon$ then exit. Here ε is a small preset positive quantity.

(7)   Use the updating formula to obtain the required approximation to the Jacobian

$$\mathbf{B}^{i+1} = \mathbf{B}^i - (\mathbf{B}^i \, \mathbf{y}^i - \mathbf{p}^i) \, (\mathbf{p}^i)^T \, \mathbf{B}^i \, / \{ (\mathbf{p}^i)^T \, \mathbf{B}^i \, \mathbf{y}^i \}$$

where $\mathbf{y}^i = \mathbf{f}^{i+1} - \mathbf{f}^i$

(8)   Set $i = i + 1$ and return to step (3)

The initial approximation to the inverse Jacobian $\mathbf{B}$ is usually taken as a scalar multiple of the unit matrix. The success of this algorithm depends on the nature of the functions to be solved and on the closeness of the initial approximation to the solution. In particular step (4) may present major problems since it may be very expensive in computer time and to avoid this t_i is sometimes set as a constant, usually one. This, however, may reduce the stability of the algorithm. Although Broyden's method does not require the calculation of derivatives the convergence is in general slower than Newton's method.

   In general the problem of solving a system of non-linear equations is a very difficult one. There is no algorithm that is guaranteed to work for all systems of equations. For large systems of equations the available algorithms tend to require large amounts of computer time to obtain accurate solutions.

2.20 CONCLUSION AND SUMMARY

The user of programs designed for solving equations will find it is an area which presents particular difficulties. It is always possible to devise or meet with problems which particular algorithms either cannot solve or take a long time to solve. For example, it is just not possible for many algorithms to find the roots of the apparently trivial problem $x^{20} = 0$ very accurately. However the algorithms described, if used with care, provide ways of solving a wide range of problems. Table 2.20.1 summarises the important practical features of these algorithms and their range of applicability.

Table 2.20.1. Summary of algorithms for finding roots of equations.

Method	Program ref	Function type	Starting values	Roots determined	Multiple roots	Complex roots
Bisection	2.2.1	General	Two	One	No	No
Iteration	None	General	One	One	No	No
Newton	2.6.1	General	One	One	No	No§
Secant	2.7.1	General	Two	One	No	No§
Mod Newton	2.9.1	General	One	One	Yes	No§
Schroder	2.9.2	General	One	One	Yes	No§
Brent	2.11.1	General	Two	One	Yes	No
Bairstow	2.13.1	Polynomial	None	All	Yes*	Yes
QD	2.14.1	Polynomial	None	All	No	Yes
Laguerre	2.15.1	Polynomial	One	All	Yes	No§
Moore	2.16.1	Polynomial	Two	All	Yes*	Yes

§ Program may be adapted to determine complex roots.
* Multiple roots may be found but convergence is usually slow.

PROBLEMS

2.1   Use Newton's method, program 2.6.1, to find a root of $x^{1.4} - \sqrt{x} + 1/x - 100 = 0$ given an initial approximation 50. Use an accuracy of 1E–4.

2.2   Use the secant method (Program 2.7.1) and bisection method (Program 2.2.1) to solve the equation given in problem 2.1 using an accuracy of 1E–4. Use initial approximations of 10 and 50 and compare your results and the number of iterations required.

2.3   Find the two real roots of $|x^3| + x - 6 = 0$ using Newton's method, program 2.6.1. Use initial approximations –1 and 1 and an accuracy 1E–4. Sketch the graph to verify the equation has only two real roots. *Hint.* Take care in finding the derivative of the function.

2.4   Explain why it is difficult to find the root of $\tan x - c = 0$ when c is a large quantity. Use the secant method, program 2.7.1, with initial approximations 1.3 and 1.4 and accuracy 1E–4, to find a root of this equation when $c = 5$ and

$c = 10$. Compare the number of iterations required in both cases. *Hint*. A sketch graph will be useful.

2.5 Solve problem 2.4 using Brent's method, program 2.11.1, and include the case where $c = 500$. Use the starting values 1.3 and 1.57 and accuracy 1E–4.

2.6 Find a root of the polynomial $x^5 - 5x^4 + 10x^3 - 10x^2 + 5x - 1 = 0$ correct to 4 decimal places by using Schroder's method (program 2.9.2) with $n = 5$ and a starting value $x_0 = 0.2$. Use Newton's method (program 2.6.1) to solve the same problem and compare the result with that obtained using Schroder's method. Explain the difference in performance between the two methods.

2.7 Use the simple iterative method to solve the equation $x^{10} = e^x$. Express the equation in the form $x = f(x)$ in different ways and start the iterations with the initial approximation $x = 1$. Compare the efficiency of the formulae you have devised and check your answer(s) using Newton's method (program 2.6.1).

2.8 Kepler's equation has the form $E - e \sin E = M$. Solve this equation for $e = 0.96727464$, the eccentricity of Halley's comet, and $M = 4.527594E–3$. Use Newton's method, program 2.6.1, with an accuracy of 0.00005 and a starting value of 1.

2.9 Compare performance of Brent and Secant programs (2.11.1 and 2.7.1 respectively) for solving $x^{11} = 0$ in the interval $(-1.5, 1)$ with an accuracy 1E–5.

2.10 The smallest positive root of the equation

$$1 - x + x^2/(2!)^2 - x^3/(3!)^2 + x^4/(4!)^2 - \ldots = 0$$

is 1.4458. By considering in turn only the first 4, 5 and 6 terms in the series show that a root of the truncated series approaches this result. Use Brent's method, program 2.11.1, using the initial values 1 and 1.5 and accuracy 1E–4.

2.11 Reduce the following system of equations to one equation in terms of x and solve the resulting equation using Newton's method, program 2.6.1.

$$e^{x/10} - y = 0$$
$$2 \log_e y - \cos x = 2$$

Use program 2.18.1 to solve these equation directly and compare your results. Use an initial approximation $x = 1$ for program 2.6.1 and approximations $x = 1$, $y = 1$ for program 2.18.1 and an accuracy 1E–4 in both cases.

2.12 Solve the pair of equations below using Newton's method for two variables, program 2.18.1, with the starting point $x = 10$, $y = -10$ and accuracy 1E–4.

$$2x = \sin[(x + y)/2]$$
$$2y = \cos[(x - y)/2]$$

2.13 Solve the two equations given below using Newton's method for two variables, program 2.18.1, with the starting point $x = 10$ and $y = 10$ and accuracy 1E–4.

$$x^3 - 3xy^2 = 1/2$$
$$3x^2y - y^3 = \sqrt{3}/2$$

2.14 The polynomial equation

$$x^4 - (13 + \varepsilon)x^3 + (57 + 8\varepsilon)x^2 - (95 + 17\varepsilon)x + 50 + 10\varepsilon = 0$$

has roots 1, 2, 5, 5 + ε. Use the methods of Bairstow, Laguerre and Moore (programs 2.13.1, 2.15.1 and 2.16.1 respectively) to find all the roots of this polynomial for $\varepsilon = 0.1, 0.01, 0.001$. What happens as ε becomes smaller? Use an accuracy 1E–5 and a starting value of 10 in Laguerre's method.

2.15 Solve problem 2.14 with $\varepsilon = 1$ using the QD method, program 2.14.1, with an accuracy of 1E–4. What happens when the method is used to solve the problem with $\varepsilon = 0.1$

2.16 Use the methods of Bairstow and Moore, programs 2.13.1 and 2.16.1 to solve the following polynomial using an accuracy requirement of 1E–4.

$$x^5 - x^4 - x^3 + x^2 - 2x + 2 = 0$$

2.17 Use Moore's method (program 2.16.1) to solve the equation

$$t^3 - 0.5 - i\sqrt{3}/2 = 0$$

Compare with the exact solution $\cos((\pi/3 + 2\pi k)/3) - i \sin((\pi/3 + 2\pi k)/3)$ for $k = 0, 1, 2$.

2.18 An outline algorithm for the Illinois method for finding a root of $f(x) = 0$ (Dowell & Jarrett, 1971) is as follows

For $k = 0, 1, 2, ...$

$$x_{k+1} = x_k - f_k / f[x_{k-1}, x_k]$$

if $f_k f_{k+1} > 0$ set $x_k = x_{k-1}$ and $f_k = \gamma f_{k-1}$

where $f_k = f(x_k)$, $f[x_{k-1}, x_k] = (f_k - f_{k-1})/(x_k - x_{k-1})$ and $\gamma = 0.5$

Program this method and use your program to solve the problem 2.8. Note that the *Regula falsi* method is similar but differs in that γ is taken as one.

2.19 The following iterative formulae can be used to solve the equation $x^2 - a = 0$:

(a) $x_{k+1} = (x_k + a/x_k)/2 \quad k = 0, 1, 2, ...$
(b) $x_{k+1} = (x_k + a/x_k)/2 - (x_k - a/x_k)^2/(8x_k) \quad k = 0, 1, 2, ...$

These iterative formulae are second and third order methods for solving this equation. Write a program to implement them and compare the number of iterations required to obtain the square root of 100.112 to 5 decimal places. For the purpose of illustration, use an initial approximation of 1000.

2.20 Modify program 2.6.1. to determine complex roots. *Hint*: In contrast to Fortran, complex quantities cannot be defined and manipulated directly in BASIC. To develop a program to find the roots of $f(z) = 0$ it is necessary to express z in the form $x + iy$ and $f(x, y)$ in the form $f_R(x, y) + i f_i(x, y)$. Thus the user defined function (FN...) cannot be used since in most versions of BASIC they are functions of one variable only. Note also that the user must supply the real and imaginary parts of the function and the real and imaginary parts of its derivative. Use the program developed to find complex roots for each of the following equations by using different starting points in the complex plane.

(a) $e^z - 3z = 0$, (b) $\sin z - 2 = 0$, (c) $z^4 - z + 1 = 0$

3

Solving linear algebraic equations

3.1 INTRODUCTION

When engineering systems are modelled, the mathematical description is frequently developed in terms of sets of algebraic simultaneous equations. Sometimes these equations are non-linear and the problems posed by them are discussed in section 2.18. In many cases, however, the equations are linear and here we describe elimination and iterative procedures for solving such systems.

Before a detailed examination of the various solution schemes is presented it is necessary to define precisely the nature of the problem. This is to find the solution of n independent linear equations in n unknowns. In matrix notation this is written

$$\mathbf{Ax} = \mathbf{b} \tag{3.1.1}$$

where $\mathbf{A}$ is an $n \times n$ square matrix of known coefficients and is usually called the coefficient matrix, $\mathbf{b}$ is a vector of n known coefficients and $\mathbf{x}$ is the vector of the n unknowns.

A system of n equations in n unknowns is said to be linearly independent if no equation in the system can be expressed as a linear combination of the others. Under these circumstances a unique solution exists. For example

$$
\begin{aligned}
2x + y - z &= 1 \\
3x - y - z &= -2 \\
x + y \quad &= 3
\end{aligned}
$$

has the unique solution $x = 1$, $y = 2$, $z = 3$. However

$$5x + y + z = 4$$
$$3x - y + z = -2$$
$$x + y \quad = 3$$

does not have a unique solution since the equations are not linearly independent; the first equation is equal to the second equation plus twice the third equation.

Although, in this chapter, we are concerned with the task of solving n equations in n unknowns it is instructive to examine briefly some variations on this problem. The discussion is limited to problems in two unknowns so that the equations can be shown graphically.

3.2 EQUATION SYSTEMS

Two independent linear equations in two unknowns are shown in Fig. 3.2.1. These equations have a unique solution which lies at the point of intersection of the two equations. In contrast, Fig. 3.2.2 shows a single linear equation in two unknowns for which there is no unique solution because every point of the infinity of points on the line $y = x + 1$ is a solution. This situation may also arise if we have a system of two linear equations that are not linearly independent. For example

$$x - y = -1$$
$$3x - 3y = -3$$

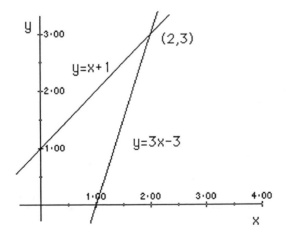

Fig 3.2.1. Geometric representation of two linear equations with a unique solution.
These distinct lines intersect.

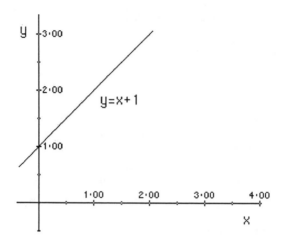

Fig 3.2.2. Geometric representaion of single linear equation equivalent two linear equations that are not independent. An infinity of solutions exists.

These two equations are equivalent to the single equation shown in Fig. 3.2.2 and no unique solution exists. In general terms, for systems of more than two equations, the same problem will arise if one or more of the equations are linear combinations of the others.

We now illustrate two further cases. Fig. 3.2.3 shows two linear equations in two unknowns but the equations are inconsistent and no solution exists. Fig. 3.2.4 shows three independent linear equations in two unknowns. The graph shows that

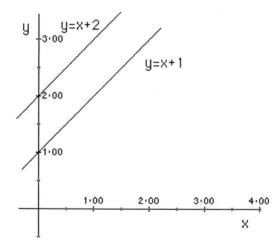

Fig. 3.2.3. Graph of two linear equations without a solution.

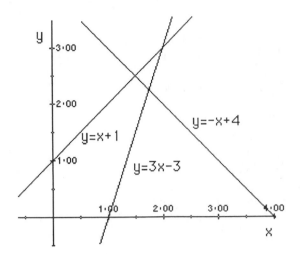

Fig. 3.2.4. Graph of three linear equations in two unknowns. These distinct lines do not intersect at a common point and consequently there is no exact solution.

the three equations do not intersect at a point and therefore no solution satisfies exactly all three equations simultaneously. This is an example of an over-determined system for which only an approximate solution can be found. This problem and its solution is discussed in section 3.20.

In Chapter 2 reference is made to the difficulty of solving non-linear equations. Sometimes a set of apparently non-linear equations can be reduced to a set of linear equations by a suitable substitution, although no general rules can be given. Two examples of this are as follows:

$$2x^2 + x + 2e^y = 16$$
$$4x^2 + 2x - e^y = 17$$

If the substitutions $X = 2x^2 + x$ and $Y = e^y$ are made then these equations become

$$X + 2Y = 16$$
$$2X - Y = 17$$

Solving these gives $X = 10$, $Y = 3$. Hence $x = 2$ or -2.5 and $y = \log_e 3 = 1.0986$.
A second example is

$$3x^2 + y = 4$$
$$4x^3 + y = 4$$

Taking the natural logarithms of these equations gives

$$\log_e 3 + 2 \log_e x = \log_e (4 - y)$$
$$\log_e 4 + 3 \log_e x = \log_e (4 - y)$$

Letting $X = \log_e x$ and $Y = \log_e (4 - y)$ gives

$$Y - 2X = 1.0986123$$
$$Y - 3X = 1.3862944$$

Thus $X = -0.2876821$ and $Y = 0.5232481$, giving $x = 0.75$ and $y = 2.3125$.

Unfortunately most sets of non-linear equations cannot be linearised in this manner and the methods of section 2.18 must be used. We will now consider further the problems that confront us when solving sets of linear equations.

3.3 DETERMINANTS AND MATRIX INVERSION

Here we will examine the limitations of using both determinants and matrix inversion to solve systems of linear equations and to overcome this difficulty an efficient, systematic elimination procedure will be introduced in section 3.4. Since we do not give the formal definition of a determinant the reader requiring further information is advised to refer to Gere and Weaver (1965) for a detailed description of determinants.

Many elementary text books cite Cramer's rule as a method for solving a set of equations. This method is grossly inefficient because to solve n equations requires the evaluation of $n + 1$ determinants, each of order n. Evaluating a determinant by expanding rows or columns requires, for large values of n, approximately $(e - 1)n!$ multiplications where $e = 2.718...$ A similar number of additions and subtractions are also required. Using this procedure and computing at a rate of one million multiplications per second, it would take about a year to evaluate a 16th order determinant and approximately 350 years to evaluate an 18th order determinant. The systematic elimination procedure for solving a set of simultaneous equations avoids the use of Cramer's rule and provides an efficient method for evaluating determinants as a by-product. In passing, it should be noted that the determinant of the coefficient matrix is a useful quantity that is sometimes required. For example, a coefficient matrix with a zero determinant is called *singular* and warns us that the equations are not independent and have no unique solution.

Let us now turn to matrix inversion. We may define the inverse of a matrix **A** as the matrix **A**$^{-1}$ where

$$\mathbf{A}^{-1}\mathbf{A} = \mathbf{A}\,\mathbf{A}^{-1} = \mathbf{I}$$

and **I** is the identity matrix. The identity matrix plays an analogous role to unity in scalar algebra. It is important to note that the inverse of a given matrix **A** does not always exist and this is directly related to whether the system of equations (3.1.1) has a solution or not. In section 3.1 we introduced matrix notation as a convenient short hand for writing down sets of simultaneous equations. This notation is also useful when manipulating equations as a complete entity, particularly when the objective is to prove or develop an analytical relationship. Having defined the inverse of a matrix we can now give an example of matrix manipulation which allows us to express the solution of (3.1.1) in matrix form.

If (3.1.1) is premultiplied by the inverse of **A**, which is written as $\mathbf{A}^{-1}$, then

$$\mathbf{A}^{-1}\mathbf{A}\mathbf{x} = \mathbf{A}^{-1}\mathbf{b}$$

Thus

$$\mathbf{x} = \mathbf{A}^{-1}\mathbf{b} \tag{3.3.1}$$

since by definition $\mathbf{A}^{-1}\mathbf{A} = \mathbf{I}$, the identity matrix, and $\mathbf{I}\mathbf{x} = \mathbf{x}$. Thus, in matrix notation, (3.3.1) is an explicit formula from which **x** may be determined. Equation (3.3.1) requires the inverse of **A**. This is formally defined as the transpose of the adjoint matrix of **A**, written adj($\mathbf{A}^T$), divided by the determinant of **A** which is a scalar quantity and is written as det(**A**). Thus

$$\mathbf{A}^{-1} = \text{adj}(\mathbf{A}^T)/\det(\mathbf{A})$$

In order to define the adjoint of a matrix **A** we must begin by considering the minors of that matrix. The minor m_{ij} is the determinant of the original matrix with the ith row and jth column removed. Thus the minors are determinants of order $n-1$. We define the cofactors c_{ij} as

$$c_{ij} = (-1)^{i+j} \, m_{ij}$$

Finally, the adjoint of a matrix is defined as the original matrix modified by replacing the original coefficients by their corresponding cofactors. Hence, we see that the adjoint matrix requires the evaluation of n^2 determinants of order $n - 1$. Solving equations using the inverse matrix computed by this method is never used in practice because it is very inefficient. In contrast we will see in section 3.8 that solving equations by the systematic elimination method is efficient. Cramer's rule solves a system of equations using a procedure which is closely related to that of finding the inverse matrix using the adjoint matrix and is also a very inefficient procedure.

If, for some purpose, the inverse of **A** is explicitly required then it may be efficiently determined as follows. Consider the equation

$$\mathbf{A}\mathbf{X} = \mathbf{I} \tag{3.3.2}$$

This equation implies that inverse of **A** is equal to **X**. This is equivalent to n equations of the form

$$\mathbf{A}\mathbf{x}_j = \mathbf{b}_j \quad j = 1, 2, \ldots, n \tag{3.3.3}$$

where $\mathbf{x}_j$ is the jth column of **X** and $\mathbf{b}_j$ is the jth column of the identity matrix **I**. Hence $\mathbf{b}_j$ is a vector of zeros except for the jth element which is equal to one. Each of the n sets of simultaneous equations, (3.3.3), can be solved efficiently by systematic elimination and from the solution of these equations **X** can be constructed. However, it must be stressed that it is only necessary to obtain the inverse of a matrix when the coefficients of the inverse have some particular physical significance which is of interest; in general a system of equations should be solved directly by systematic elimination.

We will now examine the systematic elimination procedure in detail.

3.4 GAUSSIAN ELIMINATION

This is the basic elimination procedure and is a systematic development of the simple, well known method that can be used, for example, to remove one unknown from a pair of simultaneous equations. Consider the solution of the following equations

$$\begin{aligned}
-8x_1 - 4x_2 - 17x_3 &= 85 \\
33x_1 + 16x_2 + 72x_3 &= -359 \\
-24x_1 - 10x_2 - 57x_3 &= 281
\end{aligned} \tag{3.4.1}$$

In matrix notation this becomes

$$\begin{bmatrix} -8 & -4 & -17 \\ 33 & 16 & 72 \\ -24 & -10 & -57 \end{bmatrix} \begin{bmatrix} x_1 \\ x_2 \\ x_3 \end{bmatrix} = \begin{bmatrix} 85 \\ -359 \\ 281 \end{bmatrix}$$

For the purposes of Gaussian elimination it is more convenient to write the equations in the following form:

$$\begin{bmatrix} (-8) & -4 & -17 & 85 \\ 33 & 16 & 72 & -359 \\ -24 & -10 & -57 & 281 \end{bmatrix} \begin{matrix} \text{row A1} \\ \text{row A2} \\ \text{row A3} \end{matrix}$$

Note that the terms on the right-hand side of (3.4.1) have been placed in the array. This is done because each operation of the elimination process works on a complete equation and so it is convenient to write down the coefficient matrix and augment it with the right hand side vector. This is often referred to as the "augmented array" or, alternatively, as the "detached coefficient" form. The rows have been labelled A1, A2 and A3. To eliminate x_1 from the second and third equations we divide the first equation or row, A1, by –8 to obtain row B1 as shown below. Row A1 is called the pivotal row and the coefficient –8 is the pivot. Note that at each stage the coefficient to be used as the pivot is enclosed in round brackets for clarity. We then subtract from rows A2 and A3 the coefficients of row B1 times an appropriate factor to make the coefficients of x_1 in the second and third rows equal zero. Thus row B2, for example, is formed by subtracting 33 times row B1 from row A2 as indicated on the left-hand side of the matrix below.

$$\begin{array}{c} A1/(-8) \\ A2-(33)B1 \\ A3-(-24)B1 \end{array} \begin{bmatrix} 1 & 0.5 & 2.125 & -10.625 \\ 0 & (-0.5) & 1.875 & -8.375 \\ 0 & 2 & -6 & 26 \end{bmatrix} \begin{array}{l} \text{row B1} \\ \text{row B2} \\ \text{row B3} \end{array}$$

It is seen that x_1 has been eliminated from the second and third equations. The next stage is to eliminate x_2 from the third row. Thus, using –0.5 as the pivot, we have

$$\begin{array}{c} \\ B2/(-0.5) \\ B3-(2)C2 \end{array} \begin{bmatrix} 1 & 0.5 & 2.125 & -10.625 \\ 0 & 1 & -3.75 & 16.75 \\ 0 & 0 & (1.5) & -7.5 \end{bmatrix} \begin{array}{l} \text{row B1} \\ \text{row C2} \\ \text{row C3} \end{array}$$

Note that at each stage the pivot is the coefficient on the leading diagonal and only rows below the pivotal row are manipulated. Finally we have

$$\begin{array}{c} \\ \\ C3/(1.5) \end{array} \begin{bmatrix} 1 & 0.5 & 2.125 & -10.625 \\ 0 & 1 & -3.75 & 16.75 \\ 0 & 0 & 1 & -5 \end{bmatrix} \begin{array}{l} \text{row B1} \\ \text{row C2} \\ \text{row D3} \end{array}$$

The original array of coefficients has been reduced to a triangular array with each element on the leading diagonal equal to one.

The final stage, that of determining the values of x_1, x_2 and x_3, is called back-substitution. The value of x_3 is available directly from row D3. Then x_2 can be determined from row C2 since x_3 is known. Finally, knowing x_2 and x_3, x_1 can be found from row B1.

We will now make a minor change in the detail of this elimination procedure to reduce the number of arithmetic operations that must be performed. Beginning once again with the augmented array we have

$$\begin{bmatrix} (-8) & -4 & -17 & 85 \\ 33 & 16 & 72 & -359 \\ -24 & -10 & -57 & 281 \end{bmatrix} \begin{array}{l} \text{row A1} \\ \text{row A2} \\ \text{row A3} \end{array}$$

To eliminate x_1 from the second equations we will multiply row A1 by $33/(-8)$ and subtract it from row A2 and to eliminate x_1 from the third equation we will multiply row A1 by $(-24)/(-8)$ and subtract it from row A3. Row A1 is still called the pivotal row and the coefficient -8 is the pivot. However, unlike the previously described procedure, we do not explicitly divide each coefficient of row A1 by the pivotal value and then multiply the result by -8 and 33. Instead, we multiply each coefficient by -4.125 [$= 33/(-8)$] and 3 [$=(-24)/(-8)$]. Thus we obtain

$$
\begin{array}{c}
 \\
\text{A2–(–4.125)A1} \\
\text{A3–(3)A1}
\end{array}
\begin{bmatrix}
-8 & -4 & -17 & 85 \\
0 & (-0.5) & 1.875 & -8.375 \\
0 & 2 & -6 & 26
\end{bmatrix}
\begin{array}{c}
\text{row A1} \\
\text{row B2} \\
\text{row B3}
\end{array}
$$

The next stage is to eliminate x_2 from the third row. Multiplying row B2 by $2/(-0.5)$ and subtracting it from row B3 we have

$$
\begin{array}{c}
 \\
 \\
\text{B3–(–4)B2}
\end{array}
\begin{bmatrix}
-8 & -4 & -17 & 85 \\
0 & -0.5 & 1.875 & -8.375 \\
0 & 0 & 1.5 & -7.5
\end{bmatrix}
\begin{array}{c}
\text{row A1} \\
\text{row B2} \\
\text{row C3}
\end{array}
$$

As before we can determine x_1, x_2 and x_3 by back substitution but since the coefficients on the diagonal of the matrix are not unity one extra division per row is required. This form of elimination is used in program 3.6.1, described in section 3.6.

After elimination, the determinant of the coefficient matrix can easily be calculated since it can be proved that it is equal to the product of the pivot values, see Ralston & Rabinowitz (1978). For the elimination procedure applied to (3.4.1) the pivots are -8, -0.5 and 1.5 so that

$$
\det(\mathbf{A}) = (-8)(-0.5)1.5 = 6.
$$

3.5 NUMERICAL DIFFICULTIES

In this section we consider a problem that can significantly effect the accuracy of the solution of certain systems of equations. Fortunately a systematic remedy is available for this problem and this is presented in section 3.6.

Suppose we wish to solve

$$
\begin{bmatrix}
0 & 2 & -1 \\
3 & -1 & 2 \\
1 & 3 & -5
\end{bmatrix}
\begin{bmatrix}
x_1 \\
x_2 \\
x_3
\end{bmatrix}
=
\begin{bmatrix}
1 \\
4 \\
-1
\end{bmatrix}
$$

Attempting to solve these equations using the Gaussian elimination technique described in section 3.4 will fail immediately because the term in the first row on the leading diagonal, the pivot, is zero. Thus it is impossible to divide that row by the pivot value. Clearly, this problem can be overcome by rearranging the order of the equations; for example by making the first equation the third.

As an example of a less obvious difficulty consider the equations

$$\begin{bmatrix} d & 1 \\ 2 & 1 \end{bmatrix}\begin{bmatrix} x_1 \\ x_2 \end{bmatrix} = \begin{bmatrix} 1 \\ 2 \end{bmatrix} \qquad (3.5.1)$$

where d is small. If the elimination process is followed then d is the initial pivot and x_1 and x_2 are evaluated by performing a series of algebraic manipulations. It can be readily shown that these operations are equivalent to using the following formulae:

$$x_2 = (2 - 2/d)/(1 - 2/d) \qquad (3.5.2)$$
$$x_1 = (1 - x_2)/d \qquad (3.5.3)$$

Numerical problems will arise when working to a finite precision if $2/d$ is very much greater than one; i.e. when d is very small. In this situation (3.5.2) reduces to

$$x_2 = (-2/d)/(-2/d) = 1$$

and (3.5.3) becomes

$$x_1 = (1 - 1)/d = 0$$

This is erroneous; as d tends to zero the true solution of (3.5.1) tends to

$$x_2 = 1 \text{ and } x_1 = 0.5$$

The value of d for which this problem becomes significant depends on the number of significant digits that a particular microcomputer maintains during the computation.

If the order of equations of (3.5.1) is reversed to become

$$\begin{bmatrix} 2 & 1 \\ d & 1 \end{bmatrix}\begin{bmatrix} x_1 \\ x_2 \end{bmatrix} = \begin{bmatrix} 2 \\ 1 \end{bmatrix} \qquad (3.5.4)$$

then the initial pivot is 2 and elimination is equivalent to using the following expressions to determine x_1 and x_2.

$$x_2 = (1 - d)/(1 - d/2) \qquad (3.5.5)$$
$$x_1 = (2 - x_2)/2 \qquad (3.5.6)$$

Table 3.5.1. Solution of (3.5.1) and (3.5.4) for
various values of d to seven significant digit accuracy.

d	Solution of (3.5.1) X1	X2	Solution of (3.5.4) X1	X2
.001	.5002618	.9994997	.5002501	.9994998
.0001	.500083	.99995	.500025	.99995
.00001	.500679	.999995	.5000025	.999995
.000001	.4768372	.9999995	.5000002	.9999995
.0000001	1.192093	.9999999	.5	.9999999
.00000001	0	1	.5	1
.000000001	0	1	.5	1

As d tends to zero these equations tend to the correct solutions.

Table 3.5.1 shows the solutions of equations (3.5.1), using (3.5.2) and (3.5.3), and the solution of the equivalent equations (3.5.4), using (3.5.5) and (3.5.6), for small values of d. The accuracy of the computer used for this task was seven significant digits and it is seen that solving (3.5.1) gives incorrect results if d is of order 10^{-5}; no problem is experienced in solving the equivalent equations, (3.5.4).

The conclusion to be drawn from these two examples is that it is essential to avoid using a pivot equal to zero and highly desirable to avoid a relatively small pivot. The fact that difficulties can arise even if the pivot is large in an absolute sense can be demonstrated by multiplying both equations of (3.5.1) by a very large factor. If the order of the equations is not reversed then, although the pivot is now large in an absolute sense, it is still small relative to the other coefficients. Reworking (3.5.2) and (3.5.3) with the new coefficient values shows that the problem in the elimination still persists.

3.6 PIVOTING PROCEDURES

It may be supposed that it is a simple task to rearrange, by hand, a set of linear equations so that small or zero pivots can be avoided. In practice the rearrangements are more difficult because it is possible that small or even zero coefficients may be generated during the elimination process. For example, the coefficients on the leading diagonal of the following equations are not relatively small and so it might be supposed that no numerical problems would arise.

$$\begin{bmatrix} (4) & 8 & 1 \\ 1 & 2 & 1 \\ 1 & -1 & 2 \end{bmatrix} \begin{bmatrix} x_1 \\ x_2 \\ x_3 \end{bmatrix} = \begin{bmatrix} 3 \\ 1 \\ 4 \end{bmatrix}$$

However, using the value 4 as the pivot at the first stage leads to a zero pivot (and disaster!) at the second stage of the elimination procedure.

To overcome these problems the method of "partial pivoting" is normally used. At the kth stage in the elimination, if necessary, the rows are rearranged to ensure the coefficient with the largest absolute value in the part of the kth column, in the lower triangle, is on the leading diagonal. Thus a search is made in the kth column of the coefficient matrix, beginning at row k and ending with the last row n. In solving n equations a total of $n(n + 1)/2$ coefficients must be examined.

An alternative strategy is "complete pivoting". Here, at the kth stage, all the coefficients below and to the right of the kth element on the leading diagonal are searched to find the one with the largest absolute value. Then both rows and columns are interchanged as necessary to bring this coefficient onto the leading diagonal. Complete pivoting offers little advantage over partial pivoting and it is significantly slower, requiring $n(n + 1)(2n + 1)/6$ elements to be examined in total. It is rarely used in practice and partial pivoting has been shown to be a very reliable procedure.

Program 3.6.1 performs elimination with partial pivoting. In the program the rows are not rearranged explicitly: pointers are used to indicate the next row to be manipulated and to record the order in which the rows were used.

```
100    REM -----------------------------------------
110    REM GAUSS ELIM WITH PARTIAL PIVOT JP/GL (C)
120    REM -----------------------------------------
130    PRINT "SOLUTION OF LINEAR SIMULTANEOUS EQUATIONS"
140    PRINT "GAUSS ELIMINATION & PARTIAL PIVOTING"
150    READ N
160    DIM A(N,N+1),X(N),R(N)
165    REM READ COEFFICIENTS OF MATRIX
170    FOR I=1 TO N
180        FOR J=1 TO N
190            READ A(I,J)
200        NEXT J
210    NEXT I
215    REM READ RIGHT HAND VECTOR
220    FOR I=1 TO N
230        READ A(I,N+1)
240    NEXT I
250    D=1
260    FOR K=1 TO N
265        REM CHOOSE PIVOT
270        P=0
280        FOR I=1 TO N
290            IF K=1 THEN 330
300            FOR L=1 TO K-1
310                IF I=R(L) THEN 350
320            NEXT L
330            IF ABS(A(I,K))<=ABS(P) THEN 350
340            P=A(I,K):R(K)=I
350        NEXT I
```

Progam 3.6.1. (continues).

```
360     D=D*P
370     IF P=0 THEN PRINT "ZERO PIVOT":END
380     P0=1/P
390     IF K=N THEN 490
395     REM ELIMINATION PROCEDURE
400     FOR I=1 TO N
410         FOR L=1 TO K
420             IF I=R(L) THEN 480
430         NEXT L
440         F=A(I,K)*P0
450         FOR J=K+1 TO N+1
460             A(I,J)=A(I,J)-A(R(K),J)*F
470         NEXT J
480     NEXT I
490   NEXT K
500   PRINT
505   REM BACK SUBSTITUTION
510   FOR I=N TO 1 STEP -1
520       S=A(R(I),N+1)
530       IF I=N THEN 570
540       FOR J=N TO I+1 STEP -1
550           S=S-A(R(I),J)*X(J)
560       NEXT J
570       X(I)=S/A(R(I),I)
580       PRINT"X(";I;") = ";X(I)
590   NEXT I
600   FOR K=1 TO N
610       X(R(K))=K
620   NEXT K
630   FOR I=1 TO N
640       FOR J=1 TO N-1
650           IF X(J)<=X(J+1) THEN 680
660           P=X(J):X(J)=X(J+1)
670           X(J+1)=P:D=-D
680       NEXT J
690   NEXT I
700   PRINT:PRINT "DETERMINANT = ";D
710   END
720   DATA 3 :REM NUMBER EQUATIONS
730   DATA -8,-4,-17,33,16,72,-24,-10,-57 :REM COEFFICIENTS OF MATRIX
740   DATA 85,-359,281:REM COEFFICIENTS OF RIGHT HAND VECTOR
```

Program 3.6.1.

We will now solve again the set of equations (3.4.1) to illustrate the use of partial pivoting. Elimination starts with the same augmented matrix.

$$\begin{bmatrix} -8 & -4 & -17 & 85 \\ 33 & 16 & 72 & -359 \\ -24 & -10 & -57 & 281 \end{bmatrix} \begin{matrix} \text{row A1} \\ \text{row A2} \\ \text{row A3} \end{matrix}$$

A search of the first column shows that row A2 contains the largest element and so rows A1 and A2 are interchanged to give

$$\begin{bmatrix} (33) & 16 & 72 & -359 \\ -8 & -4 & -17 & 85 \\ -24 & -10 & -57 & 281 \end{bmatrix} \begin{matrix} \text{row A2} \\ \text{row A1} \\ \text{row A3} \end{matrix}$$

Row A2 is now used as the pivotal row in the usual manner to eliminate x_1 and we have

$$\begin{matrix} \\ A1-(-8/33)A2 \\ A3-(-24/33)A2 \end{matrix} \begin{bmatrix} 33 & 16 & 72 & -359 \\ 0 & -0.12121212 & 0.45454545 & -2.03030303 \\ 0 & 1.63636364 & -4.63636364 & 19.9090909 \end{bmatrix} \begin{matrix} \text{row A2} \\ \text{row B1} \\ \text{row B3} \end{matrix}$$

At the next stage, a search of column 2 on and below the diagonal shows that the largest coefficient is in row 3 and so rows B1 and B3 are interchanged thus:

$$\begin{bmatrix} 33 & 16 & 72 & -359 \\ 0 & (1.63636364) & -4.63636364 & 19.9090909 \\ 0 & -0.12121212 & 0.45454545 & -2.03030303 \end{bmatrix} \begin{matrix} \text{row A2} \\ \text{row B3} \\ \text{row B1} \end{matrix}$$

We must now eliminate x_2. Noting that $-0.12121212/1.63636364 = -0.0740...$ we obtain

$$\begin{matrix} \\ \\ B1-(-0.0740 ...)B3 \end{matrix} \begin{bmatrix} 33 & 16 & 72 & -359 \\ 0 & 1.63636364 & -4.63636364 & 19.9090909 \\ 0 & 0 & 0.11111111 & -0.5555556 \end{bmatrix} \begin{matrix} \text{row A2} \\ \text{row B3} \\ \text{row C1} \end{matrix}$$

Back-substitution into C1, B3 and finally A2 gives the same solutions as previously determined. Again the determinant of the coefficient matrix may be found from the product of the pivots but the sign of the product must be changed if there is an odd number of row interchanges. Thus, in the previous example the product of the pivots is

$$33(1.636...)(0.111...) = 6$$

Two row interchanges were required so that the sign of the pivot product is unchanged and the determinant is equal to 6.

The above example has been solved without partial pivoting in section 3.4 and the reader can compare the steps in the two variants of the elimination procedure. However, this problem does not illustrate the need for partial pivoting since the solutions obtained are essentially identical. The need for partial pivoting is demonstrated in the solution of the following equations.

$$\begin{bmatrix} -0.375 & 0.5 & 0 & 0 & 0.125 \\ 0.5 & -1 & 0.5 & 0 & 0 \\ 0 & 0.5 & -1 & 0.5 & 0 \\ 0 & 0 & 0.5 & -1 & 0.5 \\ 0.125 & 0 & 0 & 0.5 & -0.375 \end{bmatrix} \begin{bmatrix} x_1 \\ x_2 \\ x_3 \\ x_4 \\ x_5 \end{bmatrix} = \begin{bmatrix} 1 \\ 2 \\ 3 \\ 4 \\ 5 \end{bmatrix} \quad (3.6.1)$$

A slightly modified version of program 3.6.1, allowing a solution to be determined either with or without partial pivoting, was used to solve these equations. The output from the program is shown in Table 3.6.1 and it is seen that the two estimates for the solution differ greatly. The key to this discrepancy lies in the values of the pivots used. Without partial pivoting the fourth pivot is extremely small and this introduces considerable inaccuracy into the remaining calculations. In contrast, with partial pivoting all pivots are of the same order of magnitude. Thus we see that the additional work required for partial pivoting is sometimes rewarded by significantly greater accuracy. Since there is not a reliable method of deciding whether or not partial pivoting is required it should always be used as a matter of course.

Table 3.6.1. Solution of (3.6.1) by Gaussian elimination with and without partial pivoting. The solution obtained using partial pivoting is correct.

```
SOLUTION OF LINEAR SIMULTANEOUS EQUATIONS
PARTIAL PIVOTING          │ WITHOUT PARTIAL PIVOTING
PIVOT= .5                 │ PIVOT=-.375
PIVOT= .5                 │ PIVOT=-.3333333
PIVOT= .5                 │ PIVOT=-.2499999
PIVOT= 1                  │ PIVOT= 2.384186E-07
PIVOT= .25                │ PIVOT=-4194306

X( 5 ) =  20              │ X( 5 ) =  20
X( 4 ) =  15              │ X( 4 ) =  16
X( 3 ) =  18              │ X( 3 ) =  20
X( 2 ) =  27              │ X( 2 ) =  30
X( 1 ) =  40              │ X( 1 ) =  44

DETERMINANT =  .03125     │ DETERMINANT =  .03125
```

3.7 ADDITIONAL REFINEMENTS

Several additional facilities and refinements can be introduced in the elimination process as follows:

(1) It can be shown that the accuracy of the elimination process is improved if the equations are scaled so that all the elements of the coefficient matrix are of

the same order of magnitude. The rows can be scaled by multiplying one or more of the equations by appropriate factors and the columns can be scaled by replacing the unknowns x_i by $F_i y_i$, where $i = 1, 2, \ldots ,n$. F_i may take any suitable value including unity. Unfortunately a foolproof algorithm does not exist to perform such scaling and therefore the user must consider carefully whether or not to invest the time required to scale a large set of equations since this must be done by hand.

(2) If (3.1.1) is to be solved for several different right-hand sides then the coefficient matrix can be augmented with all the right-hand side vectors and the elimination process carried out on all the vectors at the same time. Back-substitution must then be carried out separately for each of the new right-hand side vectors.

(3) If the elements of the original coefficient matrix are required for some other purpose they can be preserved instead of being overwritten as in program 3.6.1.

3.8 GAUSS-JORDAN ELIMINATION

This is a modification of the basic Gaussian elimination scheme. In this procedure the coefficients in the rows above the pivotal row are eliminated as well as those below, thereby removing the need for back-substitution. Again the procedure will be illustrated by solving equations (3.4.1). The augmented array is as before; thus:

$$
\begin{bmatrix}
(-8) & -4 & -17 & 85 \\
33 & 16 & 72 & -359 \\
-24 & -10 & -57 & 281
\end{bmatrix}
\begin{matrix}
\text{row A1} \\
\text{row A2} \\
\text{row A3}
\end{matrix}
$$

The procedure is identical to Gaussian elimination in the first stage. Eliminating x_1 from the second and third rows gives

$$
\begin{matrix}
 \\
\text{A2}-(-4.125)\text{A1} \\
\text{A3}-(3)\text{A1}
\end{matrix}
\begin{bmatrix}
-8 & -4 & -17 & 85 \\
0 & (-0.5) & 1.875 & -8.375 \\
0 & 2 & -6 & 26
\end{bmatrix}
\begin{matrix}
\text{row A1} \\
\text{row B2} \\
\text{row B3}
\end{matrix}
$$

At the next stage, x_2 is eliminated from both the first and third equations or rows. Thus, since $-4/(-0.5) = 8$ and $2/(-0.5) = -4$, we have

$$
\begin{matrix}
\text{A1}-(8)\text{B2} \\
 \\
\text{B3}-(-4)\text{B2}
\end{matrix}
\begin{bmatrix}
-8 & 0 & -32 & 152 \\
0 & -0.5 & 1.875 & -8.375 \\
0 & 0 & (1.5) & -7.5
\end{bmatrix}
\begin{matrix}
\text{row B1} \\
\text{row B2} \\
\text{row C3}
\end{matrix}
$$

In the final stage x_3 is eliminated from the first and second rows. Now since $-32/1.5 = -21.333...$ and $1.875/1.5 = 1.25$ we have

$$\begin{array}{l} \text{B1--(--21.33...)C3} \\ \text{C2--(1.25)C3} \\ \end{array} \begin{bmatrix} -8 & 0 & 0 & -8 \\ 0 & -0.5 & 0 & 1 \\ 0 & 0 & 1.5 & -7.5 \end{bmatrix} \begin{array}{l} \text{row C1} \\ \text{row C2} \\ \text{row C3} \end{array}$$

From this array it can be seen that values for x_1, x_2 and x_3 are immediately available.

Since the need for back-substitution has been avoided it may appear that this procedure is more efficient than simple Gaussian elimination. This is not the case however; more computation is required in the Gauss-Jordan than the Gauss procedure. It can be shown, Fox (1964), that the number of arithmetic operations required in each elimination process to solve n equations with a single right hand side vector is:

Gauss elimination with back substitution:
       Multiplications and divisions: $n(n^2 + 3n + 2)/3 \approx n^3/3$ for large n
       Additions and subtractions: $n(2n^2 + 3n - 5)/6 \approx n^3/3$ for large n

Gauss-Jordan elimination:
       Multiplications and divisions: $n(n^2 + 2n + 1)/2 \approx n^3/2$ for large n
       Additions and subtractions: $n(n^2 - 1)/2 \approx n^3/2$ for large n

For large values of n Gauss-Jordan elimination involves approximately 50% more arithmetic operations than does Gauss elimination.

3.9 MATRIX INVERSION BY ELIMINATION

Section 3.3 describes how the inverse of an $n \times n$ matrix can be determined by solving n sets of n linear equations. These equations can be solved by Gauss or Gauss-Jordan elimination and it has been shown, Fox (1964), that in this situation these elimination procedures involve an identical number of arithmetic operations. If unnecessary multiplications by zero and unity are avoided then to invert an $n \times n$ matrix by elimination requires a total of $n^3 + n - 1$ multiplications and divisions and a total of $n(n^2 - 2n + 1)$ additions and subtractions.

We will now examine the process of matrix inversion by Gauss-Jordan elimination. The n sets of equations to be solved, equations (3.3.3), all have identical left-hand sides and so they can be solved for all the right-hand sides at the same time. For example, consider inverting the matrix

$$\begin{bmatrix} 2 & 2 & -4 \\ 1 & -3 & 1 \\ 2 & -2 & -4 \end{bmatrix}$$

This is equivalent to solving the matrix equation

$$\begin{bmatrix} 2 & 2 & -4 \\ 1 & -3 & 1 \\ 2 & -2 & -4 \end{bmatrix} \begin{bmatrix} x_{11} & x_{12} & x_{13} \\ x_{21} & x_{22} & x_{23} \\ x_{31} & x_{32} & x_{33} \end{bmatrix} = \begin{bmatrix} 1 & 0 & 0 \\ 0 & 1 & 0 \\ 0 & 0 & 1 \end{bmatrix}$$

Thus the augmented array is

$$\begin{bmatrix} (2) & 2 & -4 & 1 & 0 & 0 \\ 1 & -3 & 1 & 0 & 1 & 0 \\ 2 & -2 & -4 & 0 & 0 & 1 \end{bmatrix} \begin{matrix} \text{row A1} \\ \text{row A2} \\ \text{row A3} \end{matrix}$$

We proceed with the usual Gauss-Jordan elimination but operate on all of the right-hand columns simultaneously. Thus stage one eliminates the coefficients of the **A** in the second and third row of the first column and reduces the coefficient on the leading diagonal of the first column to unity, giving

$$\begin{matrix} \text{A1/2} \\ \text{A2--(1)B1} \\ \text{A3--(2)B1} \end{matrix} \begin{bmatrix} 1 & 1 & -2 & 0.5 & 0 & 0 \\ 0 & (-4) & 3 & -0.5 & 1 & 0 \\ 0 & -4 & 0 & -1 & 0 & 1 \end{bmatrix} \begin{matrix} \text{row B1} \\ \text{row B2} \\ \text{row B3} \end{matrix}$$

The next stage is to eliminate the coefficients of **A** in the first and third rows of the second column and to reduce the coefficient on leading diagonal of this column to unity. This gives

$$\begin{matrix} \text{B1--(1)C2} \\ \text{B2/(--4)} \\ \text{B3--(--4)C2} \end{matrix} \begin{bmatrix} 1 & 0 & -1.25 & 0.375 & 0.25 & 0 \\ 0 & 1 & -0.75 & 0.125 & -0.25 & 0 \\ 0 & 0 & (-3) & -0.5 & -1 & 1 \end{bmatrix} \begin{matrix} \text{row C1} \\ \text{row C2} \\ \text{row C3} \end{matrix}$$

The final elimination gives

$$\begin{matrix} \text{C1--(--1.25)D3} \\ \text{C2--(--0.75)D3} \\ \text{C3/(--3)} \end{matrix} \begin{bmatrix} 1 & 0 & 0 & 0.5833 & 0.6667 & -0.4167 \\ 0 & 1 & 0 & 0.25 & 0 & -0.25 \\ 0 & 0 & 1 & 0.1667 & 0.3333 & -0.3333 \end{bmatrix} \begin{matrix} \text{row D1} \\ \text{row D2} \\ \text{row D3} \end{matrix}$$

In practice it is not necessary for the computer to store the $n \times 2n$ augmented array. At any stage only n columns (3 in this example) contain coefficients other than 0 and 1 so that only the essential n columns need be preserved. These essential columns move one place to the right at each stage.

Program 3.9.1 inverts a matrix using Gauss-Jordan elimination and also computes the value of its determinant. Only the essential columns of the augmented matrix are stored and, to maintain accuracy, partial pivoting is included in the program. Some examples of the use of this program are given in section 3.19.

```
100 REM ---------------------------------
110 REM MATRIX INV:GAUSS-JORDAN JP/GL (C)
120 REM ---------------------------------
130 PRINT "MATRIX INVERSION"
140 PRINT "GAUSS-JORDAN ELIMINATION"
150 PRINT "WITH PARTIAL PIVOTING"
160 READ N
170 DIM A(N,N),R(N),T(N)
175 REM READ COEFFICIENTS OF MATRIX
180 FOR I=1 TO N
190     FOR J=1 TO N
200         READ A(I,J)
210     NEXT J
220 NEXT I
230 D=1
240 FOR K=1 TO N
245     REM CHOOSE PIVOTS
250     P=0
260     FOR I=1 TO N
270         IF K=1 THEN 310
280         FOR L=1 TO K-1
290             IF I=R(L) THEN 330
300         NEXT L
310         IF ABS(A(I,K))<=ABS(P) THEN 330
320         P=A(I,K):R(K)=I
330     NEXT I
340     IF P=0 THEN PRINT "ZERO PIVOT":END
350     D=D*P:P0=1/P
355     REM ELIMINATION PROCEDURE
360     FOR J=1 TO N
370         A(R(K),J)=A(R(K),J)*P0
380     NEXT J
390     A(R(K),K)=P0
400     FOR I=1 TO N
410         IF I=R(K) THEN 470
420         FOR J=1 TO N
430             IF J=K THEN 450
440             A(I,J)=A(I,J)-A(I,K)*A(R(K),J)
450         NEXT J
460         A(I,K)=-A(I,K)*P0
470     NEXT I
480 NEXT K
490 FOR J=1 TO N
500     FOR I=1 TO N
510         T(I)=A(R(I),J)
520     NEXT I
530     FOR I=1 TO N
540         A(I,J)=T(I)
550     NEXT I
560 NEXT J
```

Program 3.9.1. (continues).

```
570 FOR I=1 TO N
580       FOR J=1 TO N
590             T(R(J))=A(I,J)
600       NEXT J
610       FOR J=1 TO N
620             A(I,J)=T(J)
630       NEXT J
640 NEXT I
650 FOR K=1 TO N
660       T(R(K))=K
670 NEXT K
680 FOR I=1 TO N
690       FOR J=1 TO N-1
700             IF T(J)<=T(J+1) THEN 730
710             P=T(J):T(J)=T(J+1)
720             T(J+1)=P:D=-D
730       NEXT J
740 NEXT I
750 PRINT:PRINT "INVERSE MATRIX"
760 FOR I=1 TO N
770       FOR J=1 TO N
780             PRINT A(I,J);" ";
790       NEXT J
800       PRINT
810 NEXT I
820 PRINT:PRINT "DETERMINANT = ";D
830 END
840 DATA 3 :REM SIZE OF MATRIX
850 DATA 2,2,-4,1,-3,1,2,-2,-4 :REM COEFFICIENTS OF MATRIX
```

Program 3.9.1.

3.10 ILL-CONDITIONED EQUATIONS

Consider the equations

$$2x - 2y = 2$$
$$x - Cy = 2 \qquad\qquad (3.10.1)$$

where initially we will let $C = 0.99999$. These equations can be solved by elimination to give following triangular form:

$$2x - 2y = 2$$
$$.00001y = 1$$

Thus, by back-substitution, $y = 100000$ and $x = 100001$. Table 3.10.1 shows the dramatic effect on the solution of these equations if the coefficient C is changed by a small quantity. Such equations are called "ill-conditioned" and the phenomena

Table 3.10.1. The effect of a small change in c on the solution of (3.10.1).

c	Solution		Determinant
	x	y	
.99998	50001	50000	.00004
.99999	100001	100000	.00002
1	* NO SOLUTION *		0
1.00001	-99999	-100000	-.00002
1.00002	-49999	-50000	-.00004

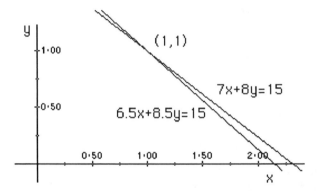

Fig. 3.10.1. Graph of two equations having an almost identical slope. Any small change in the equation coefficients will produce a large change in the point of intersection.

is illustrated for a pair of equations by Fig. 3.10.1. The gradients of the two equations are almost identical and this is the cause of the ill-conditioning. Thus a small change in the gradient or intercept in one equation causes a dramatic change in the point of intersection and thus in the solution of the equations. A formal and more general definition is that a system of linear equations is ill-conditioned if small changes in the coefficients of the system lead to far greater changes in the solution.

Ill-conditioned equations can present considerable numerical difficulties for the following reasons:

(1)   Since the solution is very sensitive to small changes in the coefficient matrix or in the right hand side vector it is essential that these coefficients are accurate. If they have been derived from experimental data then small errors will render the solution of the equations meaningless. Even if the coefficients are known exactly it may be that the number of significant digits required to specify them exceeds the number of digits that are available on a particular computer. Double precision working may be required.

(2)   In the elimination process it is necessary to compute expressions of the form
new(a_{ij}) = old$(a_{ij}) - f_{ik}$old(a_{kj}) where f_{ik} = old(a_{ik})/old(a_{kk}). If the equations are
ill-conditioned then, for   some values of i and j,   old(a_{ij}) and f_{ik}old(a_{kj}) will
be almost equal and, unless an adequate number of significant digits are carried
in the arithmetic operations, the   value calculated for the   new(a_{ij}) will be
in error. Furthermore, this poor estimate for new(a_{ij}) will then be used in
subsequent   stages of the elimination process so that the accumulation of
errors will be both rapid and inevitable.

Solving ill-conditioned equations by elimination will give a solution which must
be suspect. However, recognising such equations is difficult. It might be thought
that such equations would have a coefficient matrix with a very small determinant;
since equations with a determinant equal to zero are singular and have no solution,
(the ultimate in ill-conditioning!). It can easily be demonstrated that this is not the
case by multiplying each of the equations of (3.10.1) by 1000, thereby making the
absolute value of the determinant   equal   to 20 but leaving the equations ill-
conditioned.

3.11 MEASURING THE DEGREE OF ILL-CONDITIONING

Several methods of measuring the degree of ill-conditioning have been proposed but
they all involve considerable computation. The notion of using the determinant of
the coefficient matrix as an indicator of the condition of system of equations is useful
if each equation is   normalised so that the sum of the squares of its coefficients is
equal to unity. For a set of equations that have not been normalised in this way,
then their condition can be measured by the quantity V, where V is defined as

$$V(\mathbf{A}) = |\det(\mathbf{A})|/(p_1 p_2 \ldots p_n) \qquad (3.11.1)$$

Here **A** is the $n \times n$ coefficient matrix and

$$p_i = \sqrt{(a_{i1}^2 + a_{i2}^2 + \ldots + a_{in}^2)}$$

If $V(\mathbf{A}) = 0$ the equations are singular, if $V(\mathbf{A}) = 1$ the equations are perfectly conditioned in this case the array must be a diagonal matrix. For a system of equations
for which the system of coefficients is the identity matrix the equations represent
mutually perpendicular hyperplanes in n dimensional space. For example, for three
equations based on the identity matrix the equations represent three mutually
perpendicular planes. The closer that $V(\mathbf{A})$ is to unity, the better the condition of the
equations.

Another measure of the conditioning of a matrix is given by the value of

$$\|A\| \|A^{-1}\|$$

Here the $\|A\|$ indicates a particular norm of the matrix is taken. The norm is a measure of the size of a matrix and we can specify a number of alternative norms. For example the $\|A\|_1$ norm is defined as the maximum row sum and $\|A\|_\infty$ as the maximum column sum. Once a norm is specified the quantity $\|A\| \|A^{-1}\|$ may be calculated although in practice this is difficult because of the problem of computing A^{-1} accurately and reliably. In certain cases however this can be done and it is then possible to relate the accuracy of computation to the condition number given by $\|A\| \|A^{-1}\|$. For example if the condition number is 10^p then p decimal digits may be lost during elimination. The seriousness of this in terms of calculations will depend on how the value of p relates to the computer word length. If the condition number is greater than $1/\sqrt{\varepsilon}$, where ε is the machine accuracy, then the user should proceed with caution in any subsequent computation. Machine accuracy is discussed in Chapter 11.

It is often wrongly assumed that the accuracy of matrix inversion can be checked by verifying that $AB^* = I$ where B^* is the estimate of the inverse of A. This is not so, a poor estimate for the inverse of A can sometimes still appear to satisfy this relationship. Similarly, for the equations $Ax = b$, if the residual vector r is defined by

$$r = Ax^* - b$$

where x^* is the estimate of x, then the accuracy of the solution of a set of simultaneous equations cannot be verified by showing that r is close to a vector of zeros. The danger of this can be illustrated by the following example. Solving (3.10.1) with $C = 0.99999$ using a computer that calculates using only seven significant digits gives the following results:

$$x = 99864.38 \qquad y = 99865.38$$

These solutions are in error by approximately 0.13% but, using the same computer, the calculated residual vector was zero!

It must be stressed that the problems related to pivoting described in section 3.5 are not related to the problem of ill-conditioning. For example consider the equations (3.5.1). It was shown that as d tended to zero the solution became very inaccurate unless the equations were rearranged into the form (3.5.2), thus changing the pivot. However, using the measure of condition given by (3.11.1) we find that $V(A) = 0.89$ when $d = 0$. This indicates that the equations are quite well conditioned.

3.12 TRIANGULAR DECOMPOSITION

It is always possible to replace a square matrix A by the product of a lower triangular matrix L, a diagonal matrix D and upper triangular matrix U. Thus

$$A = LDU \qquad (3.12.1)$$

The lower and upper triangular matrices have the elements of their leading diagonal equal to one. Sometimes one of the triangular matrices is combined with the diagonal matrix to produce a new triangular matrix that does not have unit coefficients on the leading diagonal. Thus

$$A = L_1 U \qquad (3.12.2)$$

where $L_1 = LD$. The attraction of decomposition, where L_1 and U are both determined, becomes apparent if it is necessary to solve

$$Ax = b \qquad (3.12.3)$$

for many different values of b. In such case, replacing A by $L_1 U$ gives

$$L_1 Ux = b \qquad (3.12.4)$$

Writing $Ux = y$, (3.12.4) can be replaced by a pair of matrix equations

$$L_1 y = b \qquad (3.12.5)$$
$$Ux = y \qquad (3.12.6)$$

This pair of equations are easily solved because of the triangular form of L_1 and U. For each vector b, y can be determined from (3.12.5) by forward-substitution; x can be determined from (3.12.6) by back-substitution. Both of these procedures are efficient.

It is instructive to re-examine the Gauss and Gauss-Jordan elimination procedures in the light of triangular decomposition. The Gaussian elimination process effectively solves (3.12.5) to determine y from b without explicitly determining L_1 and, simultaneously, determines U explicitly. Back-substitution is, of course, the process of solving (3.12.6). Gauss-Jordan elimination solves (3.12.5) and (3.12.6) without explicitly determining either L_1 or U. For further details of triangular decomposition see Fox, (1964).

The procedure used to decompose A is similar to the elimination procedure but the order of the arithmetic operations is changed. The procedure is sometimes called compact elimination or Crout's procedure. Program 3.12.1 decomposes a matrix using this algorithm but partial pivoting has not been included in the program. For this reason this program will give inaccurate results if the pivots become very small or fail if they become zero. After decomposition the program determines a solution for any given vector **b**.

```
100   REM ----------------------------
110   REM CROUT DECOMPOSITION JP&GL (C)
120   REM ----------------------------
130   READ N
140   DIM A(N,N),B(N),X(N),Y(N)
150   PRINT "LU DECOMPOSITION & SOLUTION OF EQUATIONS"
160   PRINT:PRINT "ORIGINAL MATRIX A=LU"
165   REM READ AND PRINT COEFFICIENTS OF MATRIX
170   FOR I=1 TO N
180      FOR J=1 TO N
190         READ A(I,J)
200         PRINT TAB(5*J);A(I,J);
210      NEXT J
220      PRINT
230   NEXT I
240   D=1
245   REM DECOMPOSITION PROCESS
250   FOR K=1 TO N
260   IF K=1 THEN 340
270      FOR I=K TO N
280         S=0
290         FOR J=1 TO K-1
300            S=S+A(I,J)*A(J,K)
310         NEXT J
320         A(I,K)=A(I,K)-S
330      NEXT I
340      IF K=N THEN 450
350      FOR J=K+1 TO N
360         S=0
370         IF K=1 THEN 410
380         FOR I=1 TO K-1
390            S=S+A(K,I)*A(I,J)
400         NEXT I
410         IF A(K,K)=0 THEN PRINT "ZERO PIVOT":END
420         A(K,J)=(A(K,J)-S)/A(K,K)
430      NEXT J
440      D=D*A(K,K)
450   NEXT K
460   IF A(N,N)=0 THEN PRINT "ZERO PIVOT":END
470   D=D*A(N,N)
475   REM PRINT OUT UPPER AND LOWER TRIANGULAR MATRIX
480   PRINT:PRINT "UPPER TRIANGLE U"
490   FOR I=1 TO N
500      PRINT TAB(5*I);1;
510      IF I=N THEN 550
520      FOR J=I+1 TO N
530         PRINT TAB(5*J);A(I,J);
540      NEXT J
```

Program 3.12.1. (continues).

```
550     PRINT
560   NEXT I
570   PRINT:PRINT "LOWER TRIANGLE L"
580   FOR I=1 TO N
590     FOR J=1 TO I
600        PRINT TAB(5*J);A(I,J);
610     NEXT J
620     PRINT
630   NEXT I
640   PRINT:PRINT "DETERMINANT = ";D
650   PRINT
660   READ M
670   FOR C=1 TO M
675     REM READ RIGHT HAND SIDE VECTOR
680     FOR I=1 TO N
690        READ B(I)
700     NEXT I
705     REM SOLVE SOLVE SOLVE LY=B AND UX=Y
710     Y(1)=B(1)/A(1,1)
720     FOR I=2 TO N
730        S=B(I)
740        FOR J=1 TO I-1
750           S=S-A(I,J)*Y(J)
760        NEXT J
770        Y(I)=S/A(I,I)
780     NEXT I
790     X(N)=Y(N): PRINT "B(";N;") = ";B(N);TAB(20);"X(";N;") = ";X(N)
800     FOR I=N-1 TO 1 STEP -1
810        S=Y(I)
820        FOR J=N TO I+1 STEP -1
830           S=S-A(I,J)*X(J)
840        NEXT J
850        X(I)=S
860        PRINT "B(";I;") = ";B(I);TAB(20);"X(";I;") = ";X(I)
870     NEXT I
880     PRINT
890   NEXT C
900   END
910   DATA 3 : REM SIZE OF MATRIX
920   DATA -8,-4,-17,33,16,72,-24,-10,-57 : REM COEFFICIENTS OF MATRIX
930   DATA 1: REM NUMBER OF RIGHT HAND SIDE VECTORS
940   DATA 85,-359,281:REM RIGHT HAND SIDE VECTOR(S)
```

Program 3.12.1.

In the analysis of many physical systems sets of linear equations arise that have coefficient matrices that are both symmetric and positive definite. A positive definite matrix A is one for which $x^T A x$ is positive for any real, non-zero x. A necessary and sufficient condition for a matrix to be positive definite is that all its eigenvalues are positive. Eigenvalues are discussed in Chapter 10. If the matrix A is symmetric and positive definite it can be decomposed into the product of lower triangular matrix L and its transpose. Thus

$$A = LL^T \qquad\qquad (3.12.7)$$

To determine **L**, Cholesky's procedure is used. This is an elimination method but the details of the algorithm are not described here. In the procedure it is necessary to take the square root of the elements on the leading diagonal of the coefficient matrix. However, for a positive definite matrix the terms on its leading diagonal are positive, Fox (1964), and so no difficulty will arise when taking the square root of these terms. The method is very stable and partial pivoting is not required, nor could it be used since row interchanges would destroy the symmetry of the coefficient matrix and the procedure would be invalid.

Program 3.12.2 uses Cholesky's procedure to symmetrically decompose a symmetric matrix and then determines a solution for any given vector **b**. Table 3.12.1 shows the output from this program. The program has been used to solve a set of simultaneous equations for two different right-hand side vectors.

If the symmetric coefficient matrix is not positive definite then terms on the leading diagonal can be zero or negative and program 3.12.2 will fail. It can be shown, Wilkinson (1965), that if such a matrix can be decomposed into the form LL^T then in each column of **L**, every element will either be real or imaginary. Thus complex

Table 3.12.1. Use of Cholesky decomposition to solve a system of equations with two different right-hand sides.

```
CHOLESKI DECOMPOSITION AND SOLUTION OF EQUATIONS

ORIGINAL MATRIX
  2    3    4
  3    6    7
  4    7   10

LOWER TRIANGULAR MATRIX L
  1.414214
  2.12132      1.224745
  2.828427      .8164963      1.154701

B( 3 ) = 48          X( 3 ) = 3
B( 2 ) = 36          X( 2 ) = 2
B( 1 ) = 20          X( 1 ) = .999999

B( 3 ) = 21          X( 3 ) = 1
B( 2 ) = 16          X( 2 ) = .9999997
B( 1 ) = 9           X( 1 ) = .9999999
```

arithmetic must be used in the computation. For example

$$\begin{bmatrix} 1 & 2 & 3 \\ 2 & -5 & 9 \\ 3 & 9 & 6 \end{bmatrix} = \begin{bmatrix} 1 & 0 & 0 \\ 2 & 3i & 0 \\ 3 & -i & 2i \end{bmatrix} \begin{bmatrix} 1 & 2 & 3 \\ 0 & 3i & -i \\ 0 & 0 & 2i \end{bmatrix}$$

Generally Cholesky's procedure is only used on symmetric matrices that are positive definite.

```
100    REM -------------------------------
110    REM CHOLESKY DECOMPOSITION JP/GL (C)
120    REM -------------------------------
130    PRINT "CHOLESKY DECOMPOSITION AND SOLUTION OF EQUATIONS"
140    READ N
150    DIM A(N,N),B(N),Y(N),X(N)
155    REM READ UPPER TRIANGLE OF MATRIX COEFFICIENTS
160    FOR J=1 TO N
170       FOR I=J TO N
180          READ A(J,I)
190          A(I,J)=A(J,I)
200       NEXT I
210    NEXT J
220    PRINT:PRINT "ORIGINAL MATRIX "
230    FOR I=1 TO N
240       FOR J=1 TO N
250          PRINT A(I,J);"   ";
260       NEXT J
270       PRINT
280    NEXT I
285    REM MATRIX DECOMPOSITION
290    A(1,1)=SQR(A(1,1))
300    FOR K=2 TO N
310       A(K,1)=A(K,1)/A(1,1)
320       FOR I=2 TO K-1
330          S=0
340          FOR J=1 TO I-1
350             S=S+A(I,J)*A(K,J)
360          NEXT J
370          A(K,I)=(A(K,I)-S)/A(I,I)
380       NEXT I
390       S=0
400       FOR J=1 TO K-1
410          S=S+A(K,J)^2
420       NEXT J
430       A(K,K)=SQR(A(K,K)-S)
440    NEXT K
450    PRINT
460    PRINT "LOWER TRIANGULAR MATRIX L"
470    FOR I=1 TO N
480       FOR J=1 TO I
490          PRINT A(I,J);"     ";
500       NEXT J
510       PRINT
```

Program 3.12.2.(continues).

```
520    NEXT I
530    PRINT
535    REM READ IN NUMBER OF RIGHT HAND VECTORS
540    READ M
550    FOR C=1 TO M
560        FOR I=1 TO N
570            READ B(I)
580        NEXT I
585        REM SOLVING LY=B AND UX=B
590        Y(1)=B(1)/A(1,1)
600        FOR I=2 TO N
610            S=B(I)
620            FOR J=1 TO I-1
630                S=S-A(I,J)*Y(J)
640            NEXT J
650            Y(I)=S/A(I,I)
660        NEXT I
670        X(N)=Y(N)/A(N,N):PRINT "B(";N;")=";B(N);TAB(20);"X(";N;")=";X(N)
680        FOR I=N-1 TO 1 STEP -1
690            S=Y(I)
700            FOR J=N  TO I+1 STEP -1
710                S=S-A(J,I)*X(J)
720            NEXT J
730            X(I)=S/A(I,I)
740            PRINT "B(";I;") = ";B(I);TAB(20);"X(";I;") = ";X(I)
750        NEXT I
760        PRINT
770    NEXT C
780    END
790    DATA 3:REM SIZE OF SYMMETRIC MATRIX
800    DATA 2,3,4,6,7,10: REM COEFFICIENTS OF UPPER TRIANGLE OF MATRIX
810    DATA 2 :REM NUMBER OF RIGHT HAND VECTORS
820    DATA 20,36,48,9,16,21 :REM RIGHT HAND VECTORS
```

Program 3.12.2.

3.13 SYSTEMS WITH SPARSE AND TRIDIAGONAL MATRICES

When physical systems are modelled the resulting coefficient matrix is often sparse or tridiagonal. A sparse matrix is one in which a large proportion of the elements of the coefficient array are zero; a tridiagonal matrix is one in which all terms except those on the leading diagonal and the two diagonals adjacent to it are zero. Both types of matrix can be stored more economically than is the case for a fully populated matrix and procedures for determining the solution of a set of linear equation with these types of coefficient matrix are now presented.

Consider firstly the solution of systems of equations with a sparse coefficient matrices. Because of the importance of this problem several algorithms have been developed which either reduce the amount of memory required to store the coefficient matrix or reduce the number of calculations to obtain a solution or both. One such algorithm for sparse symmetrical matrices is that given by Jennings

(1966) and is briefly described here. The method is based on elimination but to economise on storage the coefficient matrix is stored in an unusual form. The symmetrical coefficient matrix, with an arbitrary pattern of non-zero coefficients, is stored in a manner that is best explained by an example. Suppose the coefficient matrix is as follows:

$$\begin{bmatrix} 3 & & & & & \\ 1 & 2 & & \text{symm} & & \\ 1 & 0 & 3 & & & \\ 0 & 0 & 4 & 9 & & \\ 0 & 0 & 0 & 3 & 8 & \\ 0 & 1 & 0 & 5 & 0 & 7 \end{bmatrix}$$

Only the lower triangle of this matrix is displayed since it is symmetric. In the Jennings' algorithm these coefficients are stored in a vector thus

[3 1 2 1 0 3 4 9 3 8 1 0 5 0 7]

Note that the coefficients below and including the leading diagonal are stored in sequence by rows except that all the zero coefficients to the left of the first non-zero coefficient in each row are omitted. In addition, a second vector stores the position, in the above vector, of the coefficient on the leading diagonal of the original coefficient matrix. In this example this vector would be

[1 3 6 8 10 15]

The algorithm then uses compact elimination followed by back-substitution to determine the final solutions. Program 3.13.1 is based on Jennings' algorithm. The data must be presented to the program in the manner described; this avoids the need to place a great number of zeros in the data statements. The software offers the user the choice of either inverting the coefficient matrix (by solving n sets of equations with appropriate right-hand side vectors simultaneously) or solving the set of linear equations with as many different right-hand sides as required.

```
100   REM ------------------------------------
110   REM MATRIX INVERSION: JENNINGS JP/GL (C)
120   REM ------------------------------------
130   MAX=0
140   PRINT "MATRIX INVERSION OF SYMMETRIC SPARSE MATRIX"
150   PRINT "USING JENNINGS METHOD"
160   READ U,N
170   DIM A(U),S(N)
175   REM EFFECTIVE COEFFICIENTS OF MATRIX
180   FOR I=1 TO U
190      READ A(I)
200   NEXT I
225   REM READ POSITION IN VECTOR A  OF COEFFS ON LEADING DIAGONAL
230   FOR I=1 TO N
240      READ S(I)
245      REM FIND MAXIMUM NUMBER OF EFFECTIVE COEFFICIENTS IN ANY ROW
250      IF I=1 THEN 280
260      Z1=S(I)-S(I-1)
270      IF Z1>MAX THEN MAX=Z1
280   NEXT I
290   READ T0
300   T=T0
310   IF T0=0 THEN T=N
320   DIM C(MAX),B(N,T)
330   IF T0=0 THEN 410
335   REM READ THE RIGHT HAND VECTOR(S)
340   FOR I=1 TO T
350      FOR J=1 TO N
360         READ B(J,I)
370      NEXT J
390   NEXT I
400   GOTO 470
405   REM SET ARRAY B EQUAL TO IDENTITY ARRAY FOR MATRIX INVERSION
410   FOR I=1 TO N
420      FOR J=1 TO N
430         B(I,J)=0
440      NEXT J
450      B(I,I)=1
460   NEXT I
465   REM BEGIN ELIMINATION
470   X=A(1)
480   PRINT "PIVOT = ";X
490   DET=X:R=1
500   FOR J=1 TO N
510      Q=S(J)
520      IF J=1 THEN 730
530      K=Q-R:I=J-K:P=0
540      FOR H=1 TO K
550         G=P:R=R+1:I=I+1
560         P=S(I):M=P-H+1
570         IF M<=G THEN M=G+1
580         N0=H+M-P:X=A(R)
590         GOTO 620
```

Program 3.13.1. (continues)

```
600         X=X-A(M)*C(N0)
610         M=M+1:N0=N0+1
620         IF M<P THEN 600
630         C(H)=X
640         A(R)=A(P)*X
650         IF H=K THEN 690
660         FOR M=1 TO T
670            B(J,M)=B(J,M)-B(I,M)*X
680         NEXT M
690      NEXT H
700      DET=DET*X
710      PRINT "PIVOT = ";X
730      A(Q)=1/X
740      FOR M=1 TO T
750         B(J,M)=B(J,M)*A(Q)
760      NEXT M
770   NEXT J
780   FOR J=N TO 2 STEP -1
790      Q=P-1:K=J-1:P=S(K)
800      IF P=Q THEN 870
820      FOR M=1 TO T
830         B(K,M)=B(K,M)-A(Q)*B(J,M)
840      NEXT M
850      Q=Q-1:K=K-1
860      GOTO 800
870   NEXT J
875   PRINT"SOLUTION(S)"
880   FOR I = 1 TO N
885      PRINT I;
890      FOR J = 1 TO T
900         PRINT TAB(11*J); B(I,J);
910      NEXT J
920      PRINT
930   NEXT I
940   PRINT "DETERMINANT = ";DET
950   END
960   DATA 15:REM NUMBER OF REQUIRED COEFFICIENTS
970   DATA 6:REM NUMBER OF ROWS
980   DATA 3: REM REQUIRED COEFFICIENTS OF ROW 1
990   DATA 1,2:REM REQUIRED COEFFICIENTS OF ROW 2
1000  DATA 1,0,3:REM ETC
1010  DATA 4,9
1020  DATA 3,8
1030  DATA 1,0,5,0,7:REM REQUIRED COEFFICIENTS OF LAST ROW
1040  DATA 1,3,6,8,10,15:REM REFERENCE NO OF COEFF ON LEADING DIAGONAL
1050  DATA 1:REM NUMBER OF RIGHT HAND SIDE VECTORS(ZERO FOR INVERSION)
1060  DATA 12,-14,-23,18,-76,23:REM COEFFICIENTS OF RHS  VECTOR
```

Program 3.13.1.

Turning now to the tridiagonal system, the most efficient method of storage is to place the elements of the leading diagonal in one vector and the elements of the diagonals immediately above and below the leading diagonal in two other vectors. Gaussian elimination and back-substitution can then be used to solve the equations. This is done in program 3.13.2. Examples of the use of these programs and the savings in computer storage and time that they offer are given in section 3.19.

When solving systems of equations with a sparse or tridiagonal coefficient matrix, iterative methods can sometimes be used to advantage. These methods are introduced in the next section.

```
100   REM -----------------------------
110   REM TRIDIAGONAL EQUATIONS JP/GL (C)
120   REM -----------------------------
130   PRINT "SOLUTION OF TRIDIAGONAL"
140   PRINT "EQUATIONS BY ELIMINATION"
150   READ N
160   DIM A(N),B(N),D(N),C(N)
165   REM READ LEADING DIAGONAL COEFFICIENTS
170   FOR I=1 TO N
180       READ D(I)
190   NEXT I
195   REM READ UPPER OFF-DIAGONAL COEFFICIENTS
200   FOR I=1 TO N-1
210       READ A(I)
220   NEXT I
225   REM READ LOWER OFF-DIAGONAL COEFFICIENTS
230   FOR I=1 TO N-1
240       READ B(I)
250   NEXT I
255   REM READ RIGHT HAND VECTOR
260   FOR I=1 TO N
270       READ C(I)
280   NEXT I
285   REM PERFORM ELIMINATION
290   FOR I=2 TO N
300       D(I)=D(I)-B(I-1)*A(I-1)/D(I-1)
310       C(I)=C(I)-B(I-1)*C(I-1)/D(I-1)
320   NEXT I
330   C(N)=C(N)/D(N)
340   PRINT:PRINT "X(";N;")" = ";C(N)
350   FOR I=N-1 TO 1 STEP -1
360       C(I)=(C(I)-A(I)*C(I+1))/D(I)
370       PRINT "X(";I;")" = ";C(I)
380   NEXT I
390   END
400   DATA 4 :REM NUMBER OF EQUATIONS
410   DATA 4,4,3,4 :REM DIAGONAL COEFFICIENTS
420   DATA 1,1,.5 :REM UPPER OFF-DIAGONAL COEFFICIENTS
430   DATA 2,1,2 :REM LOWER OFF-DIAGONAL COEFFICIENTS
440   DATA 3,0,5.5,8:REM RIGHT HAND VECTOR
```

Program 3.13.2.

3.14 INTRODUCTION TO ITERATIVE METHODS

A set of linear equations may be solved using iteration as an alternative to elimination. Theoretically, elimination methods produce "exact" solutions although in practice, because of the build up of rounding errors, the solutions produced by the elimination method are of limited accuracy. In contrast and in the absence of rounding errors iterative methods can only give an exact solution after an infinite number of iterations. In practice, however, a solution with an acceptable degree of accuracy can often be determined after a relatively small number of iterations.

3.15 JACOBI AND GAUSS-SEIDEL ITERATION

A simple iterative scheme is now presented. Consider the simultaneous equations

$$
\begin{aligned}
a_{11}x_1 + a_{12}x_2 + a_{13}x_3 + \ldots + a_{1n}x_n &= b_1 \\
a_{21}x_1 + a_{22}x_2 + a_{23}x_3 + \ldots + a_{2n}x_n &= b_2 \\
&\cdots\cdots\cdots\cdots\cdots\cdots\cdots\cdots \\
a_{n1}x_1 + a_{n2}x_2 + a_{n3}x_3 + \ldots + a_{nn}x_n &= b_n
\end{aligned}
\tag{3.15.1}
$$

We will begin by rearranging them to give

$$
\begin{aligned}
x_1 &= (b_1 - a_{12}x_2 - a_{13}x_3 - \ldots - a_{1n}x_n)/a_{11} \\
x_2 &= (b_2 - a_{21}x_1 - a_{23}x_3 - \ldots - a_{2n}x_n)/a_{22} \\
&\cdots\cdots\cdots\cdots\cdots\cdots\cdots\cdots \\
x_n &= (b_n - a_{n1}x_1 - a_{n2}x_2 - a_{n3}x_3 - \ldots - a_{n,n-1}x_{n-1})/a_{nn}
\end{aligned}
\tag{3.15.2}
$$

If initial (approximate) values of $x_1, x_2, x_3, \ldots, x_n$ are substituted in the right-hand side of these equations a new set of values for $x_1, x_2, x_3, \ldots, x_n$ can be determined. Thus, denoting the new value of x_i by x_i^* the iterative scheme is

$$
\begin{aligned}
x_1^* &= (b_1 - a_{12}x_2 - a_{13}x_3 - \ldots - a_{1n}x_n)/a_{11} \\
x_2^* &= (b_2 - a_{21}x_1 - a_{23}x_3 - \ldots - a_{2n}x_n)/a_{22} \\
&\cdots\cdots\cdots\cdots\cdots\cdots\cdots\cdots \\
x_n^* &= (b_n - a_{n1}x_1 - a_{n2}x_2 - a_{n3}x_3 - \ldots - a_{n,n-1}x_{n-1})/a_{nn}
\end{aligned}
\tag{3.15.3}
$$

Once we have computed all the new values of x_i these can be used in the right-hand side of (3.15.3) to compute a new set of values for x_i. This Jacobi or simultaneous iteration procedure can be terminated when the absolute differences between the old and new values of x_i are small.

The Jacobi iterative procedure can be expressed in matrix notation as follows. If the set of simultaneous equations $\mathbf{Ax} = \mathbf{b}$ are adjusted so that the terms on the leading diagonal of the coefficient matrix are all unity then the coefficient matrix $\mathbf{A}$

can be replaced by $L + I + U$ where L is a lower triangular matrix with zeros on its leading diagonal, U is an upper triangular matrix with zeros on its leading diagonal and I is the identity matrix. This process is called normalisation and it will also change the right-hand side of the equations. Letting $c_i = b_i/a_{ii}$, then

$$(L + I + U)x = c$$

Hence

$$Ix = c - (L + U)x$$

For iteration this becomes

$$x^* = c - (L + U)x \qquad (3.15.4)$$

where x^* is the new estimates for x. This matrix equation is equivalent to (3.15.3).

From (3.15.3) or its equivalent, (3.15.4), it is seen that the new estimates for x are computed from the old estimates and only when *all* the new estimates have been determined are they used in the right-hand side of the equation to perform the next iteration. An alternative procedure is to make use of the new estimates in the right-hand side of the equation as soon as they become available. This is called Gauss-Seidel or cyclic iteration and (3.15.3) becomes

$$x_1^* = (b_1 - a_{12}x_2 - a_{13}x_3 - \dots - a_{1n}x_n)/a_{11}$$
$$x_2^* = (b_2 - a_{21}x_1^* - a_{23}x_3 - \dots - a_{2n}x_n)/a_{22} \qquad (3.15.5)$$
$$\dots\dots\dots\dots\dots\dots$$
$$x_n^* = (b_n - a_{n1}x_1^* - a_{n2}x_2^* - a_{n3}x_3^* - \dots - a_{n,n-1}x_{n-1}^*)/a_{nn}$$

Similarly, (3.15.4) becomes

$$x^* = c - Lx^* - Ux \qquad (3.15.6)$$

Generally, Gauss-Seidel iteration converges more rapidly than the Jacobi iteration. It should be noted that once the equations are in the form of (3.15.2), reordering them may have an effect on the rate of convergence of the Gauss-Seidel iteration but will have no effect on the Jacobi iteration.

3.16 CONVERGENCE CRITERIA

The Jacobi and Gauss-Seidel methods do not always converge. The sufficient but not necessary conditions for convergence are as follows.

(1)   The matrix **A** should not contain a $p \times q$ submatrix of zeros, where $p + q = n$.

(2)   For each row of **A**, the absolute magnitude of the diagonal element should be at least as large as the sum of the absolute magnitudes of the other elements in the row and, in at least one row, larger than the sum.

A set of equations that satisfy condition (2) is said to be diagonally dominant. If the set of equations is not diagonally dominant it is often possible to make it so by rearranging the order of the equations or by forming new equations that are linear combinations of the original ones. Both iterative procedures converge very slowly if the equations are ill-conditioned.
    Programs 3.16.1 and 3.16.2 solve linear equations by Jacobi and Gauss-Seidel iteration respectively.

```
100    REM -------------------------
110    REM JACOBI ITERATION JP/GL (C)
120    REM -------------------------
130    PRINT "SOLUTION OF SIMULTANEOUS"
140    PRINT "EQUATIONS BY JACOBI ITERATION"
150    PRINT:PRINT "MAX CHANGE IN VECTOR";
160    INPUT EP
170    READ N
180    DIM A(N,N+1),X(N),Y(N)
185    REM READ IN THE COEFFICIENTS OF MATRIX
190    FOR I=1 TO N
200        FOR J=1 TO N
210            READ A(I,J)
220        NEXT J
230    NEXT I
235    REM READ IN COEFFICIENTS OF RIGHT HAND SIDE
240    FOR I=1 TO N
250        READ A(I,N+1)
260        X(I)=0
270    NEXT I
280    IT=0
285    REM PERFORM JACOBI ITERATIONS
290    FLAG=0:IT=IT+1
300    FOR I=1 TO N
310        Y(I)=A(I,N+1)
320        F=A(I,I): A(I,I)=0
330        FOR J=1 TO N
340            Y(I)=Y(I)-A(I,J)*X(J)
350        NEXT J
360        Y(I)=Y(I)/F: A(I,I)=F
370        IF ABS(Y(I)-X(I))>EP THEN FLAG=1
380    NEXT I
390    FOR I=1 TO N
400        X(I)=Y(I)
410    NEXT I
```

Program 3.16.1.(continues).

```
420   IF FLAG=1 THEN 290
430   PRINT
440   FOR I=1 TO N
450      PRINT"X(";I;") = ";Y(I)
460   NEXT I
470   PRINT:PRINT "NUMBER OF ITERATIONS = ";IT
480   END
490   DATA 3:REM NUMBER OF EQUATIONS
500   DATA 2,1,1,2,3,1,1,1,3: REM COEFFICIENTS OF MATRIX
510   DATA 5,9,6: REM RIGHT HAND VECTOR
```

Program 3.16.1.

```
100   REM --------------------------------
110   REM  GAUSS-SEIDEL ITERATION JP/GL (C)
120   REM --------------------------------
130   PRINT "SOLUTION OF SIMULTANEOUS EQUATIONS"
140   PRINT "USING GAUSS-SEIDEL ITERATION"
150   PRINT:PRINT "MAX CHANGE IN VECTOR";
160   INPUT EP
170   READ N
180   DIM A(N,N+1),X(N)
185   REM READ COEFFICIENT MATRIX
190   FOR I=1 TO N
200      FOR J=1 TO N
210         READ A(I,J)
220      NEXT J
230   NEXT I
235   REM READ RIGHT HAND VECTOR
240   FOR I=1 TO N
250      READ A(I,N+1)
260      X(I)=0
270   NEXT I
280   IT=0
285   REM START ITERATION
290   FLAG=0:IT=IT+1
300   FOR I=1 TO N
310      S=A(I,N+1): F=A(I,I): A(I,I)=0
320      FOR J=1 TO N
330         S=S-A(I,J)*X(J)
340      NEXT J
350      S=S/F :A(I,I)=F
360      IF ABS(S-X(I))>EP THEN FLAG=1
370      X(I)=S
380   NEXT I
390   IF FLAG=1 THEN 290
400   PRINT
410   FOR I=1 TO N
420   PRINT "X(";I;") = ";X(I)
430   NEXT I
440   PRINT:PRINT "NUMBER OF ITERATIONS =";IT
450   END
460   DATA 3 :REM NUMBER OF EQUATIONS
470   DATA 2,1,1,2,3,1,1,1,3 :REM MATRIX OF COEFFICIENTS
480   DATA 5,9,6 :REM RIGHT HAND VECTOR
```

Program 3.16.2.

Consider the application of these programs to the solution of

$$\begin{bmatrix} 1 & 1 & 3 \\ 2 & 1 & 1 \\ 2 & 3 & 1 \end{bmatrix} \begin{bmatrix} x_1 \\ x_2 \\ x_3 \end{bmatrix} = \begin{bmatrix} 6 \\ 5 \\ 9 \end{bmatrix} \qquad (3.16.1)$$

Iteration of these equations fails because the criteria for convergence is not met. However, rearranging the equations gives

$$\begin{bmatrix} 2 & 1 & 1 \\ 2 & 3 & 1 \\ 1 & 1 & 3 \end{bmatrix} \begin{bmatrix} x_1 \\ x_2 \\ x_3 \end{bmatrix} = \begin{bmatrix} 5 \\ 9 \\ 6 \end{bmatrix} \qquad (3.16.2)$$

These equation now meet the convergence criteria; in the first two equations the term on the diagonal is equal to the sum of the other terms and in the third equation the diagonal term is greater than the sum of the other terms. Tables 3.16.1 and 3.16.2 show steps in Jacobi and Gauss-Seidel iterations respectively to solve (3.16.2). These tables were produced using programs 3.16.1 and 3.16.2. The programs given in the text were adjusted to output the estimates of the solutions at stages in the iteration. It is seen that to obtain 3 digit accuracy requires about 8 iterations using the Gauss-Seidel method. In contrast approximately 70 iterations are required using Jacobi iteration.

Table 3.16.1. Solution of the linear equations (3.16.2) by Jacobi iteration.

```
SOLUTION OF SIMULTANEOUS
EQUATIONS BY JACOBI ITERATION

MAX CHANGE IN VECTOR?  1E-4

ITERATION          X(1)              X(2)              X(3)
   10            .5433145          1.527117          .6534365
   20            .8520728          1.847768          .8879785
   30            .9522139          1.950826          .9638135
   40            .9845637          1.984115          .9883108
   50            .9950136          1.994869          .996224
   60            .9983892          1.998342          .9987803
   70            .9994796          1.999465          .9996059
   80            .9998318          1.999827          .9998726
   90            .9999456          1.999944          .9999588

X( 1 ) =  .9999567
X( 2 ) =  1.999955
X( 3 ) =  .9999672

NUMBER OF ITERATIONS =  92
```

Table 3.16.2. Solution of the linear equations (3.16.2) by Gauss-Seidel iteration.

```
SOLUTION OF SIMULTANEOUS EQUATIONS
USING GAUSS-SEIDEL ITERATION

MAX CHANGE IN VECTOR?  1E-4

ITERATION       X(1)            X(2)            X(3)
    1           2.5             1.333333        .7222221
    2           1.472222        1.777778        .9166667
    3           1.152778        1.925926        .9737654
    4           1.050154        1.975309        .9915123
    5           1.016589        1.99177         .9972137
    6           1.005508        1.997256        .9990785
    7           1.001833        1.999086        .999694
    8           1.00061         1.999695        .9998981
    9           1.000203        1.999899        .9999661
   10           1.000068        1.999966        .9999887
   11           1.000023        1.999989        .9999962

X( 1 ) =  1.000023
X( 2 ) =  1.999989
X( 3 ) =  .9999962

NUMBER OF ITERATIONS =  11
```

Consider the equations

$$\begin{bmatrix} 2 & 1 & 1 \\ 2 & 2 & 1 \\ 1 & 1 & 3 \end{bmatrix} \begin{bmatrix} x_1 \\ x_2 \\ x_3 \end{bmatrix} = \begin{bmatrix} 5 \\ 9 \\ 6 \end{bmatrix} \tag{3.16.3}$$

According to the criteria given above convergence of the iteration for these equations is not guaranteed since, in the second equation, the coefficient on the diagonal is less than the sum of the other terms. In practice, with a starting vector [0,0,0] Jacobi iteration diverges but Gauss-Seidel iteration converges: a demonstration of the fact that the criteria for convergence given are sufficient but not necessary.

Gauss-Seidel iteration can be used to solve a set of equations with a banded or sparse coefficient matrix and an example of this is given in section 3.19. However, for maximum efficiency, the procedure should make use of the fact that the majority of the coefficients of the system matrix are zero and these should be ignored rather than multiplied by the current estimate of the solution.

3.17 SUCCESSIVE OVER-RELAXATION

A useful modification to the Gauss-Seidel iteration is the "successive over-relaxation" or SOR method. If x is added and subtracted from the right-hand side of (3.15.6) then

$$x^* = x + (c - Lx^* - Ux - x) \tag{3.17.1}$$

This is, of course, still the standard Gauss-Seidel scheme. However, (3.17.1) can be modified by the introduction of a constant ω so that

$$x^* = x + \omega(c - Lx^* - Ux - x) \tag{3.17.2}$$

or

$$x^* = (1 - \omega)x + \omega(c - Lx^* - Ux) \tag{3.17.3}$$

It can be formally proved that convergence can be obtained for values of ω in the range $0 < \omega < 2$. Comparing SOR with the Gauss-Seidel method a large reduction in the number of iterations can be achieved, given an efficient choice of ω. In practice ω should be chosen in the range $1 < \omega < 2$ but the precise choice of ω is a major problem. Finding the optimum value for ω depends on the particular problem and often requires careful analysis. Except in one or two special cases no simple formula exists to help in this choice and the rate of convergence is extremely sensitive to small changes in ω. A detailed study of this problem can be found in Isaacson and Keller (1966).

Program 3.17.1 solves simultaneous equations by the method of SOR. The user is required to choose a value of ω before iteration begins. Table 3.17.1 shows the number of iterations required in the solution of (3.16.3) for various values of ω; it is seen that the number of iterations required is sensitive to the value of ω.

Table 3.17.1. Number of iterations required to solve (3.16.3). Iteration stopped when the maximum change in vector is 10^{-5}.

ω	Iterations
1.0	18
1.1	14
1.2	10
1.3	13
1.4	16
1.5	20
1.6	30
1.7	59
1.8	367

```
100    REM ------------------------------------
110    REM SUCCESSIVE OVER RELAXATION JP/GL (C)
120    REM ------------------------------------
130    PRINT "SOLUTION OF SIMULTANEOUS EQUATIONS"
140    PRINT "BY SUCCESSIVE OVER RELAXATION"
150    PRINT:PRINT "MAX CHANGE IN VECTOR";
160    INPUT EP
170    PRINT "W FACTOR (1<W<2)";
180    INPUT W
190    READ N
200    DIM A(N,N+1),X(N),Y(N)
205    REM READ COEFFICIENTS OF MATRIX
210    FOR I=1 TO N
220        FOR J=1 TO N
230            READ A(I,J)
240        NEXT J
250    NEXT I
255    REM READ RIGHT HAND VECTOR
260    FOR I=1 TO N
270        READ A(I,N+1)
280        X(I)=0
290    NEXT I
300    IT=0
305    REM ITERATION STARTS
310    FLAG=0:IT=IT+1
320    FOR I=1 TO N
330    S=A(I,N+1): F=A(I,I):A(I,I)=0
340        FOR J=1 TO N
350            S=S-A(I,J)*X(J)
360        NEXT J
370        S=S/F: A(I,I)=F
375        REM OVER RELAXATION STEP
380        S=S*W+(1-W)*X(I)
390        IF ABS(S-X(I))>EP THEN FLAG=1
400        X(I)=S
410    NEXT I
420    IF FLAG=1 THEN 310
430    PRINT
440    FOR I=1 TO N
450        PRINT "X(";I;") = ";X(I)
460    NEXT I
470    PRINT:PRINT "NUMBER OF ITERATIONS = ";IT
480    END
490    DATA 3: REM NUMBER OF EQUATIONS
500    DATA 2,1,1,2,2,1,1,1,3 :REM COEFFICIENTS OF MATRIX
510    DATA 5,7,6 : REM RIGHT HAND VECTOR
```

Program 3.17.1

3.18 SYSTEMS WITH COMPLEX COEFFICIENTS

Problems can arise in science and engineering which lead to sets of simultaneous equations with complex coefficients. Let C be an $n \times n$ matrix of complex coefficients, **d** be a vector of complex coefficients and **z** be a vector of unknowns. It would be prudent to assume that these unknowns are also complex. Thus we are required to solve

$$\mathbf{Cz} = \mathbf{d} \qquad (3.18.1)$$

Let

$$\mathbf{C} = \mathbf{A} + i\mathbf{B}; \quad \mathbf{d} = \mathbf{f} + i\mathbf{h} \text{ and } \mathbf{z} = \mathbf{x} + i\mathbf{y}$$

where **A, B, f, h, x** and **y** are real. Substituting into (3.18.1) we have

$$(\mathbf{A} + i\mathbf{B})(\mathbf{x} + i\mathbf{y}) = \mathbf{f} + i\mathbf{h}$$

Hence

$$\mathbf{Ax} - \mathbf{By} + i(\mathbf{Ay} + \mathbf{Bx}) = \mathbf{f} + i\mathbf{h}$$

Equating the real and imaginary parts gives

$$\mathbf{Ay} + \mathbf{Bx} = \mathbf{h}$$
$$-\mathbf{By} + \mathbf{Ax} = \mathbf{f}$$

and combining these two equations into a single matrix equation we have

$$\begin{bmatrix} \mathbf{A} & \mathbf{B} \\ -\mathbf{B} & \mathbf{A} \end{bmatrix} \begin{bmatrix} \mathbf{y} \\ \mathbf{x} \end{bmatrix} = \begin{bmatrix} \mathbf{h} \\ \mathbf{f} \end{bmatrix} \qquad (3.18.2)$$

Thus we obtain $2n$ equations with real coefficient. For example, suppose we wish to solve the following complex equations.

$$\begin{bmatrix} 3 - 2i & 1 + i \\ 2 + i & 1 - 2i \end{bmatrix} \begin{bmatrix} z_1 \\ z_2 \end{bmatrix} = \begin{bmatrix} 1 - i \\ 2 - i \end{bmatrix} \qquad (3.18.3)$$

These equations can be rearranged thus

$$\begin{bmatrix} 3 & 1 & -2 & 1 \\ 2 & 1 & 1 & -2 \\ 2 & -1 & 3 & 1 \\ -1 & 2 & 2 & 1 \end{bmatrix} \begin{bmatrix} y_1 \\ y_2 \\ x_1 \\ x_2 \end{bmatrix} = \begin{bmatrix} -1 \\ -1 \\ 1 \\ 2 \end{bmatrix} \qquad (3.18.4)$$

Using elimination to solve these equations (program 3.6.1) the following results were obtained

$$y_1 = -0.288, \ y_2 = 0.184$$
$$x_1 = 0.416, \ x_2 = 0.512$$

Thus the solution of the original equations is

$$z_1 = 0.416 - i0.288$$
$$z_2 = 0.512 + i0.184$$

An alternative method that can be used to convert n complex equations to $2n$ real ones is to transform each complex coefficient into a 2 x 2 sub-matrix with real elements. Thus the complex scalar quantity $r = p + iq$ is represented by the matrix

$$\begin{bmatrix} p & q \\ -q & p \end{bmatrix}$$

When complex numbers are expressed in this manner it can be shown that, for the basic mathematical operations, the rules of complex algebra are replaced by the rules of matrix algebra. This is a useful technique to use when writing programs involving complex numbers in those enhanced versions of BASIC that have matrix arithmetic but not complex arithmetic as part of the language. Applying this transform to (3.18.1) the complex coefficients of **C**, **d** and **z** are replaced by 2 x 2 sub-matrices and the problem is again converted into one of $2n$ equations with real coefficients. However, since **d** and **z** now become $2n \times 2$ arrays we will obtain two vector solutions for this problem. When these are transformed back into complex notation it is found that they contain identical information that is merely arranged differently. For example, rearranging (3.18.3) using the above procedure we derive the following system of equations

$$\begin{bmatrix} 3 & -2 & 1 & 1 \\ 2 & 3 & -1 & 1 \\ 2 & 1 & 1 & -2 \\ -1 & 2 & 2 & 1 \end{bmatrix} \begin{bmatrix} x_1 & y_1 \\ -y_1 & x_1 \\ x_2 & y_2 \\ -y_2 & x_2 \end{bmatrix} = \begin{bmatrix} 1 & -1 \\ 1 & 1 \\ 2 & -1 \\ 1 & 2 \end{bmatrix}$$

3.19 SOME EXAMPLES

We will now consider some examples of systems of equations and obtain their solution using some of the methods we have considered. The examples selected vary in difficulty and some have been chosen to illustrate the effect of ill-conditioning.

Example 1 : To solve **Ax** = **b** given that **A** is defined by

$$a_{ij} = 360360/(i + j - 1) \quad i, j = 1, 2, \dots, 7$$

and that **b** is defined by

$$b_i = 1 \quad \text{For } i = 1, 2, \dots, 7$$

A is part of the Hilbert matrix which is defined by

$$a_{ij} = 1/(i + j - 1) \quad i, j = 1, 2, \dots, n$$

and the factor 360360 has been introduced to in order to ensure that the coefficients of **A** are all integer numbers. This means that the coefficients of **A** are all accurately defined; there are no rounding errors at this stage. The seven simultaneous equations have been solved using Gaussian elimination (program 3.6.1) in single and double precision as shown in Table 3.19.1. The exact solution for this set of equations is known (Wilkinson 1967) and is

$$\mathbf{x}^T = [56 \;\; -1512 \;\; 12600 \;\; -46200 \;\; 83160 \;\; -72072 \;\; 24024]/360360$$

Table 3.19.1 gives the results for this problem. The single precision solution is hopelessly inaccurate but the accuracy of the double precision results are relatively good.

Table 3.19.1. Solution of *Example* 1 using single and double precision arthmetic. The variables X(1) ... X(7) must be divided by 360360.

```
SOLUTION OF LINEAR SIMULTANEOUS EQUATIONS
GAUSS ELIMINATION & PARTIAL PIVOTING

PIVOTS                          PIVOTS
  180180                          180180
  12012                           12012
  -858.0059                       1001
  102.381                         78
  2.267498                        -3.404761906937
  -.1137736                       .0714285692617
  1.66595E-05                     -.0027056242764

X( 7 ) =  2444646              X( 7 ) =  24024.027122142
X( 6 ) = -8105682              X( 6 ) = -72072.088950906
X( 5 ) =  1.044985E+07         X( 5 ) =  83160.113181238
X( 4 ) = -6574079              X( 4 ) = -46200.07007499
X( 3 ) =  2064994              X( 3 ) =  12600.021579891
X( 2 ) = -291757.2             X( 2 ) = -1512.002973468
X( 1 ) =  13105.78             X( 1 ) =  56.00012929118

DETERMINANT =  8.171125E+08    DETERMINANT =  111192826942.16
```

Example 2 : Consider the inversion of the Hilbert matrix defined above. The inverse of this matrix can be determined analytically and, if $\mathbf{G} = \mathbf{A}^{-1}$ then

$$g_{11} = n$$
$$g_{ij} = -(i+j-2)(n-j+1)(n+j-1)\, g_{ij-1}\,/[(i+j-1)(j-1)^2]$$

$$i, j = 2, 3, \dots, n-1$$

Gauss-Jordan elimination with partial pivoting was used to invert this matrix. Program 3.9.1 was modified so that the inverted matrix was not output but instead the maximum percentage error in the calculated coefficients of the inverted matrix was determined as shown in Table 3.19.2. Working with seven figure accuracy it is seen that for values of n less than 6 the percentage error is acceptably small but for larger values of n the error increases rapidly and for $n = 7$ the calculated inverted matrix is quite meaningless. Working in double precision however gives adequate results at least up to $n = 8$. For larger values of n the Hilbert matrix generates a set of notoriously ill-conditioned equations and inverting it presents a most severe test of the accuracy of any computing machine.

Examples 1, 2 are closely related examples of very ill-conditioned equations requiring double precision computation. We will now consider two examples where the condition of the equations does not pose a problem.

Table 3.19.2. Errors arising from the inversion of the Hilbert matrix.

MATRIX INVERSION
GAUSS-JORDAN ELIMINATION WITH PARTIAL PIVOTING

N	MAX ERROR (%)	CALC DET	EXACT DET
2	7.947287E-06	8.333334E-02	8.333334E-02
3	1.716614E-04	4.629636E-04	4.62963E-04
4	3.574916E-04	1.653436E-07	1.653439E-07
5	4.770431E-02	3.750651E-12	3.749295E-12
6	4.539602	5.603089E-18	5.3673E-18
7	80.54825	1.618997E-24	4.835803E-25
8	120.5405	1.092682E-31	2.73705E-33

N	MAX ERROR (%)	CALC DETERMINANT	EXACT DETERMINANT
2	0	.08333333333333	.0833333333333333
3	1.4333333333333D-10	4.6296296296356D-04	4.62962962962962D-04
4	1.8714285714286D-09	1.6534391534103D-07	1.65343195343915D-07
5	1.6029523809524D-07	3.7492951272354D-12	3.749295I3D-12
6	4.5029563492063D-06	5.3672996668705D-18	5.36729989D-18
7	1.0171375291375D-04	4.8357985337786D-25	4.83580262D-25
8	.001277858214702	2.7370215939779D-33	2.7370501ID-33

Example 3 : Consider the solution of equations $Ax = b$ where the coefficients of A and b are shown in Table 3.19.3. It is seen that the coefficient matrix is sparse and symmetric. The equations have been solved using Gauss elimination with partial pivoting, program 3.6.1 and also by Jennings' method, program 3.13.1. A comparison of the performance of the two programs running on a typical microcomputer is as follows

> Standard Gauss: Storage, 420 coefficients
> Computing time, 37 seconds
> Jennings :   Storage, 92 coefficients plus 20 integers
> Computing time = 7 seconds.

These times include reading data but not outputting results. In this well-conditioned problem with a sparse symmetric coefficient matrix the advantage in using Jennings' method is overwhelming.

Table 3.19.3. Coefficient matrix and right-hand side vector.

$$
\begin{bmatrix}
5 & -1 & -1 & 0 & 0 & 0 & 0 & 0 & 0 & 0 & 0 & 0 & 0 & 0 & 0 & 0 & 0 & 0 & 0 & 0 \\
-1 & 3 & 0 & -3 & 0 & 0 & 0 & 0 & 0 & 0 & 0 & 0 & 0 & 0 & 0 & 0 & 0 & 0 & 0 & 0 \\
-1 & 0 & 7 & 1 & 2 & -1 & 0 & 0 & 0 & 1 & 0 & 0 & 0 & 0 & 0 & 0 & 0 & 0 & 0 & 0 \\
0 & -3 & 1 & 2 & -1 & 1 & 2 & 0 & 1 & 0 & 0 & 0 & 0 & 0 & 0 & 0 & 0 & 0 & 0 & 0 \\
0 & 0 & 2 & -1 & 1 & 0 & 0 & -6 & 0 & 3 & 0 & 0 & 0 & 0 & 0 & 0 & 0 & 0 & 0 & 0 \\
0 & 0 & -1 & 1 & 0 & 3 & 7 & 0 & 0 & 1 & 0 & 0 & 0 & 0 & 0 & 0 & 0 & 0 & 0 & 0 \\
0 & 0 & 0 & 2 & 0 & 7 & 12 & 3 & 2 & -2 & 3 & 0 & 1 & 0 & 0 & 0 & 0 & 0 & 0 & 0 \\
0 & 0 & 0 & 0 & -6 & 0 & 3 & 18 & -1 & 4 & 1 & 2 & 2 & 0 & 0 & 0 & 0 & 0 & 0 & 0 \\
0 & 0 & 0 & 1 & 0 & 0 & 2 & -1 & 4 & 0 & 2 & 0 & -1 & 0 & 0 & 0 & 0 & 0 & 0 & 0 \\
0 & 0 & 1 & 0 & 3 & 1 & -2 & 4 & 0 & 9 & -1 & 1 & 0 & 3 & 0 & 1 & 0 & 0 & 0 & 0 \\
0 & 0 & 0 & 0 & 0 & 0 & 3 & 1 & 2 & -1 & 3 & 3 & 1 & 0 & 1 & 0 & 0 & 0 & 0 & 0 \\
0 & 0 & 0 & 0 & 0 & 0 & 2 & 0 & 1 & 3 & 7 & 0 & 1 & 2 & 0 & 0 & 0 & 0 & 0 & 0 \\
0 & 0 & 0 & 0 & 0 & 0 & 1 & 2 & -1 & 0 & 1 & 0 & 12 & -1 & 1 & 0 & 7 & 3 & 0 & 0 \\
0 & 0 & 0 & 0 & 0 & 0 & 0 & 0 & 0 & 3 & 0 & 1 & -1 & 9 & 0 & 3 & 0 & 0 & 0 & 0 \\
0 & 0 & 0 & 0 & 0 & 0 & 0 & 0 & 0 & 0 & 1 & 2 & 1 & 0 & 3 & 0 & 1 & 4 & 2 & 0 \\
0 & 0 & 0 & 0 & 0 & 0 & 0 & 0 & 0 & 1 & 0 & 0 & 0 & 3 & 0 & 12 & 2 & 0 & 1 & 0 \\
0 & 0 & 0 & 0 & 0 & 0 & 0 & 0 & 0 & 0 & 0 & 0 & 7 & 0 & 1 & 2 & 18 & 1 & -1 & 2 \\
0 & 0 & 0 & 0 & 0 & 0 & 0 & 0 & 0 & 0 & 0 & 0 & 3 & 0 & 4 & 0 & 1 & 4 & 1 & -1 \\
0 & 0 & 0 & 0 & 0 & 0 & 0 & 0 & 0 & 0 & 0 & 0 & 0 & 2 & 1 & -1 & 1 & 9 & 1 \\
0 & 0 & 0 & 0 & 0 & 0 & 0 & 0 & 0 & 0 & 0 & 0 & 0 & 0 & 2 & -1 & 1 & 7
\end{bmatrix}
\begin{bmatrix}
12 \\ -14 \\ -23 \\ 18 \\ -76 \\ 23 \\ 105 \\ 191 \\ 40 \\ 77 \\ 30 \\ 38 \\ -5 \\ 18 \\ 35 \\ 10 \\ 36 \\ 33 \\ 10 \\ 15
\end{bmatrix}
$$

Example 4 : A tridiagonal system of equations has been solved using the standard elimination procedure, program 3.6.1, the elimination procedure modified for a tridiagonal coefficient matrix, program 3.13.2, and Gauss-Seidel iteration, program 3.16.2. This latter program does not take advantage of the tridiagonal nature of the coefficient matrix but it could easily be modified to do so. A comparison of the performance of the three programs running on a typical microcomputer is as follows

Standard Gauss :   Storage, 420 coefficients
                     Computing time, 37 seconds
Tridiagonal: Storage, 80 coefficients
                     Computing time, 2 second
Iteration :   Storage, 420 coefficients
[Max change=10^{-4}]:   Computing time, 18 seconds
[Max change=10^{-6}]:   Computing time, 23 seconds

The problem is well conditioned and the advantage of using a program designed to take advantage of the tridiagonal nature of the coefficient matrix is apparent.

A large number of matrices with known inverses and systems of equations with known solutions are given by Gregory & Karney, (1969). These examples have varying condition numbers and are invaluable for testing the reliability of procedures.

3.20 OVER-DETERMINED SYSTEMS

Over-determined systems are described in section 3.2. Such a system has more linearly independent equations than unknowns. For example

$$\begin{bmatrix} 1 & 1 \\ 3 & -1 \\ 3 & -4 \\ 1 & -4 \\ 2 & 7 \\ 1 & 2 \end{bmatrix} \begin{bmatrix} x_1 \\ x_2 \end{bmatrix} = \begin{bmatrix} 4 \\ 6 \\ 0 \\ -4 \\ 14 \\ 6 \end{bmatrix} \tag{3.20.1}$$

This set of six equations are shown in Fig. 3.20.1. The system has only two unknowns and no solution satisfies all the equations exactly.

In general we have

$$\mathbf{Cx} = \mathbf{p} \tag{3.20.2}$$

Given m equations and n unknowns then $\mathbf{C}$ is an $m \times n$ array. If we define the residual vector $\mathbf{r}$ as

$$\mathbf{r} = \mathbf{Cx} - \mathbf{p} \tag{3.20.3}$$

then we can find an approximate solution that minimises this residual vector according to some criteria. A frequently used criteria is that the sum of the squares of the components of the residual vector should be a minimum. This is known as the "least squares criteria" and we shall have reason to refer to this and other approximation criteria in Chapter 8. A detailed explanation of its development is given in that chapter.

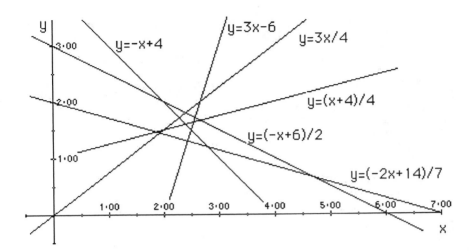

Fig. 3.20.1. Graph of six linear equations in two unknowns.

The least squares approximation is equivalent to solving

$$\mathbf{C^T C x = C^T p} \tag{3.20.4}$$

Because (3.20.2) represents m equations in n unknowns, $\mathbf{C}$ is an m x n matrix and $\mathbf{C^T}$ is an n x m matrix. Thus the product of these matrices is an n x n matrix. Writing $\mathbf{A = C^T C}$ and $\mathbf{b = C^T p}$ then (3.20.4) becomes

$$\mathbf{Ax = b} \tag{3.20.5}$$

This equation represents n equations in n unknowns and Gaussian elimination can be used to solve such a system of equations. Program 3.20.1 implements this method. Table 3.20.1 shows the result of using the program to solve (3.20.1).

Table 3.20.1. Solution of the system of equations (3.20.1).

```
APPROXIMATE SOLUTION OF AN OVER DETERMINED SYSTEM OF EQUATIONS.
LEAST SQUARES, GAUSS ELIMINATION & PARTIAL PIVOTING.

X( 2 ) =  1.475818
X( 1 ) =  2.198066

DETERMINANT =  2171
```

```
100  REM ----------------------------------
101  REM OVER DETERMINED SYSTEMS  JP/GL (C)
102  REM ----------------------------------
105  PRINT "APPROXIMATE SOLUTION OF AN OVERDETERMINED SYSTEM OF EQUA-
TIONS."
110  PRINT "LEAST SQUARES, GAUSS ELIMINATION & PARTIAL PIVOTING."
115  READ M,N
120  DIM C(M,N),A(N,N+1),X(N),R(N),V(M)
122  REM READ COEFFICIENTS OF MATRIX
125  FOR I=1 TO M
130      FOR J=1 TO N
135          READ C(I,J)
140      NEXT J
145  NEXT I
150  FOR I=1 TO N
155      FOR J=1 TO N
160          S=0
165          FOR K=1 TO M
170              S=S+C(K,I)*C(K,J)
175          NEXT K
180          A(I,J)=S
185      NEXT J
190  NEXT I
192  REM READ RIGHT HAND VECTOR
195  FOR I=1 TO M
200      READ V(I)
205  NEXT I
210  FOR I=1 TO N
215      S=0
220      FOR K=1 TO M
225          S=S+C(K,I)*V(K)
230      NEXT K
235      A(I,N+1)=S
240  NEXT I
242  REM USE GAUSSIAN ELIMINATION, PROG 3.6.1, TO SOLVE EQUATIONS
244  REM BY APPENDING PROGRAM 3.6.1, LINES 250-710 HERE.

715  REM DATA STATEMENTS
720  DATA 6 :REM NUMBER EQUATIONS
730  DATA 2:REM NUMBER OF UNKNOWNS
740  DATA 1,1,3,-1,3,-4,1,-4,2,7,1,2 :REM COEFFICIENTS OF MATRIX
750  DATA 4,6,0,-4,14,6 :REM COEFFICIENTS OF RIGHT HAND VECTOR
```

Program 3.20.1.

3.21. SUMMARY AND CONCLUSIONS

We have introduced a wide range of procedures for solving linear equation systems and inverting matrices. The programs given enable the user to solve most problems of this type which arise naturally. However, we have given examples of extreme and difficult cases which can only be solved using exceptionally high precision calculations. Fortunately, such problems are rarely encountered. Table 3.21.1 gives a useful guide to the methods we have described.

Table 3.21.1. Summary of algorithms for solving systems of equations.

Method	Program	Form of coefficient matrix	Class of problem
Gaussian elimination	2.2.1	General	Sol'n of equations
Gauss-Jordan	None	General	Matrix inversion
Crout	2.6.1	General	Sol'n of equations
Cholesky	2.7.1	Symmetric	Sol'n of equations
Jennings	2.9.1	Sparse	Sol'n of equations
Gaussian elimination	2.9.2	Tridiagonal	Sol'n of equations
Jacobi iteration	2.11.1	Diagonally dominant	Sol'n of equations
Gauss-Seidel iteration	2.13.1	Diagonally dominant	Sol'n of equations
SOR	2.14.1	Diagonally dominant	Sol'n of equations
Gaussian elimination	2.15.1	Rectangular	Sol'n of over determined system

PROBLEMS

3.1   Use Gaussian elimination, program 3.6.1, to solve the following system of equations:

$$2x_1 + 8x_2 + 4x_3 = 9$$
$$4x_1 - 5x_2 - 16x_3 = 7$$
$$10x_1 - 5x_2 + x_3 = 4$$

3.2   Rearrange the above system into the diagonal dominant form and solve it using Jacobi and Gauss-Seidel iteration, programs 3.16.1 and 3.16.2 respectively. Compare the number of iterations required by the two methods to obtain the solutions correct to 3 and 6 decimal places.

3.3   Solve the following system of equations using the successive over-relaxation method, program 3.17.1.

$$2x_1 + x_2 + 5x_3 = 5$$
$$2x_1 + 2x_2 + 3x_3 = 7$$
$$x_1 + 3x_2 + 3x_3 = 6$$

Use an accuracy of 0.00005. By experiment, find a value for the parameter ω which provides more rapid convergence than the Gauss-Seidel method.

3.4   Use the Gauss-Seidel method with an accuracy of 0.000005, program 3.16.2, to solve the system of equations $\mathbf{Ax} = \mathbf{b}$ where the elements of $\mathbf{A}$ are defined by

$$a_{ii} = -4 \text{ and } a_{ij} = 2 \text{ if } |i - j| = 1$$
$$= 0 \text{ if } |i - j| \geq 2$$
$$\text{where } i, j = 1, 2, \dots, 10.$$

and

$$\mathbf{b}^{\mathrm{T}} = [2 \ \ 3 \ \ 4 \ \ \dots \ \ 11].$$

3.5   Solve the problem given in problem 3.4 by successive over-relaxation, program 3.17.1. Use an accuracy of 0.000005 and experiment with the value of ω.

3.6   Solve the system given in problem 3.4 using program 3.13.2 to take advantage of the tridiagonal structure of the matrix.

3.7   Use program 3.9.1 to invert the matrix $\mathbf{A}$ where

$$\text{(i) } \mathbf{A} = \begin{bmatrix} 1 & 1 & 1 & 1 \\ 1 & 2 & 4 & 8 \\ 1 & 3 & 9 & 27 \\ 1 & 4 & 16 & 64 \end{bmatrix} \quad \text{(ii) } \mathbf{A} = \begin{bmatrix} 1 & 0 & 0 \\ 1 & -1 & 0 \\ 1 & -2 & 1 \end{bmatrix} \quad \text{(iii) } \mathbf{A} = \begin{bmatrix} 4 & 2 & 4 & 1 \\ 30 & 20 & 45 & 12 \\ 20 & 15 & 36 & 10 \\ 35 & 28 & 70 & 20 \end{bmatrix}$$

The matrix of case (iii) is known as the Dekker matrix.

3.8   Show that if

$$\mathbf{A} = \begin{bmatrix} 1 & 1 \\ 1 & 1 - 1/n^2 \end{bmatrix} \text{ then } \mathbf{A}^{-1} = \begin{bmatrix} 1 - n^2 & n^2 \\ n^2 & -n^2 \end{bmatrix}$$

Then solve exactly, by hand, the system $\mathbf{Ax} = \mathbf{b}$ taking

$$\mathbf{b} = [1.2, \ 3.001]^{\mathrm{T}} \text{ for } n = 10 \text{ and } n = 100.$$

Compare these results with the results you obtain using program 3.6.1 to solve these systems of linear equations. Compute the Euclidian norm $\|\mathbf{A}\|_2$, which is the square root of the sum of the squares of the elements of $\mathbf{A}$, and determine the quantity $\|\mathbf{A}\|_2 \|\mathbf{A}^{-1}\|_2$. This is a measure of the condition of $\mathbf{A}$. Comment on its value for $n = 10$ and $n = 100$.

3.9 The Vandermonde matrix $\mathbf{V}$ is defined as follows

$$\mathbf{V} = \begin{bmatrix} 1 & x_0 & x_0^2 & \cdots & x_0^n \\ 1 & x_1 & x_1^2 & \cdots & x_1^n \\ & & \cdots\cdots\cdots \\ & & \cdots\cdots\cdots \\ 1 & x_n & x_n^2 & \cdots & x_n^n \end{bmatrix}$$

Taking $x_i = x_0 + ih$ for $i = 1, 2, \dots, n$ and $x_0 = 1$ evaluate the inverse of $\mathbf{V}$ using program 3.9.1 for $n = 5$ for $h = 0.1$. For $h = 0.1$ and $n = 5$ solve the system of linear equations

$$\mathbf{Vz} = \mathbf{y} \quad \text{where } \mathbf{y} = [1, -2.4, 5.6, 6.2, -3.4, -1]^T$$

by using the inverse of $\mathbf{V}$. Then solve this problem directly using Gauss elimination, program 3.6.1, and compare your results. How has ill-conditioning affected your answer?

3.10 The elements of the matrix $\mathbf{A}$, the Hilbert matrix, are defined by

$$a_{ij} = 1/(i + j - 1) \text{ for } i, j = 1, 2, \dots, n$$

Find the inverse of $\mathbf{A}$ and the inverse of $\mathbf{A}^T\mathbf{A}$ for $n = 3$.
Noting that $(\mathbf{A}^T\mathbf{A})^{-1} = (\mathbf{A})^{-1}(\mathbf{A}^{-1})^T$ calculate the inverse of $\mathbf{A}^T\mathbf{A}$ from the inverse of $\mathbf{A}$ directly and compare your results.

3.11 Find the approximate solution of the over-determined system using program 3.20.1.

$$\begin{aligned} 2x + 3y &= 10 \\ 4x - 7y &= 8 \\ -2x + 3y &= 6 \\ 5x + 9y &= 14 \end{aligned}$$

3.12 Iteration based on the Gauss-Seidel procedure can be used to solve a set of n non-linear equations in n variables by expressing them in the form

$$x_i = F(x_1, x_2, \dots, x_n), \quad i = 1, 2, \dots, n$$

provided the following convergence conditions are satisfied

$$\sum_i \sum_j |\partial F_i / \partial x_j| < 1$$

Write a program to solve the following pair of equations by this method using starting values $x = 2$, $y = 0.25$.

$$x^2 + y^2 - 4 = 0$$
$$xy - 1 = 0$$

3.13 Attempt to solve the following equations

$$\begin{bmatrix} 1 & -2 & 1 \\ -3 & 4 & 9 \\ -2 & 3 & 4 \end{bmatrix} \begin{bmatrix} x_1 \\ x_2 \\ x_3 \end{bmatrix} = \begin{bmatrix} -2 \\ 0 \\ 1 \end{bmatrix}$$

Comment on the results.

3.14 Solve the following system of equations using Gauss-Seidel iteration, program 3.16.2.

$$\begin{bmatrix} A & 3 & 4 \\ 1 & 2A & -1 \\ 2 & -5 & A \end{bmatrix} \begin{bmatrix} x_1 \\ x_2 \\ x_3 \end{bmatrix} = \begin{bmatrix} 21 \\ 9 \\ 18 \end{bmatrix}$$

Compare the number of iterations required to solve this problem to obtain an accuracy 1E–4 when $A = 1$, 4.5, 10, 100. Comment on your results.

3.15 Consider the following system

$$\begin{bmatrix} -4 & 2 & 2 \\ 2 & -4 & 2 \\ 2 & 2 & -4 \end{bmatrix} \begin{bmatrix} x_1 \\ x_2 \\ x_3 \end{bmatrix} = \begin{bmatrix} a \\ b \\ c \end{bmatrix}$$

Attempt to solve the system by Gaussian elimination when

(i) $a = 2$, $b = 2$, $c = -4$
(ii) $a = 1$, $b = 4$, $c = 3$

3.16 Solve the set of symmetric equations $\mathbf{A}\mathbf{x} = \mathbf{b}$ by Cholesky decomposition, program 3.12.2, where

$$\mathbf{A} = \begin{bmatrix} 9 & -36 & 30 \\ -36 & 192 & -180 \\ 30 & -180 & 180 \end{bmatrix}$$

and $\mathbf{b}$ is (i) $[1\ \ 2\ \ 3]^T$ and (ii) $[-3\ \ 2\ \ -1]^T$

4

Numerical integration

4.1 INTRODUCTION

In this chapter we will examine several numerical methods for the approximate evaluation of the general integral

$$\int_a^b f(x)\,dx \qquad\qquad (4.1.1)$$

The evaluation of such integrals is often called quadrature. Most of the techniques considered require that the function $f(x)$ is continuous and that a and b are finite. However, methods will be described for dealing with discontinuities and where the range of integration is infinite.

The definite integral (4.1.1) may be interpreted as the area under the curve $y = f(x)$ from a to b. This is illustrated in Fig. 4.1.1 and it should be noted that any areas beneath the x-axis are counted as negative. Many numerical methods for integration are based on using this interpretation of the integral to derive approximations to it by dividing the interval $[a, b]$ into a number of smaller sub-intervals. By making simple approximations to the curve $y = f(x)$ in the small sub-interval its area may be obtained and on summing all the contributions we obtain an approximation to the integral in the interval $[a, b]$. Variations of this technique are derived by taking groups of sub-intervals and fitting different degree polynomials as approximations for each of these groups. The level of accuracy obtained is dependent on a number of factors. These are:

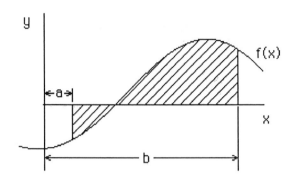

Fig.4.4.1. The value of the integral (4.1.1) is equal to the shaded area.
Areas below the x-axis are negative.

(1) the number of intervals used.
(2) the nature of the approximating function.
(3) the rounding error.

Clearly the first two factors are decided by the user but the third factor is machine dependent. We will see in Chapter 11 that different microcomputers store numerical values with different numbers of significant figures and some have double precision arithmetic while others do not. Rounding error will have a limiting effect on accuracy for different choices of the number of intervals and approximating function. Another feature of interest is the time taken to approximate the integral. This will be directly related to the number of function evaluations required. We shall now consider in detail some numerical techniques for integration.

4.2 THE TRAPEZOIDAL RULE

This rule is based on approximating the function in each sub-interval by a straight line. This is illustrated in Fig. 4.2.1. Here the straight line joining A and B approximates the curve $y = f(x)$ passing through A and B. The approximation to the area under the curve consequently constitutes a trapezoid, the area of which is given by

$$(f(x_0) + f(x_1)) (x_0 - x_1)/2$$

which if we take the interval width to be h becomes

$$h(f(x_0) + f(x_1))/2 \qquad\qquad (4.2.1)$$

The function $f(x)$ is evaluated at $n + 1$ different points, x_i, equispaced a distance

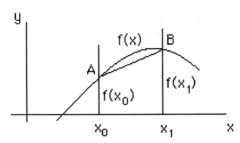

Fig.4.2.1. The trapezoidal rule. Between x_0 and x_1 the function $f(x)$ is approximated
by a straight line joining points A and B.

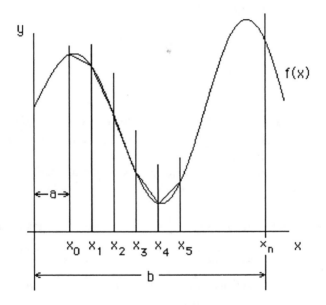

Fig 4.2.2. The trapezoidal rule for integration from a to b using n panels.

h apart in the interval $[a, b]$, where

$$x_i = x_0 + ih \quad i = 1, 2, \dots, n$$

and

$$n = (b - a)/h$$

We illustrate this situation in Fig. 4.2.2. Adding the approximations for each
interval we obtain the composite trapezoidal rule thus

$$\int_a^b f(x)dx = h[\{f(x_0) + f(x_n)\}/2 + f(x_1) + \dots + f(x_{n-1})] \qquad (4.2.2)$$

It is clear from the diagram that there will be errors arising from each approximation. It can be shown that the overall error due to the linear approximation, called the truncation error, is given by

$$E_n \approx h^3(f''(t_1) + f''(t_2) + \ldots + f''(t_n))/12$$

where each t_i lies in the interval $[x_i, x_{i-1}]$ $i = 1, 2, \ldots, n$.

Because of the computational effort required to find E_n an alternative approximation can be used that is given by

$$E_n \approx (b-a)h^2 f''(t)/12$$

where t lies between a and b. In practice we provide an upper bound for the error which is given by

$$E_n \leq (b-a)h^2 M/12 \qquad (4.2.3)$$

where M is an upper bound for $|f''(t)|$. This upper bound on the truncation error ignores the effect of rounding errors: to estimate the overall error the effect of truncation and rounding errors must be combined.

Table 4.2.1 illustrates the effect of increasing the number of intervals used in trapezoidal integration and was produced using program 4.2.1.

Table 4.2.1. Evaluation of $\int_0^1 x^9 dx$ using the trapezoidal rule.

```
EVALUATING INTEGRALS BY TRAPEZOIDAL RULE

HOW MANY DIVISIONS OF INTERVAL REQUIRED?  10
UPPER LIMIT OF INTEGRATION?  1
LOWER LIMIT OF INTEGRATION?  0

RESULTS WITH REPEATED HALVING OF INTERVAL

INTERVALS        INTEGRAL
  1              .5
  2              .2509766
  4              .1442604
  8              .1115497
 16              .102919
 32              .1007317
 64              .1001831
128              .1000458
256              .1000115
512              .1000028
1024             .1000007
```

```
100   REM ------------------------------------------
110   REM INTEGRATION BY TRAPEZOIDAL RULE   GL/JP (C)
120   REM ------------------------------------------
125   REM DEFINE FUNCTION IN STATEMENT 130
130   DEF FNA(X)=X^9
140   PRINT"EVALUATING INTEGRALS BY TRAPEZOIDAL RULE"
150   PRINT
160   PRINT"HOW MANY DIVISIONS OF INTERVAL REQUIRED";
170   INPUT D
180   PRINT"UPPER LIMIT OF INTEGRATION";
190   INPUT U
200   PRINT"LOWER LIMIT OF INTEGRATION";
210   INPUT L
220   PRINT:PRINT"RESULTS WITH REPEATED HALVING OF INTERVAL"
230   PRINT:PRINT"INTERVALS";TAB(15);"INTEGRAL"
240   FOR K=0 TO D
245       REM USE K TO DETERMINE NUMBER OF INTERVALS AND STEP SIZE H
250       N=2^K:H=(U-L)/N
260       Sum=0:YL=FNA(L)
265       REM CALCULATE FUNCTION VALUE AND SUM USING TRAPEZOIDAL RULE
270       FOR J=1 TO N
280           X=L+J*H:YH=FNA(X)
290           Sum=Sum+YL+YH
300           YL=YH
310       NEXT J
320       PRINT N;TAB(15);Sum*H/2
330   NEXT K
340   END
```

Program 4.2.1.

4.3 SIMPSON'S RULE

This rule is based on using a quadratic polynomial approximation to the function $f(x)$ over a pair of sub-intervals, and is illustrated in Fig. 4.3.1. If the polynomial passing through the points $(x_0, f(x_0))$; $(x_1, f(x_1))$; $(x_2, f(x_2))$ is derived and integrated the following formula is obtained

$$\int_a^b f(x)dx = h[f(x_0) + 4f(x_1) + f(x_2)]/3 \qquad (4.3.1)$$

This is Simpson's rule for one pair of intervals. Applying the rule to pairs of intervals in the range a to b and adding the results the following expression, known as the composite Simpson's rule, is obtained.

$$\int_a^b f(x)dx = h[f(x_0) + 4\{f(x_1) + f(x_3) + f(x_5) + \dots \}$$
$$+ 2\{f(x_2) + f(x_4) + \dots \} + f(x_{2n})]/3 \qquad (4.3.2)$$

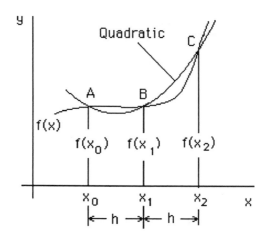

Fig.4.3.1. Simpson's rule. Between x_0 and x_2 the function $f(x)$ is approximated by a quradratic function passing through the points A, B and C.

```
100   REM --------------------------------------
110   REM INTEGRATION BY SIMPSON'S RULE  GL/JP (C)
120   REM --------------------------------------
125   REM DEFINE FUNCTION IN STATEMENT 130
130   DEF FNA(X)=X^9
140   PRINT"EVALUATING INTEGRALS BY SIMPSON'S RULE"
150   PRINT
160   PRINT"HOW MANY DIVISIONS OF INTERVAL REQUIRED";
170   INPUT D
180   PRINT"UPPER LIMIT OF INTEGRATION";
190   INPUT U
200   PRINT"LOWER LIMIT OF INTEGRATION";
210   INPUT L
220   PRINT:PRINT"RESULTS WITH REPEATED HALVING OF INTERVAL"
230   PRINT:PRINT"INTERVALS";TAB(15);"INTEGRAL"
240   FOR K=1 TO D
245      REM USE K TO DETERMINE NUMBER OF INTERVALS AND STEP H
250      N=2^K:H=(U-L)/N
260      Sum=0:YL=FNA(L)
265      REM CALCULATE FUNCTION VALUES AND FORM SUM USING SIMPSON'S RULE
270      FOR J=2 TO N STEP 2
280         X=L+(J-1)*H:YM=FNA(X)
290         X=L+J*H:YH=FNA(X)
300         Sum=Sum+YL+4*YM+YH
310         YL=YH
320      NEXT J
330      PRINT N;TAB(15);Sum*H/3
340   NEXT K
350   END
```

Program 4.3.1

Here n indicates the number of pairs of intervals and $h = (b - a)/(2n)$
    An approximation to the truncation error is given by

$$E_n \approx (b - a)h^4 f^{(iv)}(t)/180$$

where t lies between a and b and an upper bound for the error is given by

$$E_n \leq (b - a)h^4 M/180$$

where M is an upper bound for $|f^{(iv)}(t)|$. Comparing this upper bound for the truncation error with that given for the trapezoidal rule we note that it is proportional to h^4 rather than h^2.
    Program 4.3.1 evaluates definite integrals using Simpson's rule. In this program, (4.3.1) is used repeatedly to implement the composite rule.

4.4 TRUNCATION AND ROUNDING ERRORS

The expression for the truncation error for Simpson's rule involves the calculation of the fourth order derivative of $f(x)$ and because of this it is not generally used in directly calculating the error. However, the expression does show how the truncation error will decrease rapidly for values of h smaller than 1. Table 4.4.1 illustrates this. The rounding error in Simpson's rule is due to evaluating the function $f(x)$ and the subsequent multiplications and additions.

Table 4.4.1. Evaluation of $\int_0^1 x^9 dx$ using Simpson's rule.

```
EVALUATING INTEGRALS BY SIMPSON'S RULE

HOW MANY DIVISIONS OF INTERVAL REQUIRED?  8
UPPER LIMIT OF INTEGRATION?  1
LOWER LIMIT OF INTEGRATION?  0

RESULTS WITH REPEATED HALVING OF INTERVAL

INTERVALS         INTEGRAL
2                 .1679688
4                 .1086884
8                 .1006462
16                .1000421
32                .1000026
64                .1000002
128               .1
256               .1
```

4.5 NEWTON-COTES FORMULA

Both the trapezoidal rule and Simpson's rule are examples of the more general Newton-Cotes formula for integration. Other examples of these formula can be obtained by fitting higher degree polynomials through the appropriate number of points. In general we fit a polynomial of degree n through $n + 1$ points. The resulting polynomial can then be integrated to provide an integration formula. Here are some examples of Newton-Cotes formula together with estimates of their truncation errors.

For $n = 3$ we have:

$$\int_{x_0}^{x_3} f(x)dx = 3h[f(x_0) + 3f(x_1) + 3f(x_2) + f(x_3)]/8 \qquad (4.5.1)$$

The truncation error for (4.5.1) is

$$3h^5 f^{(iv)}(t)/80 \text{ where } t \text{ lies in the interval } x_0 \text{ to } x_3.$$

For $n = 4$ we have:

$$\int_{x_0}^{x_4} f(x)dx = 2h[7f(x_0) + 32f(x_1) + 12f(x_2) + 32f(x_3) + 7f(x_4)]/45 \qquad (4.5.2)$$

The truncation error for (4.5.2) is

$$8h^7 f^{(vi)}(t)/945 \text{ where } t \text{ lies in the interval } x_0 \text{ to } x_4$$

The truncation errors indicate there may be some improvement in accuracy obtained by using these rules. However, the rules are more complex; consequently greater computational effort is involved and rounding becomes a more significant problem. In most situations there is little to gain from using the higher order formula.

4.6 ROMBERG INTEGRATION

A major problem which arises with both the trapezoidal and Simpson's rule is that the number of intervals required to provide the required accuracy is initially unknown. Clearly one approach to this problem is to successively double the number of intervals used and compare the results of applying a particular rule. Romberg integration provides an organised approach to this problem and utilises the results obtained by applying the integration rule with different interval sizes to reduce the truncation error.

The Romberg integration procedure is based on the Richardson extrapolation method described in Chapter 1 and may be formulated as follows. Let I be the exact value of the integral and T_i the approximate value of the integral obtained using the trapezoidal rule with i intervals. Consequently we may write an approximation for the integral I which includes contributions from the truncation error as follows: note that the error terms are expressed in powers of h^2:

$$I = T_i + c_1 h^2 + c_2 h^4 + c_3 h^6 + \dots \tag{4.6.1}$$

Similarly if we double the number of intervals h is halved, giving

$$I = T_{2i} + c_1 (h/2)^2 + c_2 (h/2)^4 + c_3 (h/2)^6 + \dots \tag{4.6.2}$$

We can eliminate the terms in h^2 by subtracting (4.6.1) from 4 times (4.6.2), giving:

$$I = (4T_{2i} - T_i)/3 + k_2 h^4 + k_3 h^6 + \dots \tag{4.6.3}$$

Notice that the dominant or most significant term in the truncation error is now of order h^4. In general this will provide a significantly improved approximation to I. If we generate an initial set of approximations by successively halving the interval we may represent them by $T_{0,k}$ where $k = 0, 1, 2, 3, 4, \dots$ These results may be combined in a similar manner to that described in (4.6.3) by using the general formula:

$$T_{r,k} = (4^r T_{r-1,k+1} - T_{r-1,k})/(4^r - 1) \tag{4.6.4}$$

for $k = 0, 1, 2, 3, \dots$ and $r = 1, 2, 3, \dots$

Here r represents the current set of approximations we are generating. The calculations may be tabulated as follows, where in this case the interval has been halved four times to generate the first five values in the table denoted by $T_{0,k}$ the formula for $T_{r,k}$ given above is used to calculate the remaining values in the table:

$T_{0,0}$	$T_{0,1}$	$T_{0,2}$	$T_{0,3}$	$T_{0,4}$
$T_{1,0}$	$T_{1,1}$	$T_{1,2}$	$T_{1,3}$	
$T_{2,0}$	$T_{2,1}$	$T_{2,2}$		
$T_{3,0}$	$T_{3,1}$			
$T_{4,0}$				

We have chosen the above layout because it is consistent with our computer program output. A common alternative is to write the above table with the rows as columns.

At each stage the order of the truncation error is increased by two and the interval size is given by:

$$h = (b-a)/2^k \text{ for } k = 0, 1, 2, \dots \tag{4.6.5}$$

Table 4.6.1 provides the results of Romberg integration for a specific integral and was generated using program 4.6.1.

```
100    REM -------------------------------------------
110    REM INTEGRATION BY ROMBERG'S METHOD   GL/JP (C)
120    REM -------------------------------------------
125    REM DEFINE FUNCTION IN STATEMENT 130
130    DEF FNA(X)=X^9
140    PRINT"EVALUATING INTEGRALS BY ROMBERG'S METHOD"
150    PRINT
160    PRINT "UPPER LIMIT OF INTEGRATION";
170    INPUT U
180    PRINT"LOWER LIMIT OF INTEGRATION";
190    INPUT L
195    REM USER CAN CHANGE D BY CHANGING THE VALUE IN STATEMENT 200
200    D=5:D1=2^D
210    DIM T(D),Y(D1)
215    REM CALCULATE INITIAL VALUES USING THE TRAPEZOID RULE
220    FOR K=0 TO D
230        N=2^K:H=(U-L)/N
240        FOR J=0 TO N
250            X=L+J*H:Y(J)=FNA(X)
260        NEXT J
270        Sum=0
280        FOR I=1 TO N
290            Sum=Sum+Y(I-1)+Y(I)
300        NEXT I
310        T(K)=Sum*H/2
320    NEXT K
330    PRINT:PRINT"RESULTS FROM TRAPIZOIDAL RULE WITH REPEATED HALVING OF
THE INTERVAL"
340    FOR K=0 TO D
350        PRINT T(K);"   ";
360    NEXT K
370    PRINT:PRINT:PRINT"ROMBERG APPROXIMATIONS"
375    REM USE EXTRAPOLATION FORMULA TO FIND MORE ACCURATE VALUES
380    FOR P=1 TO D
390        PRINT "GROUP";P;
400        FOR K=0 TO D-P
410            Q=4^P:S=1/(Q-1)
420            T(K)=S*(Q*T(K+1)-T(K))
430            PRINT T(K);"   ";
440        NEXT K
450        PRINT
460    NEXT P
470    END
```
Program 4.6.1.

Table 4.6.1. Evaluation $\int_{0}^{1} x^{9}dx$ of using Romberg's method.

```
EVALUATING INTEGRALS BY ROMBERG'S METHOD

HOW MANY DIVISIONS OF INTERVAL REQUIRED?  5

UPPER LIMIT OF INTEGRATION?  1

LOWER LIMIT OF INTEGRATION?  0

RESULTS FROM TRAPIZOIDAL RULE WITH REPEATED HALVING OF THE INTERVAL
 .5    .2509766    .1442604    .1115497    .102919    .1007317

ROMBERG APPROXIMATIONS
GROUP  1  .1679688    .1086884    .1006462    .1000421    .1000026
GROUP  2  .1047363    .1001101    .1000019    .1
GROUP  3  .1000366    .1000002    .1
GROUP  4  .1    .1
GROUP  5  .1
```

4.7 GAUSSIAN INTEGRATION FORMULAE

The common feature of the methods considered so far is that the integrand was evaluated at equal intervals within the range of integration. In contrast Gaussian integration requires the evaluation of the integrand at specified, but unequal, intervals. The general form of the rule is:

$$\int_{-1}^{1} f(x)dx = \sum_{i=1}^{n} A_i\, f(x_i) \tag{4.7.1}$$

The parameters A_i and x_i are chosen so that for a given n the rule is exact for polynomials up to and including degree $2n - 1$. For example when $n = 2$ we wish to determine four parameters on the basis that the rule is correct for cubic and lower degree polynomials. This procedure is described in detail later in this section. It should be noticed that the range of integration is required to be from -1 to 1. This does not restrict the integrals to which Gaussian integration can be applied since if $f(x)$ is to be integrated in the range a to b then it can be replaced by the function $g(t)$ integrated from -1 to 1 where

$$t = (2x - a - b)/(b - a)$$

Note that in the above formula, when $x = a$, $t = -1$ and when $x = b$, $t = 1$.

The simplest example of Gaussian integration is the formula

$$\int_{-1}^{1} f(x)dx = A_1 f(x_1) + A_2 f(x_2)$$ (4.7.2)

Since we have four parameters that may be fixed we can expect the rule to be exact for any cubic polynomial. By ensuring the rule is exact for the polynomials 1, x, x^2, x^3 in turn, four equations are obtained as follows.

Taking $f(x) = 1$ gives

$$\int_{-1}^{1} 1\, dx = 2 = A_1 + A_2$$

Taking $f(x) = x$ gives

$$\int_{-1}^{1} x\, dx = 0 = A_1 x_1 + A_2 x_2$$

Taking $f(x) = x^2$ gives

$$\int_{-1}^{1} x^2 dx = 2/3 = A_1 x_1^2 + A_2 x_2^2$$

Taking $f(x) = x^3$ gives

$$\int_{-1}^{1} x^3 dx = 0 = A_1 x_1^3 + A_2 x_2^3$$

Solving gives

$$x_1 = -1/\sqrt{3} \qquad x_2 = 1/\sqrt{3}$$
$$A_1 = 1 \qquad\qquad A_2 = 1$$

Thus

$$\int_{-1}^{1} f(x)\, dx \approx f(-1/\sqrt{3}) + f(1/\sqrt{3})$$ (4.7.3)

Notice that this rule, like Simpson's rule, is exact for cubics but it requires fewer function evaluations.

A general procedure for obtaining the values of A_i and x_i is based on the following results:

(a)   The points in the range of integration $x_1, x_2, x_3, ... , x_n$ can be shown to be the roots of the Legendre polynomial of degree n. The Legendre polynomials $P_n(x)$ are defined by the formula:

$$P_0(x) = 1; P_1(x) = x$$

$$P_n(x) = [(2n - 1)xP_{n-1}(x) - (n - 1)P_{n-2}(x)]/n, \quad n = 2, 3, 4, ...$$

(b)   The values of A_i can be obtained from

$$A_i = 2(1 - x_i^2)/[nP_{n-1}(x_i)]^2 \tag{4.7.4}$$

Tables have been produced for the values of the x_i and A_i for various values of n. Examples are given in the Table 4.7.1 and a more extensive list may be found in Abramowitz and Stegun (1964). This book provides an excellent reference not only for these functions but for a very extensive range of mathematical functions. Program 4.7.1 performs Gaussian integration, using the coefficients of Table 4.7.1. and some results are given in Table 4.7.2.

Table 4.7.1. Abscissae and weight factors for Gaussian integration.

n	$\pm x_i$	A_i
2	0.57735 02691 89626	1.00000 00000 00000
4	0.33998 10435 84856	0.65214 51548 62546
	0.86113 63115 94053	0.34785 48451 37454
8	0.18343 46424 95650	0.36268 37833 78364
	0.52553 24099 16329	0.31370 66458 77887
	0.79666 64774 13627	0.22238 10344 53374
	0.96028 98564 97536	0.10122 85362 90376

Table 4.7.2. Evaluation of $\int_0^1 x^9 dx$ using Gauss's method.

```
GAUSSIAN INTEGRATION

UPPER LIMIT OF INTEGRATION?  1
LOWER LIMIT OF INTEGRATION?  0


   2 POINT: VALUE OF INTEGRAL IS  5.902778E-02
   4 POINT: VALUE OF INTEGRAL IS  9.989797E-02
   8 POINT: VALUE OF INTEGRAL IS  .1
  16 POINT: VALUE OF INTEGRAL IS  .1
```

```
100  REM ------------------------------
110  REM GAUSSIAN INTEGRATION  GL/JP (C)
120  REM ------------------------------
125  REM DEFINE FUNCTION IN STATEMENT 130
130  DEF FNA(X)=X^9
140  DIM X1(8),A(8)
150  PRINT" GAUSSIAN INTEGRATION"
160  PRINT
170  PRINT"UPPER LIMIT OF INTEGRATION";
180  INPUT U
190  PRINT"LOWER LIMIT OF INTEGRATION";
200  INPUT L
210  PRINT
220  FOR C=1 TO 4
230      N=2^C
240      FOR I=1 TO N/2
250          READ A(I)
260      NEXT I
270      FOR I=1 TO N/2
280          READ X1(I)
290      NEXT I
300      Sum=0
305      REM FORM GAUSS SUMMATION
310      FOR K=1 TO N/2
320          X1=(X1(K)*(U-L)+U+L)/2: X2=(-X1(K)*(U-L)+U+L)/2
330          Y=FNA(X1)+FNA(X2): Sum=Sum+A(K)*Y
340      NEXT K
350      In=(U-L)*Sum/2
360      PRINT N;"POINT: VALUE OF INTEGRAL IS ";In
370  NEXT C
380  END
385  REM DATA FOR 2,4,8,16 POINT RULES AT 390, 400,410-420,430-460.
390  DATA 1,.577350269190
400  DATA .347854845137,.652145154863,.861136311594,.339981043585
410  DATA .101228536290,.222381034453,.313706645878,.362683783378
420  DATA .960289856498,.796666477414,.525532409916,.183434642496
430  DATA .027152459412,.062253523939,.095158511683,.124628971256
440  DATA .149595988817,.169156519395,.182603415045,.189450610455
450  DATA .989400934992,.944575023073,.865631202388,.755404408355
460  DATA .617876244403,.458016777657,.281603550779,.095012509838
```

Program 4.7.1.

4.8 INFINITE RANGES OF INTEGRATION

Other formulae of the Gauss type are available to allow us to deal with integrals having a special form and infinite ranges of integration. These are the Gauss-Laguerre and Gauss-Hermite formulae and take the following form.

(a) *Gauss-Laguerre formulae:*

$$\int_0^\infty e^{-x} g(x)\, dx = \sum_{i=1}^{n} A_i g(x_i) \qquad (4.8.1)$$

where x_i are the roots of the nth order Laguerre polynomial $L_n(x)$. We may now define function $L_n(x)$ by the equation

$$L_n(x) = e^x (d^n/dx^n)(e^{-x}x^n) \qquad (4.8.2)$$

and the coefficients A_i are calculated from

$$A_i = (n!)^2/[x_i(L'_n(x_i))^2] \qquad (4.8.3)$$

In general we will wish to evaluate integrals of the form

$$\int_0^\infty f(x)\, dx$$

We may write this integral as

$$\int_0^\infty e^{-x} [e^x f(x)]\, dx = \sum_{i=1}^{n} A_i g(x_i)$$

so that $g(x) = e^x f(x)$ and thus

$$\int_0^\infty f(x)\, dx = \sum_{i=1}^{n} A_i \exp(x_i) f(x_i) \qquad (4.8.4)$$

Table 4.8.1 gives sample values x_i, A_i and the product $A_i \exp(x_i)$. A more complete list may be found in Abramowitz and Stegun, (1964) and more accurate values are given by Davis and Rabinowitz (1956). These coefficients are used in program 4.8.1. Table 4.8.2 gives results for the evaluation of a specific integral for various values of n.

Table 4.8.1. Abscissae and weight factors for Gauss-Laguerre integration.

n	x_i	A_i	$A_i \exp(x_i)$
2	0.58578 64376 27	8.53553 390593 E-1	1.53332 603312
	3.41421 35623 73	1.46446 609407 E-1	4.45097 733505
4	0.32254 76896 19	6.03154 104342 E-1	0.83273 912384
	1.74576 11011 58	3.57418 692438 E-1	2.04810 243845
	4.53662 02969 21	3.88879 085150 E-2	3.63114 630582
	9.39507 09123 01	5.39294 705561 E-4	6.48714 508441

```
100   REM ------------------------------------
110   REM GAUSS LAGUERRE INTEGRATION  GL/JP (C)
120   REM ------------------------------------
125   REM DEFINE FUNCTION IN STATEMENT 130
130   DEF FNA(X)=LOG(1+EXP(-X))
140   DIM X1(8),A(8)
150   PRINT"GAUSS-LAGUERRE INTEGRATION BETWEEN ZERO & +INF"
160   PRINT
170   FOR C=1 TO 3
180     N=2^C
185     REM READ  DATA FOR GAUSS N POINT RULE
190     FOR I=1 TO N
200         READ A(I)
210     NEXT I
220     FOR I=1 TO N
230         READ X1(I)
240     NEXT I
250     In=0
255     REM FORM GAUSS SUMMATION
260     FOR K=1 TO N
270         Y=FNA(X1(K)):In=In+A(K)*Y
280     NEXT K
290     PRINT N;"POINT: VALUE OF INTEGRAL IS ";In
300   NEXT C
310   END
315   REM DATA FOR GAUSS TWO POINT RULE
320   DATA 1.53332603312,4.45095733505,.585786437627,3.41421356237
325   REM DATA FOR GAUSS FOUR POINT RULE
330   DATA .832739123838,2.04810243845,3.63114630582,6.48714508441
340   DATA .322547689619,1.74576110116,4.53662029692,9.39507091230
345   REM DATA FOR GAUSS EIGHT POINT RULE
360   DATA .437723410493,1.03386934767,1.66970976566,2.37692470176
370   DATA 3.20854091335,4.26857551083,5.81808336867,8.90622621529
380   DATA .170279632305,.903701776799,2.25108662987,4.26670017029
390   DATA 7.04590540239,10.7585160102,15.7406786413,22.8631317369
```

Program 4.8.1.

Table 4.8.2. Evaluation of $\int_0^\infty \log_e(1 + e^{-x}) dx$ using the Gauss-Laguerre method.

```
GAUSS-LAGUERRE INTEGRATION BETWEEN ZERO & +INF

2 POINT: VALUE OF INTEGRAL IS  .8226584
4 POINT: VALUE OF INTEGRAL IS  .8223578
8 POINT: VALUE OF INTEGRAL IS  .8224665
```

(b) *Gauss-Hermite formulae*:

$$\int_{-\infty}^\infty \exp(-x^2) g(x) dx = \sum_{i=1}^n A_i g(x_i)$$

$$\int_{-\infty}^\infty f(x) dx = \int_{-\infty}^\infty \exp(-x^2) [\exp(x^2) f(x)] dx$$

$$= \sum_{i=1}^n A_i \exp(x_i^2) f(x_i)$$

The x_i are the roots of the nth order Hermite polynomial $H_n(x)$. We may define function $H_n(x)$ by the equation

$$H_n(x) = (-1)^n \exp(x^2) (d^n/dx^n) \exp(-x^2) \tag{4.8.5}$$

$$A_i = 2^{n+1} n! \sqrt{\pi}/[H'_n(x_i)]^2 \tag{4.8.6}$$

Table 4.8.3 gives values of x_i and A_i and $A_i \exp(x_i^2)$ for various values of n. More extensive table are again given in Abramowitz and Stegun, (1964) and more accurate tables in Salzer *et al* (1952).

Table 4.8.4 gives the results of evaluating a specific integral for various values of n using program 4.8.2.

Table 4.8.3. Abscisae and weight factors for Gauss-Hermite integration.

n	$\pm x_i$	A_i	$A_i \exp(x_i^2)$
2	0.70710 67811 87	8.86226 92545 28 E-1	1.46114 11826 61
4	0.52464 76232 75	8.86226 92545 28 E-1	1.05996 44828 95
	1.65068 01238 86	8.13128 35447 25 E-1	1.24022 58176 96
8	0.38118 69902 07	6.61147 01255 82 E-1	0.76454 41286 52
	1.15719 37142 47	2.07802 32581 49 E-1	0.79289 00483 86
	1.98165 67566 96	1.70779 83007 41 E-2	0.86675 26065 63
	2.93063 74202 57	1.99604 07221 14 E-4	1.07193 01442 48

```
100    REM -----------------------------------
110    REM GAUSS HERMITE INTEGRATION  GL/JP (C)
120    REM -----------------------------------
130    DEF FNA(X)=1/(1+X*X)^2                  '
140    DIM X1(8),A(8)
150    PRINT" GAUSS-HERMITE INTEGRATION BETWEEN -INF & +INF"
160    PRINT
170    FOR C=1 TO 4
180       N=2^C
185       REM READ DATA FOR GAUSS N POINT FORMULA
190       FOR I=1 TO N/2
200          READ A(I)
210       NEXT I
220       FOR I=1 TO N/2
230          READ X1(I)
240       NEXT I
250       In=0
255       REM FORM GAUSS SUMMATION
260       FOR K=1 TO N/2
270          Y=FNA(X1(K))+FNA(-X1(K))
280          In=In+A(K)*Y
290       NEXT K
300       PRINT N;"POINT: VALUE OF INTEGRAL IS ";In
310    NEXT C
320    END
325    REM DATA FOR TWO POINT FORMULA
330    DATA 1.46114118266,.707106781187
335    REM DATA FOR FOUR POINT FORMULA
340    DATA 1.05996448290,1.24022581770,.524647623275,1.65068012389
345    REM DATA FOR EIGHT POINT FORMULA
350    DATA .764544128652,.792890048386,.866752606563,1.07193014425
370    DATA .381186990207,1.15719371245,1.98165675670,2.93083742026
375    REM DATA FOR SIXTEEN POINT FORMULA
360    DATA .547375205038,.552441957368,.563217829088,.581247275401
370    DATA .609736958256,.655755672876,.736245622278,.936874492884
380    DATA .273481046138,.822951449145,1.38025853920,1.95178799092
390    DATA 2.54620215785,3.17699916198,3.86944790486,4.68873893931
```

Program 4.8.2.

Table 4.8.4. Evaluation of $\int_{-\infty}^{\infty} dx/(1 + x^2)$ using the Gauss-Hermite method.

```
GAUSS-HERMITE INTEGRATION BETWEEN –INF & +INF

 2 POINT: VALUE OF INTEGRAL IS  1.298792
 4 POINT: VALUE OF INTEGRAL IS  1.482336
 8 POINT: VALUE OF INTEGRAL IS  1.550268
16 POINT: VALUE OF INTEGRAL IS  1.565924
```

4.9 THE CURTIS-CLENSHAW METHOD

This method approximates the integral, again assumed to be in the range –1 to 1, by a series of Chebyshev polynomials, each of which can be integrated exactly. The method has been found to be very successful for a wide range of integrals because the Chebyshev series approximation is very good in the range –1 to 1. The integrand is expressed as follows:

$$f(x) = a_0 T_0(x)/2 + a_1 T_1(x) + a_2 T_2(x) + \dots \tag{4.9.1}$$

where the T_i are the Chebyshev polynomials. Using the properties of this function, described in Chapter 7, we obtain the following algorithm:

$$I = b_0 - 2 \sum_{i=2,4,6\dots} b_i /[(i - 1)(i + 1)] \tag{4.9.2}$$

The b_i are calculated from the formula:

$$b_i = \{g_0 + g_n + 2 \sum_{k=1}^{n-1} g_k\}/n \tag{4.9.3}$$

where

$$g_k = \cos(ki\pi/n)f[\cos(k\pi/n)]$$

The value of n is set to ensure the method provides the level of accuracy required.

   This method is implemented in program 4.9.1. It has the advantage that the final terms in the series for I can be calculated first and thus a check can be made to determine if the series is converging for a given n. If it is not converging, because the final terms in the series are not sufficiently small, then n may be doubled and the procedure restarted. A detailed development of the method is given in Clenshaw & Curtis (1960).

```
100    REM -------------------------------------------------
110    REM INTEGRATION BY CURTIS CLENSHAW METHOD   GL/JP (C)
120    REM -------------------------------------------------
125    REM DEFINE FUNCTION IN STATEMENT 130
130    DEF FNA(X)=X^X : COUNT =0
140    DIM X1(16),A(16),Y1(16)
150    PRINT"EVALUATING INTEGRALS BY CURTIS CLENSHAW METHOD"
160    PRINT
170    PRINT"ENTER ACCURACY REQUIRED";
180    INPUT Acc
190    PRINT"UPPER LIMIT OF INTEGRATION";
200    INPUT U
210    PRINT"LOWER LIMIT OF INTEGRATION";
220    INPUT L
230    PI=3.141592653#
240    FOR N=2 TO 16 STEP 2
250        In=0:G=PI/N
260        GOSUB 440
270        FOR J=0 TO N STEP 2
280            Sum=0
285            REM APPROXIMATE CHEBYCHEV COEFFICIENTS
290            FOR K=0 TO N
300                Arg=K*G
310                Sum=Sum+COS(Arg*J)*Y1(K)
320            NEXT K
330            A(J)=2*(Sum-(Y1(0)+Y1(N))/2)/N
335            REM CACULATE APPROXIMATION FOR INTEGRAL
340            IF J=0 THEN 370
350            In=In-A(J)/((J-1)*(J+1))
360            GOTO 380
370            In=A(0)/2
380            IF (ABS(A(J))<Acc) AND (N>4) THEN 400
390        NEXT J
400    NEXT N
410    In=(U-L)*In
420    PRINT:PRINT"THE VALUE OF THE INTEGRAL IS ";In;"COUNT=";COUNT
430    END
435    REM SUBROUTINE TO CALC FUNCTION VALUES
440    FOR K=0 TO N
450        Arg=K*G
460        X1(K)=COS(Arg)
470        X=(X1(K)*(U-L)+U+L)/2
480        Y1(K)=FNA(X):COUNT=COUNT+1
490    NEXT K
500    RETURN
```

Program 4.9.1.

4.10 PROBLEMS IN THE EVALUATION OF INTEGRALS

The major problems which occur are as follows:-

      (1) The function is discontinuous in the range of integration.
      (2) The function has singularities.
      (3) The range of integration is infinite.

In most cases these problems cannot be dealt with directly by numerical techniques and consequently some preparation of the integrand is required before the integral can be evaluated by the appropriate numerical method. In case (1) the position of the discontinuities must be found and the integral split into a sum of two or more integrals, the ranges of which avoid the discontinuities. Case (2) can be dealt with in various ways: using a change of variable, integration by parts and splitting the integral.

    Examples of the above situations will now be given, the first two of these are taken from Fox and Mayer (1977). Consider the integral

$$\int_0^1 x^{-1/2} f(x)\, dx$$

Clearly this integral is infinite when $x = 0$. Putting $x = t^2$ the integral becomes

$$\int_0^1 2f(t^2)\, dt$$

This integral is well behaved throughout the range of integration.
    Consider the discontinuous integral

$$\int_0^1 x^{-1/2} \cos x\, dx$$

Integrating by parts gives

$$\int_0^1 x^{-1/2} \sin x\, dx + 2 \cos 1 - 2$$

The integral is now continuous throughout its range.
    With regard to integrals with infinite ranges of integration we have discussed methods that can deal directly with this situation. However, we can sometimes deal with this problem by using an initial substitution. Consider the example

$$\int_1^\infty dx/\{x^2 + \cos(1/x)\} \tag{4.10.1}$$

putting $z = 1/x$, then $dz = -dx/x^2$ and (4.10.1) may be transformed as follows:-

$$I = -\int_1^0 dz/\{1 + z^2 \cos z\} = \int_0^1 dz/\{1 + z^2 \cos z\} \tag{4.10.2}$$

4.11 A GENERAL PROBLEM IN NUMERICAL INTEGRATION

We have discussed above a number of techniques for numerical integration, some of which are very effective in providing accurate answers to difficult problems. It must be said, however, that even the best methods have difficulty with functions which change very rapidly with small changes in the independent variable. An example of this type of function is $\cos(x^6)$. In attempting to obtain a computer generated graph for this function in the range 0 to 2π it is likely to be found that the graph does not represent the function correctly, particularly if the graph is being displayed on a video display unit. The reason for this is that the computer graphics will not have sufficient resolution to represent the fast changing nature of the function, where, for even a relatively small change in x the value of the function can change from an extreme positive to an extreme negative value. Fig.4.11.1 illustrates the inadequate nature of such a graphical display by showing, together with the original graph, a second graph which is an expansion of the part indicated. It should be noted that in the range 0 to 2π the function $\cos(x^6)$ contains some 9792 cycles. The function $\cos(x^6)$ is a particularly extreme example of a function that presents computational difficulties. The consequence of this situation is that a great number of divisions of the range of integration are needed to provide the required level of accuracy. Filon's Method, which is described in section 4.13, can be used for functions of the form $f(x)\cos(kx)$ and $f(x)\sin(kx)$. For more general functions adaptive integration methods have been introduced which increase the number of intervals only in those regions where the function is changing very rapidly, thus reducing the overall number of calculations required. Patterson's method, described in section 4.14, is an example of such a technique.

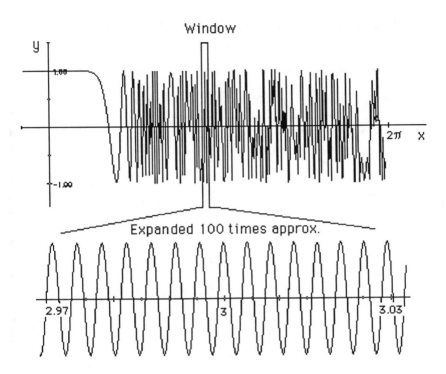

Fig. 4.11.1. Upper diagram attempts to show the function $\cos(x^6)$ in the range 0 to 2π. The lower diagram is an expanded plot in the region $x = 3$.

4.12 TEST INTEGRALS

The methods of integration previously described have been tested using several integrands. The number of function evaluations required to obtain a specific accuracy are listed in Table 4.12.1.

Several conclusions can be drawn from these results. As may be expected, Simpson's rule generally requires fewer functional evaluations than the trapezoidal rule but there are a few exceptions to this pattern. Gaussian integration is usually effective and requires only a few functional evaluations. The program given is limited to 16 intervals but this can be extended to 96 intervals if so desired using the coefficients given in Abramowitz and Stegun (1964). Finally the Curtis-Clenshaw method is generally effective although the method fails to obtain an accurate result in certain cases.

Table 4.12.1. Results obtained from various integration methods
for the evaluation of specific integrals.

Integral	Error	Number of function evaluations for:			
		Trapezoidal	Simpson	Gauss	Curtis
$\int_0^4 x^2(x-1)^2(x-2)^2(x-3)^2(x-4)^2\,dx$	10^{-2}	64	64	8	F
	10^{-4}	128	256	8	F
$\int_0^{\pi/2} \log(1+x)dx$	10^{-2}	16	4	2	15
	10^{-4}	64	8	4	15
$\int_0^1 x^x\,dx$	10^{-2}	32	16	4	15
	10^{-4}	256	64	8	48
$\int_{-1}^1 dx/(x^4+x^2+0.9)$	10^{-2}	12	4	8	15
	10^{-4}	16	4	8	63
$\int_0^1 dx/(1+x^4)\,dx$	10^{-2}	16	4	4	15
	10^{-4}	64	8	4	24
$\int_0^4 2e^{-x^2}dx/\sqrt{\pi}$	10^{-2}	4	8	8	24
	10^{-4}	8	8	8	63

4.13 FILON'S SINE AND COSINE FORMULAE

These formula can be applied to integrals of the form

$$\int_a^b f(x)\cos kx\,dx \text{ and } \int_a^b f(x)\sin kx\,dx$$

Consider $\int_0^{2\pi} f(x)\cos kx\,dx$

By the method of undetermined coefficients we can obtain an approximation to this integrand as follows. Let

$$\int_0^{2\pi} f(x) \cos x \, dx = A_1 f(0) + A_2 f(\pi) + A_3 f(2\pi) \qquad (4.13.1)$$

Requiring that this should be exact for $f(x) = 1, x$ and x^2 we have

$$0 = A_1 + A_2 + A_3$$
$$0 = A_2 \pi + A_3 2\pi$$
$$4\pi = A_2 \pi^2 + A_3 4\pi^2$$

Thus $A_1 = 2/\pi$ $A_2 = -4/\pi$ $A_3 = 2/\pi$ and hence

$$\int_0^{2\pi} f(x) \cos x \, dx = [2f(0) - 4f(\pi) + 2f(2\pi)]/\pi \qquad (4.13.2)$$

More general results can be developed as follows:-

$$\int_a^b f(x) \cos kx \, dx = h[A\{f(x_n) \sin k x_n - f(x_0) \sin k x_0\} + BC_e + DC_o]$$

$$\int_a^b f(x) \sin kx \, dx = h[A\{f(x_0) \cos k x_0 - f(x_n) \cos k x_n\} + BS_e + DS_o]$$

where $h = (b - a)/n$, $q = kh$ and

$$A = [q^2 + q \sin 2q/2 - 2 \sin^2 q]/q^3 \qquad (4.13.3)$$

$$B = 2[q (1 + \cos^2 q) - \sin 2q]/q^3 \qquad (4.13.4)$$

$$D = 4[\sin q - q \cos q]/q^3 \qquad (4.13.5)$$

$$C_o = \sum_{i=1,3,5}^{n-1} f(x_i) \cos kx_i$$

$$C_e = [f(x_0) \cos kx_0 + f(x_n) \cos kx_n]/2 + \sum_{i=2,4,6}^{n-2} f(x_i) \cos kx_i$$

It can be seen that C_o and C_e are odd and even sums of cosine terms. S_o and S_e are similarly defined with respect to sine terms.

It is important to note that Filon's method, when applied to functions of the form given above, usually gives better results than Simpson's method for the same number of intervals.

It is sometimes useful to use approximations to the expressions for A, B, C and D given in (4.13.3), (4.13.4) and (4.13.5) by expanding them in series of ascending powers of q. This leads to the following results

$$A = 2q^2(q/45 - q^3/315 + q^5/4725 + \ldots)$$

$$B = 2(1/3 + q^2/15 + 2q^4/105 + q^6/567 + \ldots)$$

$$D = 4/3 - 2q^2/15 + q^4/210 - q^6/11340 + \ldots$$

When the number of intervals becomes very large, h and q become small. As q tends to zero A tends to zero, B tends to 2/3 and D tends to 4/3. Substituting these values into the formula for Filon's method, it can be shown that it becomes equivalent to Simpson's rule. However in these circumstances the accuracy of Filon's rule may then be worse than Simpson's rule due to the additional complexity of the calculations.

Program 4.13.1 implements Filon's method for the evaluation of appropriate integrals. The program incorporates a modification to the standard Filon method in that we switch from using formula (4.13.3), (4.13.4) and (4.13.5) to the series approximation if q is less than 0.1. The justification for this is that as q becomes small, the accuracy of the trigonometric evaluation is limited by the accuracy of the particular computer, whereas the accuracy of the truncated series improves. This is illustrated in Table 4.13.1. In this table a specified integral is evaluated with and without the switching procedure described above. The results are striking; with switching the integral is evaluated to a high degree of accuracy but without switching the method fails completely.

```
100   REM -----------------------------------------
110   REM INTEGRATION BY FILON'S METHOD   GL/JP (C)
120   REM -----------------------------------------
125   REM DEFINE FUNCTION IN STATEMENT 130
130   DEF FNA(X)=EXP(-X/2)
140   PRINT"EVALUATING INTEGRALS BY FILON RULE"
150   PRINT:PRINT"FUNCTION TO BE INTEGRATED MUST BE OF FORM"
160   PRINT "                    f(X)*COS(KX)        (1)"
170   PRINT "                    f(X)*SIN(KX)        (2)"
180   PRINT"SELECT (1) OR (2)";
190   INPUT CASE
```

Program 4.13.1 (continues)

```
200   PRINT"DEFINE ONLY f(X) IN DEF STATEMENT":PRINT
210   PRINT"ENTER COEFFICIENT K";
220   INPUT K
230   PRINT"HOW MANY DIVISIONS OF INTERVAL REQUIRED";
240   INPUT D
250   PRINT"UPPER LIMIT OF INTEGRATION";
260   INPUT U
270   PRINT"LOWER LIMIT OF INTEGRATION";
280   INPUT L
290   PRINT:PRINT"INTERVALS";TAB(15);"INTEGRAL";TAB(30);" Q VALUE"
300   FU=FNA(U):FL=FNA(L)
310   FOR I=2 TO D
320       N=2^I:H=(U-L)/N
330       Q=K*H:Q2=Q*Q:Q3=Q2*Q
335       REM IF Q IS LESS THAN .1 THEN SWITCH TO SERIES APPROXIMATION
340       IF Q<.1 THEN 390
350       A=(Q2+Q*SIN(2*Q)/2-2*(SIN(Q))^2)/Q3
360       B=2*(Q*(1+(COS(Q))^2)-SIN(2*Q))/Q3
370       C=4*(SIN(Q)-Q*COS(Q))/Q3
380       GOTO 420
390       A=2*Q2*(Q/45-Q*Q2/315+Q2*Q3/4725)
400       B=2*(1/3+Q2/15+2*Q2*Q2/105+Q3*Q3/567)
410       C=4/3-2*Q2/15+Q2*Q2/210-Q3*Q3/11340
420       IF CASE=2 THEN 530
425       REM CALCULATION OF FILON APPROXIMATION IN COSINE CASE
430       Co=0:Ce=(FL*COS(K*L)+FU*COS(K*U))/2
440       FOR J=1 TO N-1 STEP 2
450           X=L+J*H: Co=Co+FNA(X)*COS(K*X)
460       NEXT J
470       FOR J=2 TO N-2 STEP 2
480           X=L+J*H: Ce=Ce+FNA(X)*COS(K*X)
490       NEXT J
500       In=H*(A*(FU*SIN(K*U)-FL*SIN(K*L))+B*Ce+C*Co)
510       PRINT N;TAB(15);In;TAB(30);Q
520       GOTO 620
525       REM  CALCULATION OF FILON APPROXIMATION IN SINE CASE
530       So=0:Se=(FL*SIN(K*L)+FU*SIN(K*U))/2
540       FOR J=1 TO N-1 STEP 2
550           X=L+J*H: So=So+FNA(X)*SIN(K*X)
560       NEXT J
570       FOR J=2 TO N-2 STEP 2
580           X=L+J*H: Se=Se+FNA(X)*SIN(K*X)
590       NEXT J
600       In=H*(-A*(FU*COS(K*U)-FL*COS(K*L))+B*Se+C*So)
610       PRINT N;TAB(15);In;TAB(30);Q
620   NEXT I
630   END
```

Program 4.13.1

Table 4.13.1. Filon's method for $\int_0^1 \dfrac{\sin x}{x}\,dx$ with and without switching.

To avoid the singularity at $x = 0$ the lower limit is taken as 10^{-10}.

INTERVALS	WITH SWITCH	WITHOUT SWITCH	Q VALUE
4	1717807	1717807	.25
8	107289.3	107289.3	.125
16	6778.847	8345.597	.0625
32	424.7422	.9461985	.03125
64	27.43611	3577.224	.015625
128	2.601753	2385.132	.0078125
256	1.049563	2385.132	3.90625E-03
512	.9525506	2385.132	1.953125E-03
1024	.9464873	-2383.003	9.765625E-04
2048	.9461082	-1191.147	4.882813E-04
4096	.9460847	0	2.441406E-04
8192	.9460832	0	1.220703E-04
Exact value	.9460831		

Table 4.13.2 compares the Filon and Simpson rules for a given integral using double precision arithmetic. The table shows that using 2048 intervals Filon's method is accurate to 8 significant digits whereas Simpson's rule provides only 2 digit accuracy.

Table 4.13.2. Evaluation of $\int_0^{2\pi} \exp(-x/2)\cos(100x)\,dx$.

INTERVALS	FILON'S RULE VALUE	SIMPSON'S RULE VALUE
4	4.552294395525306D-05	1.963214980751727
8	4.723385401248605D-05	1.917338330117465
16	4.723385401248586D-05	-.5731930365678587
32	4.766419313320027D-05	2.428016839503756D-02
64	4.777341086419405D-05	2.922634611081637D-02
128	4.783086776704809D-05	-8.744198806217476D-03
256	4.784047870077462D-05	5.551247460306683D-04
512	4.783817859104394D-05	-1.302617002085689D-04
1024	4.783811204203145D-05	4.534024288873272D-05
2048	4.783810837104364D-05	4.771479513174270D-05
Exact value	4.78381081343E-5	

4.14 PATTERSON'S METHOD

Patterson's procedure, (Patterson 1973), provides an extremely powerful integration technique. The method employs a basic integration algorithm which uses a
sequence of integration rules of increasing order. The rules start with a three point
Gauss rule and then employ $7, 15, 31, 63, 127$ and 255 point rules successively.
The points are added in an optimum way so that no integrand evaluations are wasted.
The basic integration algorithm is considered to have converged when successive
integration rules differ by less than or equal to the prescribed accuracy. If it does not
converge then the method employs further sub-division of the range of integration
and the re-application of the integration rules.
     The method is found to be successful on a wide range of integrals including those
where the integrand value fluctuates rapidly with small changes in the independent
variable. The method is also capable of dealing with singularities at the ends of the
range of integration. It should, however be noted that the applications of Patterson's
method, while providing reliable results to high accuracy, may require significant
computer time.
     To illustrate the power of this method the following integrals have been evaluated
using the program 4.14.1.

$$\int_0^1 x^{0.001} dx$$

  Exact value: $1000/1001 = 0.999000999$ recurring.

$$\int_0^1 \frac{dx}{[1 + (230x - 30)^2]}$$

  Exact value: $(\tan^{-1} 200 + \tan^{-1} 30)/230 = 0.0134924856495$

$$\int_0^4 x^2(x - 1)^2(x - 2)^2(x - 3)^2(x - 4)^2 dx$$

  Exact value: $1024/693 = 14.776334776$

     Fig. 4.14.1 shows plots of these functions over the range of integration. It
can be seen that each function, at some point, changes rapidly with small changes
of the independent variable, making such functions extremely difficult to integrate
numerically if a high degree of accuracy is required. It is of interest to note that all
of the other methods described failed to give accurate evaluations of the above
integrals; in some cases an accuracy of two significant figures could not be obtained.
     Even for Patterson's method the accuracy depends on the microcomputer used.
Table 4.14.1 illustrates the effect of single and double precision working.

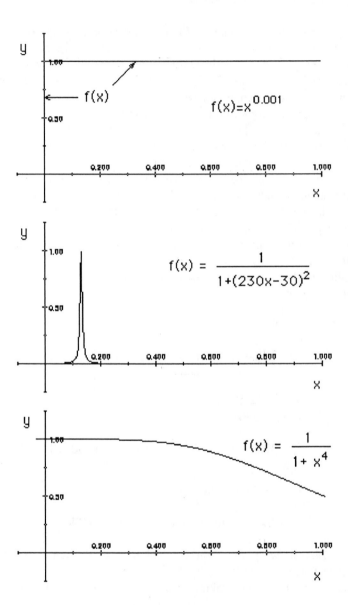

Fig. 4.14.1. The three functions integrated by Patterson's
method in the range 0 to 1.

```
100    REM ------------------------------------------------
110    REM PATTERSON'S METHOD FOR INTEGRATION GL/JP (C)
120    REM ------------------------------------------------
125    REM DEFINE FUNCTION IN STATEMENT 130
130    DEF FNA(X)=1/(1+X^4)
140    PRINT "EVALUATING INTEGRALS USING PATTERSON'S METHOD"
150    PRINT
160    PRINT "INPUT THE UPPER LIMIT FOR THE INTEGRAL";
170    INPUT U
180    PRINT "INPUT THE LOWER LIMIT FOR THE INTEGRAL";
190    INPUT L
200    PRINT
210    DIM R(9),Fu(400),P(381)
220    Ep=.000001
230    Np=0
235    REM ADAPTIVE QUADRATURE PROCEDURE
240    FOR I=1 TO 381
250        READ P(I)
260    NEXT I
270    Nm=4096
280    A=L:B=U
290    GOSUB 870
300    Sol=R(K-1)
310    Re=0
320    IF Sol<>0 THEN Re=ABS((R(K)-R(K-1))/Sol)
330    IF Ic=0 THEN 810
340    Es=ABS(Sol*Ep)
350    C1=1
360    Rh=1
370    N=1
380    H=U-L
390    Ba=1
400    Sol=0
410    Re=0
420    H=H/2
430    N=N+N
440    M1=Ba
450    M2=Ba+1:Ou=1
460    IF M1>M2 THEN 690
470    FOR J0=M1 TO M2
480        J1=J0
490        IF Rh<>1 THEN J1=M2+M1-J0
500        Al=L+H*(J1-1)
510        Be=Al+H
520        A=Al
530        B=Be
540        GOSUB 870
550        Co=ABS(R(K)-R(K-1))
560        IF Ic<>1 THEN 660
570        IF Co>=Es THEN 600
580        C1=SGN(C1)*2
590        GOTO 660
600        IF N=Nm THEN 650
610        Ba=2*J1-1
620        Rh=1
```

Program 4.14.1 (continues).

```
630      IF  J1-2*INT(J1/2)=0  THEN  Rh=0
640      GOTO  400
650      C1=-ABS(C1)
660      Sol=Sol+R(K)
670  NEXT  J0
680  Re=Re+Co
690  ON  Ou  GOTO  700,750,790
700  M1=1
710  M2=Ba-1
720  Rh=1
730  Ou=2
740  GOTO  460
750  M1=Ba+2
760  M2=N
770  Ou=3
780  GOTO  460
790  Ic=C1
800  Re=Re/ABS(Sol)
810  PRINT  "VALUE OF  INTEGRAL =";Sol
820  PRINT  "              CHECK  =";Ic
830  PRINT  "            EPSILON=";Ep
840  PRINT  " NO OF FUNCT EVALS=";Np
850  PRINT
860  END
865  REM QUADRATURE SUBROUTINE USING GAUSSIAN INTEGRATION
870  Ic=0
880  Sum=(B+A)/2
890  Dif=(B-A)/2
900  Fz=FNA(Sum)
910  R(1)=2*Fz*Dif
920  I=0
930  Io=0
940  In=1
950  Ac=0
960  FOR  K=2  TO  8
970      Io=Io+In
980      FOR  J=In  TO  Io
990          I=I+1
1000         X=P(I)*Dif
1010         Fu(J)=FNA(Sum+X)+FNA(Sum-X)
1020         I=I+1
1030         Ac=Ac+P(I)*Fu(J)
1040     NEXT  J
1050     In=Io+1
1060     I=I+1
1070     R(K)=(Ac+P(I)*Fz)*Dif
1080     IF ABS(R(K)-R(K-1))-Ep*ABS(R(K))<=0 THEN 1170
1090     IF K=8 THEN 1150
1100     Ac=0
1110     FOR  J=1  TO  Io
1120         I=I+1
1130         Ac=Ac+P(I)*Fu(J)
1140     NEXT  J
1150 NEXT  K
```

Program 4.14.1 (continues).

```
1160 Ic=1
1170 Np=In+Io+Np
1180 RETURN
1190 END
1195 REM FOR THE 381 COEFFICIENTS   FOR VARIOUS ORDER GAUSSIAN
1196 REM INTEGRATION FORMULA REFER TO PATTERSON'S PAPER
```

Program 4.14.1.

Table 4.14.1 Evaluation of the integrals of the functions A, B and C betwen 0 and 1 using Patterson's method where A = $x^{0.001}$, B = $1/(1 + x^4)$ and C = $1/[1 + (230x - 30)^2]$

	Integral A	Integral B	Integral C
Evaluation	0.9990019	0.8669729	0.01349249
Check	0	1	1
Epsilon	1E-6 Single	1E-6 Single	1E-6 Single
No of evals.	63 Precision	793 Precision	9038 Precision
Evaluation	0.9990018956447	0.86697298733908	0.013492485649448
Check	0	1	1
Epsilon	1E-6 Double	1E-6 Double	1E-6 Double
No of evals.	63 Precision	538 Precision	8987 Precision
Evaluation	0.99901187306100	0.86697298733908	0.013492485649448
Check	0	1	1
Epsilon	1E-7 Double	1E-10 Double	1E-7 Double
No of evals.	255 Precision	1562 Precision	10453 Precision
Exact	0.99900099900100	0.86697298733988	0.013492485649468

4.15 REPEATED INTEGRALS

In this section we confine ourselves to a discussion of repeated integrals using up to two variables. It is important to note that there is a significant difference between double integrals and repeated integrals. However, it can be shown that if the integrand satisfies certain requirements then double integrals and repeated integrals are equal in value. A detailed discussion of this result is given in Jeffrey (1979).

We have discussed in this chapter various techniques for evaluating single integrals. The extension of these methods to repeated integrals can present considerable programming difficulties. Furthermore the number of computations required for the accurate evaluation of a repeated integral can be enormous. For these reasons only programs for the evaluation of repeated integrals involving two variables will be presented here and an alternative method for estimating repeated integrals using random numbers is given in Chapter 9. Whilst many algorithms for the evaluation of single integrals can be extended to repeated integrals, here only extensions to the

Simpson and Gauss methods are presented. These have been chosen as the best
compromise between programming simplicity and efficiency.

An example of a repeated integral is

$$\int_{a_1}^{b_1} dx \int_{a_2}^{b_2} f(x,y)\, dy \qquad (4.15.1)$$

In this notation the function is integrated with respect to x from a_1 to b_1 and with
respect to y from a_2 to b_2. Here the limits of integration are constant but in some
applications they may be variables.

4.16 SIMPSON'S RULE FOR REPEATED INTEGRALS

We now apply Simpson's rule to the repeated integral (4.15.1) by applying the rule
first in the y direction and then in the x direction. Consider three equi-interval values
of y which are y_0, y_1 and y_2. On applying Simpson's rule, (4.3.1), to integration with
respect to y in (4.15.1) we have

$$I = \int_{x_0}^{x_2} dx \int_{y_0}^{y_2} f(x,y)\, dy \approx \int_{x_0}^{x_2} k[f(x,y_0) + 4f(x,y_1) + f(x,y_2)]/3\, dx \qquad (4.16.1)$$

where $k = y_2 - y_1 = y_1 - y_0$.

Consider now three equi-interval values of x; x_0, x_1 and x_2. Applying Simpson's rule
again to integration with respect to x, from (4.16.1) we have

$$I \approx hk[f(x_0,y_0) + f(x_0,y_2) + f(x_2,y_0) + f(x_2,y_2) +$$
$$4\{f(x_0,y_1) + f(x_1,y_0) + f(x_1,y_2) + f(x_2,y_1)\} + 16f(x_1,y_1)]/9 \qquad (4.16.2)$$

where $h = x_2 - x_1 = x_1 - x_0$

This is an "operator" for Simpson's rule in two variables and it is shown diagrammatically in Fig. 4.16.1 By applying this operator to each group of 9 points on
the surface $f(x,y)$ the composite Simpson's rule is obtained. The form of the
coefficients of this rule are easier to comprehend when presented diagrammatically,
rather than as a formula, and they are shown in Fig. 4.16.2.

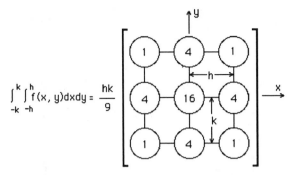

$$\int_{-k}^{k}\int_{-h}^{h} f(x, y)\,dx\,dy = \frac{hk}{9}$$

Fig. 4.16.1. Simpson's rule. Operator for two dimensions.

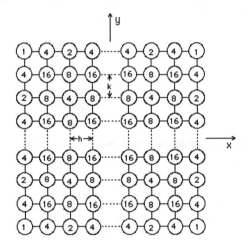

Fig. 4.16.2. Simpson's rule. Pattern of coefficients for the composite operator in two dimensions.

In Fig.4.16.2 h and k are defined by

$$h = (b_1 - a_1)/n \quad \text{and} \quad k = (b_2 - a_2)/m$$

where n and m must be even and are the number of steps in the x and y directions respectively.

When developing a program to evaluate a single integral using Simpson's method the basic rule was used repeatedly rather than employing the composite rule directly. In contrast program 4.16.1 evaluates repeated integrals in two variables by making direct use of the composite rule. Table 4.16.1 gives the results of evaluating

$$\int_{0}^{1} dx \int_{0}^{1} e^{-xy}\,dy \tag{4.16.3}$$

Table 4.16.1. Evaluation of the integral of (4.16.3).

```
TWO DIMENSIONAL INTEGRATION USING SIMPSON'S RULE

NO OF DIVISIONS OF INTERVAL REQUIRED?  5

ENTER RATIO OF INTERVALS IN X/INTERVALS IN Y?  1
ENTER UPPER LIMIT FOR Y?  1
ENTER LOWER LIMIT FOR Y?  0
ENTER UPPER LIMIT FOR X?  1
ENTER LOWER LIMIT FOR X?  0

X   (DIVISIONS)   Y          INTEGRAL
2                 2          .7966927
4                 4          .7966054
8                 8          .7966002·
16                16         .7965998
32                32         .7965996
```

```
100   REM ------------------------------------------------------------
110   REM TWO DIMENSIONAL INTEGRATION USING SIMPSON'S RULE JP/GL (C)
120   REM ------------------------------------------------------------
130   PRINT"TWO DIMENSIONAL INTEGRATION USING SIMPSON'S RULE"
140   PRINT:PRINT"NO OF DIVISIONS OF INTERVAL REQUIRED";
150   INPUT D
160   PRINT"ENTER RATIO OF INTERVALS IN X/INTERVALS IN Y";
170   INPUT R
180   PRINT"ENTER UPPER LIMIT FOR Y";
190   INPUT YU
200   PRINT"ENTER LOWER LIMIT FOR Y";
210   INPUT YL
220   PRINT"ENTER UPPER LIMIT FOR X";
230   INPUT XU
240   PRINT"ENTER LOWER LIMIT FOR X";
250   INPUT XL
260   PRINT:PRINT " X     (DIVISIONS)     Y";TAB(26);"INTEGRAL"
270   FOR P=1 TO D
280      IF R>1 THEN F=INT(R+.5):YN=2^P:XN=F*YN:GOTO 310
290      IF R<1 THEN F=INT(1/R+.5): XN=2^P:YN=F*XN:GOTO 310
300      YN=2^P:XN=YN
310      K=(YU-YL)/YN:H=(XU-XL)/XN
320      T=0
325      REM USES SIMPSON'S ON X AND Y VARIABLES
```

Program 4.16.1 (continues)

```
330      FOR I=0 TO YN
340          Y=YL+I*K
350          FOR J=0 TO XN
360              X=XL+J*H
370              GOSUB 500
380              IF I=0 OR I=YN THEN MI=1:GOTO 410
390              IF I/2=INT(I/2)THEN MI=2:GOTO 410
400              MI=4
410              IF J=0 OR J=XN THEN MJ=1:GOTO 440
420              IF J/2=INT(J/2)THEN MJ=2:GOTO 440
430              MJ=4
440              T=T+MI*MJ*F
450          NEXT J
460      NEXT I
470      PRINT XN;TAB(15);YN;TAB(25);T*K*H/9
480  NEXT P
490  END
495  REM DEFINE FUNCTION IN TERMS OF X AND Y AT STATEMENT 500
500  F=EXP(-X*Y)
510  RETURN
```

Program 4.16.1.

This integral is equal to 0.796599599 and it is seen that to obtain an accurate result a large number of computations must be performed. It can be proved, see Salvadori and Baron (1961), that when Simpson's rule is adapted to evaluate repeated integrals the error is still of order h^4 and thus it is possible to use an extrapolation scheme similar to the Romberg method of section 4.6.

Consider a more general repeated integral given by

$$I = \int_a^b dx \int_{\phi_1(x)}^{\phi_2(x)} f(x,y)\, dy \qquad (4.16.4)$$

where $\phi_1(x)$ and $\phi_2(x)$ are single valued functions. Applying Simpson's rule to estimate the integration of the function with respect to y between $\phi_1(x)$ and $\phi_2(x)$ gives

$$A_i \approx k_i\{f(x_i,y_0) + 4[f(x_i,y_1) + f(x_i,y_3) + \dots]$$

$$+ 2[f(x_i,y_2) + f(x_i,y_4) + \dots] + f(x_i,y_m)\}/3$$
$$i = 0, 1, \dots ,n \qquad (4.16.5)$$

where $k_i = [\phi_2(x_i) - \phi_1(x_i)]/m$, m is the number of intervals which must be even. Applying the composite rule to estimate the integration of the function with respect to x between a and b gives

$$I \approx h[A_0 + 4(A_1 + A_3 + \dots) + 2(A_2 + A_4 + \dots) + A_n]/3 \qquad (4.16.6)$$

where $h = (b-a)/n$; n is the number of interval and must be even. Program 4.16.2 evaluates repeated integrals in two variables of the type defined by (4.16.4) using (4.16.5) and (4.16.6). For example, the evaluation of

$$I = \int_1^2 dx \int_{x^2}^{x^4} x^2 y \, dy \qquad (4.16.7)$$

is shown in Table 4.16.2. The best estimate compares well with the exact value, 83.974026.

```
100    REM -----------------------------------------------------------------
110    REM TWO DIMENSIONAL INTEGRATION WITH VARIABLE LIMITS:SIMPSON'S RULE
120    REM -----------------------------------------------------------------
125    REM DEFINE FUNCTION LIMITS IN STATEMENTS 130 AND 140
130    DEF FNU(X)=X^4
140    DEF FNL(X)=X^2
150    PRINT"TWO DIMENSIONAL INTEGRATION USING SIMPSON'S RULE"
160    PRINT" LIMITS OF INTEGRATION IN Y ARE FUNCTIONS OF X"
170    PRINT:PRINT"NO OF DIVISIONS OF INTERVAL REQUIRED";
180    INPUT D
190    PRINT"ENTER RATIO OF INTERVALS IN X/INTERVALS IN Y";
200    INPUT R
210    PRINT"ENTER UPPER LIMIT FOR X";
220    INPUT XU
230    PRINT"ENTER LOWER LIMIT FOR X";
240    INPUT XL
250    PRINT
260    PRINT " X         DIVISIONS      Y ";TAB(26);"INTEGRAL"
270    FOR P=1 TO D
280        IF R>1 THEN F=INT(R+.5)  : YN=2^P:XN=F*YN:GOTO 310
290        IF R<1 THEN F=INT(1/R+.5):XN=2^P:YN=F*XN:GOTO 310
300        XN=2^P:YN=XN
310        H=(XU-XL)/XN
320        T=0
325        REM CALCULATE FUNCTION VALUES AND USE SIMPSON'S RULE
330        FOR I=0 TO XN
340            X=XL+I*H
350            S=0
360            YU=FNU(X)
370            YL=FNL(X)
380            K=(YU-YL)/YN
390            FOR J=0 TO YN
400                Y=YL+J*K
410                GOSUB 540
```

Program 4.16.2 (continues).

```
420              IF J=0 OR J=YN THEN S=S+F:GOTO 450
430              IF J/2=INT(J/2) THEN S=S+2*F:GOTO 450
440              S=S+4*F
450          NEXT J
460          S=S*K/3
470          IF I=0 OR I=XN THEN T=T+S:GOTO 500
480          IF I/2=INT(I/2) THEN T=T+2*S:GOTO 500
490          T=T+4*S
500      NEXT I
510      PRINT XN;TAB(15);YN;TAB(25);T*H/3
520  NEXT P
530  END
535  REM DEFINE FUNCTION IN SUBROUTINE AT STATEMENT 540
540  F=X^2*Y
550  RETURN
```

Program 4.16.2.

Table 4.16.2. Evaluation of the integral (4.16.7)

```
TWO DIMENSIONAL INTEGRATION USING SIMPSON'S RULE
LIMITS OF INTEGRATION IN Y ARE FUNCTIONS OF X

NO OF DIVISIONS OF INTERVAL REQUIRED? 5
ENTER RATIO OF INTERVALS IN X/INTERVALS IN Y? 1
ENTER UPPER LIMIT FOR X? 2
ENTER LOWER LIMIT FOR X? 1


X    DIVISIONS    Y         INTEGRAL
2                 2         95.4248
4                 4         84.88371
8                 8         84.03429
16                16        83.97787
32                32        83.97527
```

4.17 GAUSSIAN INTEGRATION FOR REPEATED INTEGRALS

The Gaussian method can developed to evaluate repeated integrals with constant limits of integration. In section 4.7 it was shown that for single integrals the integrand must be evaluated at specified points. Thus, if

$$I = \int_{-1}^{1} dx \int_{-1}^{1} f(x,y) dy$$

then

$$I \approx \int_{-1}^{1} \left[\sum_{i=1}^{n} A_i f(x,y_i) \right] dx$$

$$I \approx \sum_{i=1}^{n} \sum_{j=1}^{n} A_i A_j \, f(x_j, y_i)$$

The rules for calculating x_j, y_i and A_i are given in section 4.7. Program 4.17.1 evaluates integrals using this technique. Because the values of x and y are chosen on the assumption that the integration takes place in the range -1 to 1 the program includes the necessary manipulations to adjust the function so as to accommodate an arbitrary range of integration. Table 4.17.1 shows that output of program 4.17.1 when evaluating the integral defined by (4.16.3). Comparison with Table 4.16.1 shows Gaussian integration to be greatly superior to Simpson's rule.

Table 4.17.1. Evaluation of the integral (4.16.3) using the Gaussian integration.

```
TWO DIMENSIONAL GAUSSIAN INTEGRATION

NO OF DIVISIONS OF INTERVAL REQUIRED <=4?  4
ENTER UPPER LIMIT FOR Y?  1
ENTER LOWER LIMIT FOR Y?  0
ENTER UPPER LIMIT FOR X?  1
ENTER LOWER LIMIT FOR X?  0

    STEPS                        INTEGRAL
      2                          .7965382
      4                          .7965996
      8                          .7965997
     16                          .7965996
```

The integral of (4.16.7) cannot be evaluated using program 4.17.1 since the limits of integration of y are functions of x. However the limits of integration can be changed to limits -1 and 1 by letting

$$y = [(x^4 - x^2)s + (x^4 + x^2)]/2$$

Thus

$$dy = (x^4 - x^2)ds/2$$

and the integral becomes

$$\int_{1}^{2} dx \int_{-1}^{1} x^2 [(x^4 - x^2)s + (x^4 + x^2)](x^4 - x^2) \, ds/4 \qquad (4.17.1)$$

```
100    REM --------------------------------------------------
110    REM TWO DIMENSIONAL GAUSSIAN INTEGRATION JP/GL (C)
120    REM --------------------------------------------------
130    DIM A(16),X1(16)
140    PRINT"TWO DIMENSIONAL GAUSSIAN INTEGRATION"
150    PRINT
160    PRINT"NO OF DIVISIONS OF INTERVAL REQUIRED <=4";
170    INPUT D
180    PRINT"ENTER UPPER LIMIT FOR Y";
190    INPUT YU
200    PRINT"ENTER LOWER LIMIT FOR Y";
210    INPUT YL
220    PRINT"ENTER UPPER LIMIT FOR X";
230    INPUT XU
240    PRINT"ENTER LOWER LIMIT FOR X";
250    INPUT XL
260    PRINT
270    PRINT "STEPS";TAB(25);"INTEGRAL"
280    FOR P=1 TO D
290        N=2^P
295        REM READ DATA FOR N POINT GAUSS FORMULA
300        FOR I=1 TO N/2
310            READ A(I)
320            A(N+1-I)=A(I)
330        NEXT I
340        FOR I=1 TO N/2
350            READ X1(I)
360            X1(N+1-I)=-X1(I)
370        NEXT I
380        T=0
385        REM FORM GAUSS SUMMATION
390        FOR I=1 TO N
400            X=(X1(I)*(XU-XL)+XU+XL)/2
410            FOR J=1 TO N
420                Y=(X1(J)*(YU-YL)+YU+YL)/2
430                GOSUB 590
440                T=T+F*A(I)*A(J)
450            NEXT J
460        NEXT I
470        IN=(XU-XL)*(YU-YL)*T/4
480        PRINT N;TAB(25);IN
490    NEXT P
500    END
505    REM DATA FOR GAUSSIAN INTEGRATION RULES
510    DATA 1,.577350269#
520    DATA .347854845,.652145155#,.861136312#,.339981044#
530    DATA .101228536,.222381034#,.313706646#,.362683783#
540    DATA .960289856,.796666477#,.5255324099999999#,.183434642#
550    DATA .027152460,.062253524#,.095158512#,.124628971#
560    DATA .149595989,.169156519#,.182603415#,.18945061#
570    DATA .989400935,.944575023#,.865631202#,.755404408#
580    DATA .617876244,.458016778#,.281603551#,9.501250999999999D-02
585    REM DEFINE FUNCTION AT STATEMENT 590
590    F=EXP(-X*Y)
600    RETURN
```

Program 4.17.1.

This integral can be evaluated using program 4.17.1 and the results are shown in Table 4.17.2. Again Gaussian integration is shown to be superior to Simpson's rule.

Table 4.17.2. Evaluation of integral (4.17.1) using Gauss's method. This is equivalent to evaluating (4.16.7).

```
TWO DIMENSIONAL GAUSSIAN INTEGRATION

NO OF DIVISIONS OF INTERVAL REQUIRED <=4?  4
ENTER UPPER LIMIT FOR Y?  1
ENTER LOWER LIMIT FOR Y?  -1
ENTER UPPER LIMIT FOR X?  2
ENTER LOWER LIMIT FOR X?  1


STEPS                     INTEGRAL
  2                       76.52557
  4                       83.97291
  8                       83.97404
 16                       83.974
```

4.18 SUMMARY

In this chapter we have presented a range of techniques for evaluating single and repeated integrals with finite, and in the case of single integrals, infinite limits. Here we give in table 4.18.1 a summary of the most appropriate use of the techniques described to provide an easy reference for the reader.

Table 4.18.1. Summary of integration methods.

Method	Program	Type of function
Trapezoidal	4.2.1	General
Simpson	4.3.1	General
Romberg	4.6.1	General
Gaussian	4.7.1	General
Gauss-Laguerre	4.8.1	Infinite upper limit
Gauss-Hermite	4.8.2	Infinite limits
Curtis-Clenshaw	4.9.1	General
Filon	4.13.1	$f(x)\cos x$ and $f(x)\sin x$
Patterson	4.14.1	For difficult integrals
Simpson repeated	4.16.1/2	Two variables
Gauss repeated	4.17.1	Two variables

PROBLEMS

4.1 Use the trapezoidal rule, program 4.2.1, to evaluate the integral

$$\int_{\varepsilon}^{1} \sqrt{t}\ (1 + \log_e t)\, dt$$

using 1024 intervals. Take values of $\varepsilon = 1E{-}2$, $1E{-}4$ and $1E{-}8$. Compare your answers with the value of the integral when $\varepsilon = 0$ which is 2/9. Note that integrating this problem numerically with $\varepsilon = 0$ presents difficulties since the program must evaluate $\log_e t$ when $t = 0$. When $t = 0$ the function tends to minus infinity.

4.2 Solve problem 4.1 using the Simpson rule, program 4.3.1.

4.3 Use Simpson's rule, program 4.3.1, to evaluate the Fresnel integrals

$$C(1) = \int_{0}^{1} \cos(\pi\, t^2/2)\, dt \ \text{ and } \ S(1) = \int_{0}^{1} \sin(\pi\, t^2/2)\, dt$$

Use 32 intervals. The exact values, to 7 decimal places, are

$$C(1) = 0.7798934 \text{ and } S(1) = 0.4382591.$$

How does the accuracy of your answer compare with the order of the theoretically predicted error.

4.4 The Elliptic integral of the second kind is given

$$\int_{0}^{t} \sqrt{(1 - \sin^2 \beta \sin^2 \theta)}\ d\theta$$

Taking β as $\pi/6$ and t as $\pi/2$ evaluate this integral using Romberg's rule.(program 4.6.1). Compare your answer with the exact value to seven decimal places which is 1.46746221.

4.5 The integral

$$\text{li}(x) = \int_{0}^{x} dt\, /\log_e t$$

can be used to estimate the number of prime numbers less than x. Use Romberg's method, program 4.6.1, to evaluate this integral in order to estimate the number of primes less than 1000. Take the lower limit of integration to be 1E–4. The number of primes less than 1000 is 168. The exact value of this integral is Ei($\log_e 1000$) which is approximately 177.6.

4.6   Use Filon's method, program 4.13.1, with 64 intervals, to evaluate the integral

$$\int_0^\pi \sin x \cos kx \; dx$$

for $k = 0, 4$ and 100. Compare your results with the exact answer, which equals $2/(1 - k^2)$. Compare the result you obtained for $k = 100$ with result of using Simpson's rule, program 4.3.1.

4.7   Solve the following integral using the Gauss-Laguerre method, program 4.8.1.

$$\int_0^\infty e^{-x} dx \; /(x + 100)$$

Compare your answer with the exact solution, 9.9019419×10^{-3}, (103/10402).

4.8   Solve the following integral using the Gauss-Hermite method, program 4.8.2:

$$\int_{-\infty}^\infty \exp(-x^2) \cos x \; dx$$

Compare your answer with the exact solution $\sqrt{\pi} \exp(-1/4)$.

4.9   Evaluate the following integrals, using Simpson's rule for repeated integrals (program 4.16.1), using 64 divisions in each direction

$$\text{(i)} \quad \int_{-1}^1 dy \int_{-\pi}^\pi x^4 y^4 dx \qquad \text{(ii)} \quad \int_{-1}^1 dy \int_{-\pi}^\pi x^{10} y^{10} dx$$

4.10 Evaluate, using Simpson's rule for repeated integrals (program 4.16.2), using 64 divisions in each direction

$$\text{(i)} \quad \int_0^3 dx \int_1^{\sqrt{x/3}} \exp(y^3) dy \qquad \text{(ii)} \quad \int_0^2 dx \int_0^{2-x} (1 + x + y)^{-3} dy$$

4.11 Evaluate part (ii) of problems 4.9 and 4.10 using Gaussian integration, program 4.17.1. *Note*: To use this program the range of integration must be constant.

4.12 A definition of the Bessel function of the first kind of order zero, $J_0(z)$ is given by

$$J_0(z) = (1/\pi) \int_0^\pi \cos(z \sin q) \, dq$$

Evaluate this function using Simpson's rule, program 4.3.1 using 64 intervals for $z = 1, 3$ and 5

4.13 A definition of the error function erf(z) is given by

$$\text{erf}(z) = (2/\sqrt{\pi}) \int_0^z \exp(-t^2) \, dt$$

Evaluate this function using Simpson's rule, program 4.3.1 using 64 intervals for values of $z = 0.2, 0.76$ and 1.35.

4.14 A definition of the sine integral Si(z) is given by

$$\text{Si}(z) = \int_0^z \frac{\sin t}{t} \, dt$$

Evaluate this integral using the Gauss method for values of $z = 0.5, 1$ and 2. Why does the Gaussian method work and yet the Simpson and Romberg methods fail?

4.15 Evaluate

$$\int_0^\infty \frac{dx}{(1 + x^3 + x^{-x})}$$

by (i) using the Gauss-Laguarre method, program 4.8.1, and (ii) noting the value of the integral

$$\int_0^\infty \frac{dx}{(1 + x^3)} = \frac{\pi}{3 \sin(\pi/3)}$$

and subtracting this value from the difference of the integrals

$$\int_0^5 \frac{dx}{(1 + x^3 + x^{-x})} \qquad \int_0^5 \frac{dx}{(1 + x^3)}$$

evaluated numerically using Simpsons rule. *Note.* This technique avoids the infinite upper limit and is based on the fact that these functions converge for values 5 and above.

5

Numerical differentiation

5.1 INTRODUCTION

In this section we introduce numerical methods for finding approximations to the derivatives of given functions. This approach has an important application in the development of those scientific computer programs which require the calculation of derivatives. Since numerical approximations require only function values, the user of a program employing such approximations need only provide the function and not the derivatives. These methods are usually based on some kind of polynomial approximation to the function. The nature of differentiation is such that the errors in the approximating functions tend to be magnified and care must be taken to minimise these errors.

There are a number of alternative procedures for obtaining approximating formulae for derivatives we shall now consider a simple method based on the use of Taylor's series.

5.2 APPROXIMATING FORMULA

Consider the function

$$y = f(x)$$

Assuming that we have a tabulated set of equi-spaced values for x and y given by (x_i, y_i) for $i = 0, 1, 2, \ldots$ where $x_i = x_0 + ih$ and $y_i = y(x_0 + ih)$ then the derivative of y at the point x_i can be written as dy/dx at $x = x_i$

This can be written in shortened form as y'_i where the dash denotes differentiation once and the subscript i denotes that x_i is the point at which the derivative is obtained.

If we expand $y(x_0 + ih)$ about x_0 we obtain

$$y(x_0 + ih) = y_i + y'_i h + y''_i h^2/2! + y'''_i h^3/3! + \dots \qquad (5.2.1)$$

Similarly

$$y(x_0 - ih) = y_i - y'_i h + y''_i h^2/2! - y'''_i h^3/3! + \dots \qquad (5.2.2)$$

If we subtract (5.2.2) from (5.2.1) we obtain

$$y'_i = [y(x_0 + ih) - y(x_0 - ih)]/(2h) + O(h^2) \qquad (5.2.3)$$

The term $O(h^2)$ indicates that the remaining terms are of order h^2 or above. We may write (5.2.3) in the form

$$y'_i = (y_{i+1} - y_{i-1})/(2h) + O(h^2) \qquad (5.2.4)$$

Fig. 5.2.1 illustrates how this provides an approximation for the derivative. We can

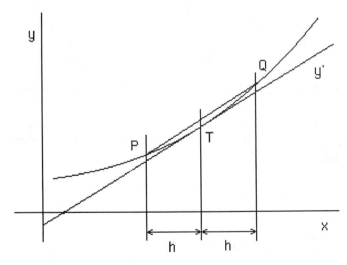

Fig. 5.2.1. The diagram shows how the tangent at T
may be well approximated by the secant PQ.

Table 5.2.1. Approximation to the first derivative of
x^9 at the point $x = 1$ for various values of h.

h	First derivative
.1	9.852638
.01	9.008389
.001	9.000242
.0001	9.001494
.00001	9.012223
.000001	8.851291
.0000001	10.72884

see the slope of the line joining P and Q is approximately equal to the slope of the tangent at T. Since the slope of the tangent at T is y'_i this gives the relation

$$y'_i \approx (y_{i+1} - y_{i-1})/(2h)$$

which is essentially the same as (5.2.4).

We can see from both Fig 5.2.1 and the order of the error in (5.2.4) that the smaller h becomes the better the approximation given by (5.2.4). However Table 5.2.1, which was generated by a 7-digit accuracy computer, shows how the accuracy of this approximation varies for different values of h in practice, when applied to finding the derivative of $y = x^9$ at the point $x = 1$. From the results given in Table 5.2.1 we can see that at first, as h is decreased, the approximate value of the derivative improves. The best result is accurate to four significant places. However, for values of h less than 0.001 the approximation becomes less accurate because of errors introduced by the limitation of machine accuracy. This causes some difficulty in the choice of h since it is not simply a matter of taking it as small as possible.

5.3 APPROXIMATIONS FOR HIGHER ORDER DERIVATIVES

Having derived an approximating formula for the first order derivative of a function we now obtain similar formulae for approximating higher order derivatives. If we add equations (5.2.1) and (5.2.2) we obtain

$$y_{i+1} + y_{i-1} = 2y_i + y''_i h^2/2 + O(h^4)$$

or

$$y''_i = (y_{i+1} - 2y_i + y_{i-1})/h^2 + O(h^2) \tag{5.3.1}$$

Here again the derivative is expressed only in terms of the value of the function at specified points. The remaining terms are, as before, of order h^2. This means that the error in both the first and second order derivatives is of order h^2 and will, theoretically, decrease as h decreases.

We can also obtain an approximation for third order derivatives. This requires consideration of further terms in the Taylor's series expansions as follows

$$y_{i+2} = y_i + y'_i 2h + y''_i (2h)^2/2! + y'''_i (2h)^3/3! + O(h^4) \qquad (5.3.2)$$

$$y_{i-2} = y_i - y'_i 2h + y''_i (2h)^2/2! - y'''_i (2h)^3/3! + O(h^4) \qquad (5.3.3)$$

Subtracting (5.3.3) from (5.3.2) gives

$$y_{i+2} - y_{i-2} = 4y'_i h + (8/3)y'''_i h^3 + O(h^5) \qquad (5.3.4)$$

Now if we subtract equation (5.2.2) from (5.2.1) we have

$$y_{i+1} - y_{i-1} = 2y'_i h + y'''_i h^3/3 + O(h^5) \qquad (5.3.5)$$

Taking twice (5.3.5) from (5.3.4) gives on rearrangement

$$y'''_i = (y_{i+2} - 2y_{i+1} + 2y_{i-1} - y_{i-2})/(2h^3) + O(h^2) \qquad (5.3.6)$$

We can continue in this way to derive expressions for higher order derivatives. In each case the approximation will involve only function values evaluated at various points. Any order derivative can be obtained by proceeding in this way, but the algebra becomes increasingly lengthy. We give below, without the details of its derivation, an expression for the fourth order derivative, which is useful in a number of applications.

$$y^{(iv)}_i \approx (y_{i+2} - 4y_{i+1} + 6y_i - 4y_{i-1} + y_{i-2})/h^4 \qquad (5.3.7)$$

Table 5.3.1 gives the results of using the above expressions to obtain approximations for the first, second and third order derivatives for a specific function and for decreasing values of h. The function to be differentiated is $y = x^7$ and its derivatives are to be evaluated at the point $x = 1$. The values of the first, second and third derivatives are exactly 7, 42, 210 respectively. The results show that accuracy improves at first as h decreases and then the approximation becomes worse as h becomes smaller. The best result for the first order derivative is obtained when $h = 0.001$. The best results for the higher order derivatives are obtained when $h = 0.01$. It should be noted that as h becomes smaller than 0.01 the approximations to the higher order derivatives worsen dramatically.

Table 5.3.1. Values of the first, second and third order
derivatives of x^7 at $x = 1$ for various values of h.

	Derivative order		
h	First	Second	Third
.1	7.352103	42.70141	216.3128
.01	7.003495	42.00697	209.9872
.001	7.000119	42.43851	-149.0116
.0001	7.001162	47.68373	-208616.3
.00001	7.009507	0	0
.000001	6.884337	-476837.3	4.172327E+11
.0000001	8.344651	0	-2.086163E+14

5.4 OBTAINING HIGHER ACCURACY DERIVATIVES

By using combinations of various Taylor's series expansions we can obtain alternative expressions for the different order derivatives with smaller errors. Examples of
these are:

$$y'_i = (-y_{i+2} + 8y_{i+1} - 8y_{i-1} + y_{i-2})/(12h) + O(h^4) \tag{5.4.1}$$

$$y''_i = (-y_{i+2} + 16y_{i+1} - 30y_i + 16y_{i-1} - y_{i-2})/(12h^2) + O(h^4) \tag{5.4.2}$$

Table 5.4.1 gives the results of using the above approximations to differentiate x^7 at
$x = 1$ and it may be compared with the results obtained from using (5.2.4) and (5.3.1)
given in Table 5.3.1. We can see from Table 5.4.1 that formulae (5.4.1) and (5.4.2)
enable us to obtain more accurate approximations for the first and second order
derivatives. As before, it is important to make a careful choice of h; too small a value
of h leads to very inaccurate values for the derivative even with the more accurate
formulae. The limited results given above seem to indicate that for the higher order
methods a larger value of h should be taken.

These numerical studies show that by using a carefully chosen value of h we can
obtain   approximations for the various order derivatives that we require. How
satisfactory these approximations are will depend on the nature of the function being
differentiated. For functions whose values and derivatives change rapidly for small
changes of the independent variable it may be difficult to obtain the accuracy
required.

We now consider a method which allows continuous adjustment to the interval
value h so that improved approximations for derivatives can be obtained.

Table 5.4.1. Values of the first and second derivative of x^7 for various values of h using the more exact formulae of (5.4.1) and (5.4.2).

	Derivative order:	
x	First	Second
.1	6.991581	41.99442
.01	6.999995	42.00082
.001	7.000164	42.52295
.0001	7.00161	47.18702
.00001	7.009507	0
.000001	6.804864	-566244.3
.0000001	8.493663	-5463760

5.5 APPLYING THE RICHARDSON EXTRAPOLATION

Richardson extrapolation is used in the calculations for Romberg integration, see section 4.6, to obtain improved approximations to the value of an integral by successively halving the interval. The extrapolation method can also be applied to the process of numerical differentiation and its application is very similar to that used for integration.

Consider again formula (5.2.4)

$$y'_i = (y_{i+1} - y_{i-1})/(2h) + O(h^2)$$

For a specific function $y = f(x)$ we can obtain a sequence of approximations for its derivative by using an initial value of h, halving it a specific number of times, and for each new h value finding the approximate derivative using formula (5.2.4). Let these approximations be $A_1, A_2, \dots , A_k$. We can now use Richardson's extrapolation of Chapter 1 to calculate a new sequence of approximations $B_1, B_2, \dots ,B_{k-1}$. These results will be a significant improvement on the previous ones if we use sufficient accuracy, which for certain problems requires working to double precision. The calculations are performed using the formula

$$B_i = (A_{i+1}4^m - A_i)/(4^m - 1) \quad \text{for } i = 1, 2, 3, \dots ,k - m$$

Initially m is taken as 1 and then the process is repeated restarting with the latest improved values and increasing the value of m by one each time until the required accuracy is reached or until m equals $k - 1$. Table 5.5.1 gives the results of applying the extrapolation method to find the first derivative of x^7. It should be noted that this table was produced on the Macintosh computer with double precision arithmetic: this precision is required to avoid the improvement in results being obscured by rounding errors. It is interesting to compare these results with those obtained on the Macintosh computer with single precision which are given in Table 5.5.2. The method is easily extended to give improved approximations for second order derivatives. The only change required is that the initial sequence of approximations must be derived from a formula for the second order derivative such as (5.3.1). Program 5.5.1 implements the extrapolation technique using a fixed number of divisions of the interval to estimate the first or second order derivative.   The program requires the user to define the function for which the derivative is to be determined.

Table 5.5.1. Very accurate results are obtained using the extrapolation method for the derivative of x^7 when $x = 1$.

```
EVALUATION OF DERIVATIVES USING EXTRAPOLATION METHOD WITH DOUBLE PRECISION

IF YOU REQUIRE FIRST ORDER DERIVATIVE ENTER 1
                    OR IF SECOND ORDER ENTER 2?  1
ENTER VALUE OF INTERVAL H?  .1
ENTER VALUE AT WHICH DERIVATIVE IS TO BE EVALUATED? 1
RESULTS USING CENTRAL DIFFERENCES WITH REPEATED HALVING OF INTERVALS

APPROXIMATIONS:

GROUP  1
 6.999474687468681    6.999967182615232    6.999997949142334 6.99999987182496
GROUP  2
 7.000000015625003    7.000000000244141    7.000000000003809
GROUP  3
 7   6.999999999999995
GROUP  4
 6.999999999999995
```

Table 5.5.2. Single precision computation does not allow the full potential of the extrapolation method to be exploited in the evaluation of the derivative of x^7 at $x = 1$.

```
EVALUATION OF DERIVATIVES USING EXTRAPOLATION METHOD WITH SINGLE PRECISION

IF YOU REQUIRE FIRST ORDER DERIVATIVE ENTER 1
                    OR IF SECOND ORDER ENTER 2?  1
ENTER VALUE OF INTERVAL H?  .1
ENTER VALUE AT WHICH DERIVATIVE IS TO BE EVALUATED? 1
RESULTS USING CENTRAL DIFFERENCES WITH REPEATED HALVING OF INTERVALS

 APPROXIMATIONS:
 GROUP  1    6.999468    6.999958    7.000017    7.000027
 GROUP  2    6.999991    7.000021    7.000028
 GROUP  3    7.000022    7.000028
 GROUP  4    7.000028
```

```
100  REM ------------------------------------------------------------
110  REM EVALUATING DERIVATIVES USING EXTRAPOLATION METHOD GL/JP(C)
120  REM ------------------------------------------------------------
140  DIM A(5),B(5)
145  REM DEFINE FUNCTION IN STATEMENT 150
150  DEF FNA(X) = X^7
160  PRINT"EVALUATION OF DERIVATIVES USING EXTRAPOLATION METHOD WITH
SINGLE PRECISION"
170  PRINT
180  PRINT "IF YOU REQUIRE FIRST ORDER DERIVATIVES ENTER 1"
190  PRINT "                    OR IF SECOND ORDER ENTER 2";
200  INPUT OD
210  PRINT "ENTER VALUE OF INTERVAL H";
220  INPUT H
230  PRINT "ENTER VALUE AT WHICH DERIVATIVE IS TO BE EVALUATED";
240  INPUT X
250  PRINT "RESULTS USING CENTRAL DIFFERENCES WITH REPEATED HALVING OF
INTERVALS"
255  REM INITIAL VALUES CALCULATED BY CENTRAL DIFFERENCE FORMULA
260  FOR I=1 TO 5
270     IF OD =2 THEN 300
280     A(I)= (FNA(X+H)-FNA(X-H))/(2*H)
290     GOTO 320
300     A(I)= (FNA(X+H)-2*FNA(X)+FNA(X-H))/(H*H)
310     PRINT A(I);
320     H=H/2
330  NEXT I
335  REM IMPROVED VALUES CALCULATED BY EXTRAPOLATION METHOD
340  PRINT  : PRINT"APPROXIMATIONS:"
350  FOR M=1 TO 4
360     PRINT "GROUP";M;
370     FOR I=1 TO 5-M
380        B(I)=(A(I+1)*4^M-A(I))/(4^M-1)  : A(I)=B(I)
390        PRINT B(I);
400     NEXT I
410  PRINT
420  NEXT M
430  END
```

Program 5.5.1.

5.6 USE OF APPROXIMATING FUNCTIONS WITH OBSERVED DATA

When we wish to determine the rates of change relating to experimental data derived from observations the methods described above can often lead to gross errors. These errors are caused by inaccuracies in the measurement process and are additional to the truncation errors that have already been discussed. For example, let y^e_{i-1} be the experimental observation of the true value y_{i-1} so that

$$y^e_{i-1} = y_{i-1} + \delta_{i-1} \text{ and } y^e_{i+1} = y_{i+1} + \delta_{i+1}$$

The worst case situation for the calculation of the first derivatives arises when δ_{i-1} and δ_{i+1} have opposite signs. For simplicity let us assume that $\delta_{i-1} = -\delta$ and $\delta_{i+1} = \delta$. Then from (5.2.4) we have

$$y^{e\prime}_i = (y^e_{i+1} - y^e_{i-1})/(2h)$$

$$= (y_{i+1} - y_{i-1} + \delta_{i+1} - \delta_{i-1})/(2h) \approx y'_i + \delta/h$$

Since h is small the value of δ/h may be large and in the case of experimental observations δ/h may also be large compared to rounding errors.

To avoid these difficulties experimental data is usually smoothed or an approximating function is fitted to the data and this function is differentiated analytically. The use of approximating functions is described in detail in Chapter 8. Another method for approximating derivatives is based on the cubic spline function and this is also described in detail in Chapter 8. The spline function is also particularly useful when we wish to find the rate of change of accurate tabulated data which is well separated so that h is large and cannot be adjusted.

5.7 CONCLUSIONS AND SUMMARY

In this chapter we have described simple methods for obtaining the approximate derivatives of various orders for specified functions at given values of the independent variable. The results indicate these methods, although easy to program, are very sensitive to small changes in key parameters and should be used with considerable care.

PROBLEMS

5.1   Use formulae (5.2.4) and (5.3.1) to find the first and second derivatives of the function $x^2\cos x$ at $x = 1$ using $h = 0.1$ and $h = 0.01$.

5.2   Repeat problem 5.1 using Richardson extrapolation of the interval with $h = 0.1$.

5.3   Evaluate the first derivative of $\cos x^6$ when $x = 1$ and $x = 2$ using $h = 0.001$ by formula (5.2.4) and compare with the Richardson's extrapolation using $h = 0.1$.

5.4   Write a program to evaluate derivatives using approximations (5.4.1) and (5.4.2). Use your program to find the first and second derivatives of (i) $\cos x^6$ and (ii) $x^2\cos x$ at $x = 1$ using $h = 0.1, 0.01$.

5.5 Illustrate that small changes of x can lead to dramatic changes in the gradient of a specific function by finding the gradient of $y = \cos x^6$ at $x = 3.1, 3.01, 3.001$ and 3 using (5.2.4) with $h = 0.0001$. Compare your results with the exact value of the gradient.

5.6 The approximations for partial derivatives may be defined as follows:

$$\partial f/\partial x = [f(x + h,y) - f(x - h,y)]/2h$$

$$\partial f/\partial y = [f(x,y + h) - f(x,y - h)]/2h$$

Use these formulae, with $h = 0.001$, to find the first order partial derivatives of $\exp(x^2 + y)$ at $x = 2, y = 1$.

5.7 Use Richardson's extrapolation method, program 5.6.1, to find the derivative of $x - \tan x$ for $x = 1, 2, ... , 5$ and $h = 0.1$. Compare with the exact results obtained by direct differentiation.

5.8 An approximate expression for the Gamma function (Abramowitz & Stegun, 1964) is as follows:

$$\Gamma(x + 1) = 1 + a_1 x + a_2 x^2 + a_3 x^3 + a_4 x^4 + a_5 x^5 + \varepsilon(x) \quad 0 \le x \le 1$$

where $a_1 = -0.57486\ 46$ $a_4 = 0.42455\ 49$
          $a_2 = 0.95123\ 63$ $a_5 = -0.10106\ 78$
          $a_3 = -0.69985\ 88$

and $|\varepsilon(x)| \le 5 \times 10^{-5}$

Use formula (5.2.4) to find an estimate of the derivative of $\Gamma(x + 1)$ at $x = 0.5$ with $h = 0.01$, using the given approximation to obtain the values of the Gamma function you require. Check your estimate by differentiating the polynomial approximation directly.

6

The solution of differential equations

6.1 INTRODUCTION

Many mathematical models derived in the study of physical systems involve instantaneous rates of change that are given mathematically by the derivatives of one variable with respect to another. For example, the rate of change of the current i in a given electrical circuit at a given instant of time t would be given by

$$di/dt$$

Differential equations relate the derivatives of variables and functions of these variables. For example, consider the circuit diagram given in Fig. 6.1.1. The diagram represents a simple electrical circuit involving a resistance R in parallel with an inductance L. The voltage applied across the circuit is E. Initially at time $t = 0$ the switch is open and the current is consequently zero. A physical analysis of this circuit leads to the following differential equation giving the current at any time t.

$$E - L\, di/dt = iR \qquad (6.1.1)$$

together with the initial condition that $i = 0$ when $t = 0$.

This differential equation is known as a first order differential equation because the highest order derivative it contains is first order. An example of a second order differential equation is given by

$$m\,d^2x/dt^2 = -kx$$

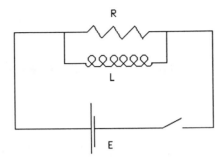

Fig. 6.1.1. A simple electrical circuit.

This equation describes the motion of a body of mass m when it moves in a straight line under the action of a force directed towards a fixed point which is proportional to the distance x of the body from the fixed point. This type of motion is known as simple harmonic motion. Specific conditions must be imposed on the initial nature of the motion. For example the motion may start from the fixed point with velocity v at time $t = 0$. This would lead to the initial conditions

$$dx/dt = v \text{ and } x = 0 \text{ at time } t = 0$$

The solutions of these equations are the functions of the independent variable t that satisfy the differential equation and the given initial conditions. For these differential equations we can find specific functional relationships between the dependent and independent variable but this is not always the case. For many problems it is impossible to determine a functional relationship between the variables and for others the algebraic procedure for determining these relationships can be extremely lengthy and complex. Thus we require numerical procedures which allow us to obtain approximate solutions to a wide range of problems.

The problems considered so far are called initial value problems because initial values are given for the dependent variable and its derivatives which allow a specific solution to be found. We shall discuss another class of differential equations called boundary value problems in a later section of this chapter.

6.2 THE INITIAL VALUE PROBLEM

In the following sections we shall examine numerical methods for the solution of the first order initial value problem of the form

$$dy/dx = f(x,y) \tag{6.2.1}$$

with the initial condition

$$y = y_0 \text{ when } x = x_0$$

If this problem could be solved by analytical methods we would be able to find a solution in the form

$$y = y(x)$$

Using a numerical approach to the solution of this problem will provide a series of discrete points (x_i, y_i) where

$$x_i = x_0 + ih \text{ and } y_i = y(x_i)$$

An appropriate value for h is decided by the user. These points provide a discrete approximation to the continuous curve $y = y(x)$ as shown in Fig. 6.2.1.

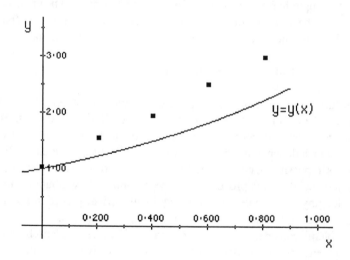

Fig.6.2.1. A comparison of the discrete numerical approximation,
indicated by •, with the exact solution of y=y(x).

6.3 THE TAYLOR'S SERIES METHOD

This is a particularly simple approach for solving the initial value problem and is based on obtaining a Taylor's series approximation for the exact solution of the differential equation. The Taylor's series expansion of the function $y(x)$ about the point x_0 is given by

$$y(x) = y_0 + (x-x_0)y'_0 + (x-x_0)^2 y''_0/2! + \ldots + (x-x_0)^n y^{(n)}_0/n! + \ldots \quad (6.3.1)$$

where y'_0 represents the first derivative, y''_0 the second derivative and $y^{(n)}_0$ the nth derivative of y with respect to x evaluated at $x = x_0$. The coefficients of the Taylor series expansion are easily found since we know y_0 and so y'_0 can be found from the differential equation

$$y'(x) = f(x,y) \quad (6.3.2)$$

By repeated differentiation of (6.3.2) we can find the derivatives y''_0, y'''_0 and so on. The result of this process is an approximation for the function $y(x)$ in an ascending series of powers of x. The approximation in terms of x can then be used to calculate y for any range of values of x that we require. The method can be easily extended to deal with any order differential equation. As a simple example the method is used to solve the differential equation

$$y' = y$$

with initial conditions $y_0 = 1$ when $x_0 = 0$. By repeated differentiation we have, in this example, $y^{(n)}_0 = \ldots = y''_0 = y'_0 = 1$. Consequently the Taylor's series approximation is

$$y = 1 + x + x^2/2! + x^3/3! + \ldots$$

The major disadvantage of this method is that the initial work to find the approximation in terms of x must be performed by the user and cannot be practically implemented on the computer. Since this constitutes the bulk of the work the Taylor's series method is not suitable for computer implementation and for this reason the method will not be discussed further. Two useful references for these techniques are given in Redish (1961) and Kreyzig (1967). The first of these gives a number of examples of the use of Taylor's series to solve first order differential equations while the second provides a theoretical discussion of power series methods for the solution of differential equations in general.

We shall now consider methods that provide a direct numerical approach to the problem.

6.4 EULER'S METHOD

One of the simplest approaches to obtaining the numerical solution of a differential equation is the method of Euler. Like the previous method this employs Taylor's series but uses only the first two terms of the expansion. Consider the following form of Taylor's series in which the third term is called the remainder term and represents the contribution of all the terms not included in the series.

$$y(x_0 + h) = y(x_0) + y'(x_0)h + y''(t)h^2/2 \qquad (6.4.1)$$

where t lies in the interval (x_0, x_1) and $x_1 = x_0 + h$. For small values of h we may neglect the terms in h^2 and (6.4.1) leads to the formula

$$y_1 = y_0 + hy'_0$$

and in general

$$y_{n+1} = y_n + hy'_n \text{ for } n = 0, 1, 2 \ldots$$

By virtue of (6.2.1) this may be written

$$y_{n+1} = y_n + h f(x_n, y_n) \text{ for } n = 0, 1, 2 \ldots \qquad (6.4.2)$$

This is known as Euler's method and it is illustrated geometrically in Fig. 6.4.1. From (6.4.1) we can see that the truncation error is of order h^2.

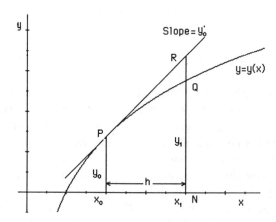

Fig. 6.4.1. The Euler approximation is the length of the line RN, the true value is QN.

```
100   REM ------------------------------------------------------------
110   REM  SOLUTION OF DIFFERENTIAL EQUATIONS: EULER'S METHOD GL/JP (C)
120   REM ------------------------------------------------------------
130   PRINT"SOLUTION OF A FIRST ORDER"
140   PRINT"DIFFERENTIAL EQUATION USING EULER'S METHOD":PRINT
150   PRINT"ASSUMING YOUR DIFFERENTIAL EQUATION IS"
160   PRINT"OF THE FORM dy/dx=f(x,y)"
170   PRINT
180   PRINT"ENTER INITIAL AND FINAL VALUES OF INDEPENDENT VARIABLE x";
190   INPUT X0,XN
200   PRINT"ENTER INCREMENTAL VALUE h";
210   INPUT H
220   NM=(XN-X0)/H:NM=NM+1
230   PRINT"ENTER THE INITIAL VALUE FOR y";
240   INPUT Y0:X=X0:Y=Y0
250   PRINT "x";TAB(13);"y"
255   REM EULER'S METHOD IS EXECUTED FOR THE STEPS REQUIRED
260   FOR N=1 TO NM
270      GOSUB 330
280      PRINT X;TAB(10);Y
290      Y1=Y+H*FXY
300      X=X+H : Y=Y1
310   NEXT N
320   END
325   REM SUBROUTINE TO EVALUATE F(X,Y) DEFINE FUNCTION IN STATEMENT 330
330   FXY=Y
340   RETURN
```
Program 6.4.1.

Table 6.4.1. The solution of $10 - 5\mathrm{d}i/\mathrm{d}t = i$, where $i = 0$ when $t = 0$ for
various step sizes using Euler's method.

t	Values of i for step value:			
	h=1	h=0.5	h=0.1	Exact value
0	0	0	0	0
1	2	1.9	1.829272	1.812692
2	3.6	3.439	3.32392	3.2968
3	4.88	4.68559	4.545156	4.511884
4	5.904	5.695328	5.542995	5.506711
5	6.7232	6.513216	6.358303	6.321206

The method is simple to program and is implemented by program 6.4.1. Applying
this program to the differential equation (6.1.1) with $E = 10$ volt, $R = 1$ ohm and
$L = 5$ henry gives Table 6.4.1 which illustrates how the approximate solution varies
for different values of h. Table 6.4.2 compares true solution values with approximations given by Euler's method for various values of h for the equation

$$\mathrm{d}y/\mathrm{d}x = y \quad \text{where } y_0 = 1 \text{ when } x_0 = 0$$

The true solution is $y = e^x$

Table 6.4.2. Solution of $y' = y$ where $y = 1$ when $x = 0$ for small step sizes.

x	Value of y for step size value :			
	h=0.1	h=0.01	h=0.001	Exact y
0	1	1	1	1
1	2.593743	2.704814	2.71692	2.718282
2	6.7275	7.316019	7.381675	7.389056
3	17.4494	19.78847	20.05544	20.08554
4	45.25925	53.5241	54.4891	54.59815
5	117.3909	144.7727	148.0427	148.4132

The Euler method, although simple, requires a very small step h to provide reasonable levels of accuracy. If the differential equation is to be solved for a wide range of values of x, the method becomes very expensive in terms of computer time because of the very large number of small steps required to span the interval of interest. In addition, the errors made at each step may accumulate in an unpredictable way. A simple modification of Euler's method that leads to a significant improvement is discussed below.

6.5 THE MODIFIED EULER METHOD

The basis of the modified Euler method is to use the average of the derivatives at two successive points instead of the derivative at the current point. The geometrical interpretation of this step, shown in Fig. 6.5.1, provides an illuminating insight into the nature of the method. Since the derivatives represent the slopes of the curve at the given points, it should be better to take the average of the slopes at the beginning and end of the interval rather than at one end to calculate the change in y over the interval.

These considerations lead to the following algorithm.

1) Start with n set at zero where n indicates the iteration.

2) Calculate $y^{(1)}_{n+1} = y_n + hf(x_n, y_n)$

3) Calculate $f(x_{n+1}, y^{(1)}_{n+1})$ where $x_{n+1} = x_n + h$ $\hspace{2cm}$ (6.5.1)

4) For $k = 1, 2, \ldots$ calculate

$$y^{(k+1)}_{n+1} = y_n + h\{f(x_{n+1}, y^{(k)}_{n+1}) + f(x_n, y_n)\}/2$$

At step 4, when the difference between successive values of y_{n+1} are sufficiently small increment n by 1 and repeat steps 2, 3 and 4.

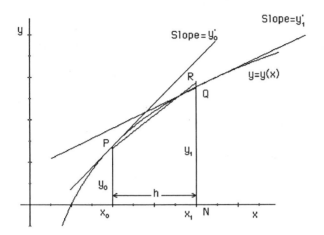

Fig. 6.5.1. In Euler's modified method the approximation to y is given by RN and the true value is QN. The slope PR is the average of the slopes at P and Q.

By considering the Taylor's series expansion of y_{n+1} we can obtain the order of the error in terms of the step size h.

$$y_{n+1} = y_n + hy'_n + h^2 y''_n /2! + h^3 y'''_n(t)/3! \qquad (6.5.2)$$

where t lies in the interval (x_n, x_{n+1}). It can be shown that y''_n may be approximated by

$$y''_n = (y'_{n+1} - y'_n)/h + O(h) \qquad (6.5.3)$$

Substituting this expression for y''_n in (6.5.2) gives

$$y_{n+1} = y_n + hy'_n + h(y'_{n+1} - y'_n)/2! + O(h^3)$$

$$= y_n + h(y'_{n+1} + y'_n)/2! + O(h^3)$$

This shows the local truncation error is of order h^3 so there is a significant improvement in accuracy over the basic Euler method which has a truncation error of order h^2. The problem of accuracy still remains however since errors will accumulate from step to step. In particular, since the function $f(x,y)$ calculated repeatedly from the values of y which include accumulated errors, these errors may grow in an unpredictable way. We will return to this important problem of error growth in a later section.

```
100    REM ---------------------------------------------------------------
110    REM SOL'N OF DIFFERENTIAL EQUATIONS:MODIFIED EULER'S METHOD GL/JP(C)
120    REM ---------------------------------------------------------------
130    PRINT"SOLUTION OF A FIRST ORDER"
140    PRINT"DIFFERENTIAL EQUATION USING  MODIFIED EULER'S METHOD":PRINT
150    PRINT"ASSUMING YOUR DIFFERENTIAL EQUATION IS"
160    PRINT"OF THE FORM dy/dx=f(x,y)"
170    PRINT
180    PRINT"ENTER INITIAL AND FINAL VALUES OF INDEPENDENT VARIABLE x";
190    INPUT X0,XN
200    PRINT"ENTER INCREMENTAL VALUE h";
210    INPUT H
220    NM=(XN-X0)/H:NM=NM+1
230    PRINT"ENTER THE INITIAL VALUE FOR y";
240    INPUT Y0
250    X=X0:Y=Y0
260    ACC=.0005
270    PRINT
280    PRINT "x";TAB(13);"y"
290    FOR N=1 TO NM
300       GOSUB 400: F1=F
310       PRINT X;TAB(10);Y
320       Y=Y0+H*F1
330       X=X+H
335       REM USE MODIFIED EULER TO GENERATE NEW VALUES
340       GOSUB 400: F2=F
350       Y2=Y0+H*(F1+F2)/2
360       IF ABS(Y-Y2)>ACC THEN Y=Y2:GOTO 340
370       Y=Y2:Y0=Y2
380    NEXT N
390    END
395    REM SUBROUTINE TO CALCULATE f(x,y). DEFINE FUNCTION AT STATEMENT 400
400    F=Y
410    RETURN
```

Program 6.5.1.

Program 6.5.1 implements this method. The results from the program when applied to specific problems are given in Table 6.5.1 and Table 6.5.2. Table 6.5.1 shows how improved accuracy may be obtained by using the modified Euler method for the same problem solved by Euler's method in Table 6.4.1. Table 6.5.2 shows that this method may still produce poor results for some problems even when a relatively small step size is used. This error tends to increase as the computation proceeds.

The modified Euler method is classified as a predictor-corrector method. This means that in the case of the modified Euler method the initial value is given by the formula

$$y_n^{(1)} = y_n + hf(x_n, y_n)$$

The modified Euler method

Table 6.5.1. The modified Euler method used to solve $10 - 5di/dt = i$ given that when $t = 0$, $i = 0$. Step size $h = 0.5$.

t	i	Exact value
0	0	0
1	1.814048	1.812692
2	3.299019	3.2968
3	4.514609	4.511884
4	5.509686	5.506711
5	6.32425	6.321206

Table 6.5.2. The modified Euler method used to solve $y' = y$ given that when $x = 0$, $y = 1$. Step size $h = 0.1$.

x	y	Exact value
0	1	1
.1	1.105263	1.105171
.2	1.221605	1.221403
.3	1.350194	1.349859
.4	1.492319	1.491825
.5	1.649405	1.648721
.6	1.823025	1.822119
.7	2.014921	2.013753
.8	2.227017	2.225541
.9	2.461438	2.459603
1	2.720535	2.718282

which is called the predictor and this is corrected by the repeated application of the formula

$$y^{(k+1)}_{n+1} = y_n + h\{f(x_{n+1}, y^{(k)}_{n+1}) + f(x_n, y_n)\}/2 \qquad k = 0, 1, 2, \dots$$

which is called the corrector.

There are many predictor-corrector formulae and some provide much greater accuracy than the relatively simple modified Euler method. These methods, however, require accurate estimates for a number of initial y values before they can be initiated. In sections 6.8 and 6.9 we shall consider a number of predictor-corrector methods but in the next section we consider a method which can provide approximations at each step without requiring a number of initial values for y.

6.6 RUNGE-KUTTA METHODS

An important group of methods which allow us to obtain greater accuracy at each step and yet require only the initial value of y to be given with the differential equation are called the Runge-Kutta methods. Any degree of accuracy may be obtained at each step but the better the accuracy the greater the number of function evaluations required. An intelligent choice has to be made combining the level of accuracy and the amount of computational effort at each step. We shall now describe some typical Runge-Kutta methods. Consider the following particular type of Runge-Kutta formula.

$$y_{n+1} = y_n + a_1k_1 + a_2k_2 + a_3k_3 + a_4k_4 \qquad (6.6.1)$$

where

$$k_1 = hf(x_n, y_n)$$
$$k_2 = hf(x_n + b_1h, y_n + b_1k_1)$$
$$k_3 = hf(x_n + b_2h, y_n + b_2k_2)$$
$$k_4 = hf(x_n + b_3h, y_n + b_3k_3)$$

The values of the constants a_1, a_2, a_3, a_4 and b_1, b_2, b_3, b_4 will be given explicitly for a particular method. The derivation of these constants is lengthy but straightforward. They are obtained from the above equations by inserting the expressions for k_1, k_2, k_3 and k_4 in the first equation for y_n and expanding both sides using the Taylor's series expansion. The values of the constants are then derived from the system of linear equations produced by equating the coefficients of the same powers of h on both sides of (6.6.1). On equating coefficients up to and including h^4 we obtain a system of equations where the number of variables exceeds the number of equations. This allows a number of degrees of freedom in obtaining the coefficients we require and provides a number of alternative Runge-Kutta methods. All of these methods we derive from formula (6.6.1) are called fourth order and each has an error term of order h^5.

Theoretically, any order Runge-Kutta method can be derived and used to solve differential equations. However, for high order methods the computational effort becomes excessive. In the section below we consider a number of methods which have been used practically with some success.

6.7 SOME USEFUL RUNGE-KUTTA TECHNIQUES

We give the equations for two fourth order Runge-Kutta methods. The classical method has the form

$$y_{n+1} = y_n + (k_1 + 2k_2 + 2k_3 + k_4)/6 \qquad (6.7.1)$$

where

$$k_1 = hf(x_n, y_n)$$
$$k_2 = hf(x_n + h/2, y_n + k_1/2)$$
$$k_3 = hf(x_n + h/2, y_n + k_2/2)$$ (6.7.2)
$$k_4 = hf(x_n + h, y_n + k_3)$$

The next Runge-Kutta method is a variation on the formula (6.6.1) and is due to Gill, (1951) and takes the form

$$y_{n+1} = y_n + (k_1 + (2 - \sqrt{2})k_2 + (2 + \sqrt{2})k_3 + k_4)/6 \qquad (6.7.3)$$

where

$$k_1 = hf(x_n, y_n)$$
$$k_2 = hf(x_n + h/2, y_n + k_1/2)$$ (6.7.4)
$$k_3 = hf(x_n + h/2, y_n + (\sqrt{2} - 1)k_1/2 + (2 - \sqrt{2})k_2/2)$$
$$k_4 = hf(x_n + h, y_n - \sqrt{2}k_2/2 + (1 + \sqrt{2}/2)k_3)$$

A number of other forms of the Runge-Kutta method have been derived which have particularly advantageous properties. The equations for these methods will not be given but their important features are as folows:–

(1) *The Runge-Kutta-Merson method.* (Merson, 1957) This method has an error term of order h^5 and in addition allows an estimate of the local truncation error to be obtained at each step in terms of known values.

(2) *The Ralston-Runge-Kutta method.* (Ralston,1962) As was indicated earlier we have some degree of freedom in assigning the coefficients for a particular Runge-Kutta method in this formula the values of the coefficients are chosen so as to minimise the truncation error.

(3) *The Butcher-Runge-Kutta method.* (Butcher, 1964) This method provides higher accuracy at each step, the error being of order h^6.

Program 6.7.1 implements all the methods together since the principle is the same for each method, only the coefficients of the individual equations differ and these are placed in data statements at the end of the program. The individual methods are each preceded by a remark statement indicating the particular method and by examining the program the user can see the structure of the equations for each. Results from this program are given in Table 6.7.1.

The disadvantage of the Runge-Kutta methods is that they require a considerable number of function evaluations at each step. In the case of the fourth order methods considered above each step needs four function evaluations. In the next section we shall consider a number of predictor-corrector methods, some of which give the same order truncation error as Runge-Kutta methods at each step but may require fewer function evaluations.

```
100   REM --------------------------------------------------------------
110   REM SOLUTION OF DIFFERENTIAL EQUATIONS:RUNGE KUTTA METHOD GL/JP (C)
120   REM --------------------------------------------------------------
130   PRINT"SOLUTION OF A FIRST ORDER"
140   PRINT"DIFFERENTIAL EQUATION USING  RUNGE KUTTA METHOD":PRINT
150   PRINT"ASSUMING YOUR DIFFERENTIAL EQUATION IS"
160   PRINT"OF THE FORM dy/dx=f(x,y)"
170   PRINT:PRINT"ENTER INITIAL & FINAL VALUES OF INDEPENDENT VARIABLE x";
180   INPUT X0,XN
190   PRINT"ENTER STEP SIZE";
200   INPUT H
210   NM=(XN-X0)/H
220   PRINT"ENTER THE INITIAL VALUE FOR y";
230   INPUT Y0:X=X0:Y=Y0
240   FOR ME=1 TO 5
250      Y=Y0:X=X0:Y(ME)=Y
260      READ OD,MT$
270      PRINT:PRINT MT$;" METHOD":PRINT:PRINT" X";TAB(12);" Y"
275      REM READ DATA FOR RUNGE KUTTA METHOD
280      FOR I=1 TO OD
290         READ A(I)
300      NEXT I
310      FOR I=1 TO OD
320         FOR J=1 TO OD
330            READ B(I,J)
340         NEXT J
350      NEXT I
360      FOR I=1 TO OD
370         READ C(I)
380      NEXT I
390      FOR N=1 TO NM+1
395         REM CALCULATE STEP USING RUNGE KUTTA FORMULA
400         PRINT X;TAB(8);Y(ME)
410         FOR I=1 TO OD
420            XN=X+A(I)*H
430            YN=Y
440            FOR J=1 TO I-1
450               YN=YN+B(I,J)*K(J)
460            NEXT J
470            GOSUB 550
480            K(I)=H*FXY
490            Y(ME)=Y(ME)+K(I)*C(I)
500         NEXT I
510         X=X+H:Y=Y(ME)
520      NEXT N
530   NEXT ME
540   END
545   REM SUBROUTINE FOR FUNCTION DEFINED IN STATEMENT 550 AS FXY=F(XN,YN)
550   FXY=YN
560   RETURN
```

Program 6.7.1 (continues).

```
565   REM DATA FOR CLASSICAL METHOD
570   DATA 4,"CLASSICAL"
580   DATA 0,0.5,0.5,1
590   DATA 0,0,0,0
600   DATA 0.5,0,0,0
610   DATA 0,0.5,0,0
620   DATA 0,0,1,0
630   DATA 0.166666667,0.333333333,0.333333333,0.166666667
635   REM DATA FOR GILL METHOD
640   DATA 4,"GILL"
650   DATA 0,0.5,0.5,1
660   DATA 0,0,0,0
670   DATA 0.5,0,0,0
680   DATA 0.20710678,0.29289319,0,0
690   DATA 0,-0.707106781,1.707106781,0
700   DATA 0.166666667,.097631073#,.56903559#,.166666667#
705   REM DATA FOR MERSON METHOD
710   DATA 5, "MERSON"
720   DATA 0, 0.333333333,0.3333333333,0.5,1
730   DATA 0,0,0,0,0
740   DATA 0.333333333,0,0,0,0
750   DATA 0.1666666667,0.1666666667,0,0,0
760   DATA 0.125,0,0.375,0,0
770   DATA 0.5,0,-1.5,2,0
780   DATA 0.1666666667,0,0,0.666666667,0.1666666667
785   REM DATA FOR BUTCHER METHOD
790   DATA 6,"BUTCHER"
800   DATA 0,0.25,0.25,0.5,0.75,1
810   DATA 0,0 ,0,0 ,0,0
820   DATA 0.25,0,0,0,0,0
830   DATA 0.125,0.125,0,0,0,0
840   DATA 0,-0.5,1,0,0,0
850   DATA 0.1875,0,0,0.5625,0,0
860   DATA -0.428571429,0.285714286,1.71428571,-1.71428571,1.14285714,0
870   DATA 0.0777777778,0,0.355555556,0.133333333,0.355555556,0.0777777778
875   REM DATA FOR RALSTON METHOD
880   DATA 4,"RALSTON"
890   DATA 0,0.4,0.45573725,1
900   DATA 0,0,0,0
910   DATA 0.4,0,0,0
920   DATA 0.2969776,0.15875964,0,0
930   DATA 0.2181004,-3.05096516,3.83286476,0
940   DATA 0.17476028,-0.55148066,1.20553560,0.17118478
```

Program 6.7.1

Table 6.7.1. The solution of $y' = y$ given that when $x = 0$, $y = 1$. The step size is $h = 1$. The Butcher and Merson methods give the best results.

X	Classical	Gill	Merson	Butcher	Ralston	Exact value
0	1	1	1	1	1	1
1	2.708333	2.708333	2.715278	2.718229	2.708333	2.718282
2	7.33507	7.33507	7.372733	7.38877	7.33507	7.389056
3	19.86581	19.86581	20.01902	20.08437	19.86581	20.08554
4	53.80325	53.80324	54.3572	54.59392	53.80325	54.59815
5	145.7171	145.7171	147.5949	148.3988	145.7171	148.4132
6	394.6505	394.6505	400.7611	403.3819	394.6506	403.4288
7	1068.845	1068.845	1088.178	1096.485	1068.845	1096.633
8	2894.789	2894.788	2954.705	2980.496	2894.79	2980.958
9	7840.055	7840.052	8022.845	8101.672	7840.057	8103.084
10	21233.48	21233.47	21784.26	22022.2	21233.49	22026.46

6.8 PREDICTOR-CORRECTOR METHODS

The modified Euler method, which has already been described, is a simple example of a predictor-corrector method with a truncation error of order h^3. The predictor-corrector methods we shall consider now have much smaller truncation errors. As an initial example we will consider Milne's method (Milne 1949) which is based on the following equations

$$y_{n+1}^{(1)} = y_{n-3} + 4h(2y'_n - y'_{n-1} + 2y'_{n-2})/3$$

$$y'_{n+1} = f(x_{n+1}, y_{n+1}^{(1)}) \qquad (6.8.1)$$

$$x_{n+1} = x_n + h$$

and for $k = 1, 2, 3, ...$

$$y_{n+1}^{(k+1)} = y_{n-1} + h(y'_{n-1} + 4y'_n + y'_{n+1}{}^{(k)})/3$$

$$y'_{n+1}{}^{(k+1)} = f(x_{n+1}, y_{n+1}^{(k+1)}) \qquad (6.8.2)$$

The equations (6.8.1) constitute the predictor and the equations (6.8.2) constitute the corrector. The corrector is iterated for the values of k indicated until convergence has been achieved. After convergence for this step the independent variable x_n is incremented by h, n is incremented by one and the process repeated until the differential equation has been solved in the range of interest. The method is started with $n = 3$ and consequently the values of y_3, y_2, y_1 and y_0 must be known before this method can be applied. These values must be obtained by some other self starting procedure such as one of the Runge-Kutta methods described in section 6.6. The self starting procedure chosen must have the same order truncation error as the predictor-corrector method.

By using Taylor's series expansions on both sides of the predictor equation the truncation error for this equation, EP, can be found for each step. This may be estimated as

$$EP = 14\, y^{(v)}(t)h^5/45 \qquad\qquad (6.8.3)$$

Similarly the local truncation error for the predictor, EC, may be estimated as

$$EC = -y^{(v)}(t)h^5/90 \qquad\qquad (6.8.4)$$

where t lies in the interval (x_{n-1}, x_{n+1}). We see that the calculation of these error terms involves calculating high order derivatives, this can be avoided by noting that if Y is the true value of y_{n+1} and P is the predicted value of y_{n+1} then, assuming the rounding error is negligible, we have by virtue of (6.8.3)

$$Y = P + 14y^{(v)}(t)h^5/45 \qquad\qquad (6.8.5)$$

Similarly if C is the corrected value of y_{n+1} by virtue of (6.8.4) we have

$$Y = C - y^{(v)}(t)h^5/90 \qquad\qquad (6.8.6)$$

Subtracting (6.8.5) from (6.8.6) gives

$$C - P = (1/90 + 14/45)y^{(v)}(t)h^5$$

and therefore our approximation to the corrector is

$$EC = (P - C)/29 \qquad\qquad (6.8.7)$$

and similarly we obtain an approximation to the error of the predictor as

$$EP = 28(C - P)/29 \qquad\qquad (6.8.8)$$

For small values of h the method of Milne would appear to give high accuracy at each step. Comparing it with fourth order Runge-Kutta methods we see that the order of the truncation error is the same. However, the Runge-Kutta methods require four function evaluations while the Milne method may, with careful choice of step size, require only two for each step.

Table 6.8.1. Milne's method used to solve $y' = y$ given that
when $x = 0$, $y = 1$. Step size $h = 1$.

x	y	Exact value
0	1	1
1	2.708333	2.718282
2	7.33507	7.389056
3	20.17144	20.08554
4	54.40176	54.59815
5	148.5352	148.4132
6	405.8738	403.4288
7	1108.818	1096.633
8	3029.384	2980.958
9	8276.402	8103.084
10	22611.57	22026.46

Apart from the need for initial starting values the Milne method would appear to be generally superior to the Runge-Kutta method. For a true comparison of these methods, however, it is necessary to consider how they behave over a whole range of problems since applying any method to some differential equations results at each step in a growth of error that ultimately swamps the calculation. Table 6.8.1 illustrates the behaviour of Milne's method when applied to a specific problem. These results are produced by taking a large step size. Although not as accurate as the Runge-Kutta method they are comparable. Taking a step size of $h = 0.1$ would lead to greatly increased accuracy.

Table 6.8.2 provides the results of solving a different problem using Milne's method. The results of Table 6.8.2 after the initial steps fail to come near to the true values. In this case errors inherent to the method swamp the true value. We shall discuss this problem in greater detail in section 6.10. The method is implemented by program 6.8.1.

Table 6.8.2. Use of Milne's method to solve $y' = -10y$
where $y = 1$ when $x = 0$. A step size of $h = 0.05$ gives poor results.

x	y	Exact value
0	1	1
.1	.3681709	.3678795
.2	.1353248	.1353353
.3	4.966646E-02	4.978707E-02
.4	1.812655E-02	1.831563E-02
.5	6.473925E-03	6.737944E-03
.6	2.114263E-03	2.47875E-03
.7	4.078527E-04	9.118811E-04
.8	-3.630132E-04	3.354623E-04
.9	-8.452527E-04	1.234096E-04
1	-1.298348E-03	4.539989E-05

```
100   REM ----------------------------------------------------------------
110   REM SOLUTION OF A DIFFERENTIAL EQUATION: MILNE'S METHOD GL/JP (C)
120   REM ----------------------------------------------------------------
130   PRINT"SOLUTION OF A FIRST ORDER"
140   PRINT"DIFFERENTIAL EQUATION USING  MILNE METHOD"
150   PRINT
160   PRINT"ASSUMING YOUR DIFFERENTIAL EQUATION IS"
170   PRINT"OF THE FORM dy/dx=f(x,y)"
180   PRINT
190   PRINT"ENTER INITIAL AND FINAL VALUES OF INDEPENDENT VARIABLE x";
200   INPUT X0,XN
210   PRINT"ENTER STEP SIZE";
220   INPUT H
230   NM=(XN-X0)/H
240   PRINT"ENTER THE INITIAL VALUE FOR y";
250   INPUT Y0:X=X0:Y=Y0:Y(0)=Y:YN=Y
260   PRINT:PRINT" X";TAB(12);" Y"
270   FOR N=1 TO 3
280       XN=X:YN=Y
290       PRINT X;TAB(8);YN
300       GOSUB 640
305       REM CLASSICAL RUNGE KUTTA METHOD USED TO GIVE INITIAL VALUES
310       K1=H*FXY
320       XN=X+H/2:YN=Y+K1/2
330       GOSUB 640
340       K2=H*FXY:   YN=Y+K2/2
350       GOSUB 640
360       K3=H*FXY
370       YN=Y+K3:XN=X+H
380       GOSUB 640
390       K4=H*FXY
400       Y(N)=Y+(K1+2*K2+2*K3+K4)/6
410       X=X+H:Y=Y(N)
420   NEXT N
430   PRINT X;TAB(8);YN
440   FOR N=4 TO NM
445       REM NOW USE MILNE PREDICTOR CORRECTOR METHOD
450       XN=X-2*H:YN=Y(1)
460       GOSUB 640
470       F1=FXY
480       XN=X-H:YN=Y(2)
490       GOSUB 640
500       F2=FXY
510       XN=X:YN=Y(3)
520       GOSUB 640
530       F3=FXY
540       Y4=Y(0)+4*H/3*(2*F3-F2+2*F1)
550       XN=X+H:YN=Y4
560       GOSUB 640
570       F4=FXY
580       YN=Y(2)+H/3*(F2+4*F3+F4)
590       IF ABS(YN-YC)>.000001 THEN YC=YN:GOTO 560
600       X=X+H:Y(0)=Y(1):Y(1)=Y(2):Y(2)=Y(3):Y(3)=YN
610       PRINT X;TAB(8);YN
```

Program 6.8.1.(continues).

```
620    NEXT N
630    END
635    REM SUB TO CALC FUNC'N DEFINED IN STATEMENT 640 AS FXY=F(XN,YN)
640    FXY=-10*YN
650    RETURN
```

Program 6.8.1.

Another predictor-corrector method is that of Adams. This method is given by the predictor-corrector equations

$$y_{n+1} = y_n + h(55y'_n - 59y'_{n-1} + 37y'_{n-2} - 9y'_{n-3})/24 \qquad (6.8.9)$$

$$y_{n+1} = y_n + h(9y'_{n+1} + 19y'_n - 5y'_{n-1} + y'_{n-2})/24 \qquad (6.8.10)$$

Equation (6.8.9) is the predictor and (6.8.10) the corrector. The procedure is started with $n = 3$ and consequently the method requires the starting values y_3, y_2, y_1 and y_0. The corrector is iterated until convergence is achieved. It can be shown that each step of this procedure has a truncation error of order h^5 and that the method is generally more stable than Milne's method.

The method is implemented by program 6.8.2 and Table 6.8.3 illustrates the behaviour of this method for a specific example. The problem solved here is the same as that solved in Table 6.8.2. However the results are much better. Although there are clearly errors in the solution using Adam's method the growth of the error is stable.

Table 6.8.3. Using Adams' method with $h = 0.05$ to solve $y' = -10y$ where $y = 1$ when $x = 0$ gives better results than Milne's method.

x	y	Exact value
0	1	1
.1	.3681709	.3678795
.2	.1353081	.1353353
.3	4.965726E-02	4.978707E-02
.4	1.822289E-02	1.831563E-02
.5	6.687511E-03	6.737944E-03
.6	2.453567E-03	2.47875E-03
.7	8.999989E-04	9.118811E-04
.8	3.309633E-04	3.354623E-04
.9	1.21718E-04	1.234096E-04
1	4.4762E-05	4.539989E-05

```
100    REM -------------------------------------------------------------
110    REM SOLUTION OF A DIFFERENTIAL EQUATION: ADAMS METHOD GL/JP (C)
120    REM -------------------------------------------------------------
130    PRINT"SOLUTION OF A FIRST ORDER"
140    PRINT"DIFFERENTIAL EQUATION USING  ADAMS' METHOD"
150    PRINT
```

Program 6.8.2. (continues)

```
160    PRINT"ASSUMING YOUR DIFFERENTIAL EQUATION IS"
170    PRINT"OF THE FORM dy/dx=f(x,y)"
180    PRINT
190    PRINT"ENTER INITIAL AND FINAL VALUES OF INDEPENDENT VARIABLE x";
200    INPUT X0,XN
210    PRINT"ENTER STEP SIZE";
220    INPUT H
230    NM=(XN-X0)/H
240    PRINT"ENTER THE INITIAL VALUE FOR y";
250    INPUT Y0:X=X0:Y=Y0:Y(0)=Y:YN=Y
260    PRINT:PRINT" X";TAB(12);" Y"
270    FOR N=1 TO 3
280        XN=X:YN=Y
290        PRINT X;TAB(8);YN
300        GOSUB 660
305        REM CLASSICAL RUNGE KUTTA METHOD USED TO GIVE INITIAL VALUES
310        K1=H*FXY
320        XN=X+H/2:YN=Y+K1/2
330        GOSUB 660
340        K2=H*FXY:   YN=Y+K2/2
350        GOSUB 660
360        K3=H*FXY
370        YN=Y+K3:XN=X+H
380        GOSUB 660
390        K4=H*FXY
400        Y(N)=Y+(K1+2*K2+2*K3+K4)/6
410        X=X+H:Y=Y(N)
420    NEXT N
430    PRINT X;TAB(8);YN
440    FOR N=4 TO NM
450        XN=X-3*H:YN=Y(0):GOSUB 660
460        F0=FXY
470        XN=X-2*H:YN=Y(1)
480        GOSUB 660
490        F1=FXY
500        XN=X-H:YN=Y(2)
510        GOSUB 660
520        F2=FXY
530        XN=X:YN=Y(3)
540        GOSUB 660
550        F3=FXY
560        Y4=Y(3)+H/24*(55*F3-59*F2+37*F1-9*F0)
570        XN=X+H:YN=Y4
580        GOSUB 660
590        F4=FXY
600        YN=Y(3)+H/24*(F1-5*F2+19*F3+9*F4)
610        IF ABS(YN-YC)>.000005 THEN YC=YN:GOTO 580
620        X=X+H:Y(0)=Y(1):Y(1)=Y(2):Y(2)=Y(3):Y(3)=YN
630        PRINT X;TAB(8);YN
640    NEXT N
650    END
655    REM SUB TO CALC FUNCTION DEFINED IN STATEMENT 660 AS FXY=F(XN,YN)
660    FXY=YN
670    RETURN
```

Program 6.8.2.

6.9 HAMMING'S METHOD AND THE USE OF ERROR ESTIMATES

The method of Hamming, (1959), is based on the following pair of predictor-corrector equations

$$y_{n+1} = y_{n-3} + 4h(2y'_n - y'_{n-1} + 2y'_{n-2})/3$$

$$y_{n+1} = (9y_n - y_{n-2} + 3h(y'_{n+1} + 2y'_n - y'_{n-1}))/8$$

$$(6.9.1)$$

The first equation is used as the predictor and the second as the corrector. To obtain a further improvement in accuracy at each step in the predictor and corrector we modify these equations by using expressions for the local truncation errors. We have seen how approximations can be obtained for the local truncation errors of Milne's method by using the predicted and corrected values of the current approximation to y. We can obtain similar approximations for the local truncation errors in Hamming's method. This leads to the equations

$$y_{n+1} = y_{n-3} + 4h(2y'_n - y'_{n-1} + 2y'_{n-2})/3 \qquad (6.9.2)$$

$$y^M_{n+1} = y_{n+1} - 112(P - C)/121 \qquad (6.9.3)$$

In this equation P and C represent the predicted and corrected value of y at the nth step.

$$y^*_{n+1} = (9y_n - y_{n-2} + 3h(y'^M_{n+1} + 2y'_n - y'_{n-1}))/8 \qquad (6.9.4)$$

In this equation y'^M_{n+1} is the value of y'_{n+1} calculated using the modified value of y_{n+1} which is y^M_{n+1}.

$$y_{n+1} = y^*_{n+1} + 9(y_{n+1} - y^*_{n+1})/121 \qquad (6.9.5)$$

Equation (6.9.2) is the predictor and (6.9.3) modifies the predicted value by using an estimate of the truncation error. Equation (6.9.4) is the corrector which is modified by (6.9.5) using an estimate of the truncation error. The equations in this form are each used only once before n is incremented and the steps repeated again.

The choice of h must be made carefully so that the error does not increase without bound. The details of how this choice should be made will be discussed in section 6.10. Table 6.9.1 illustrates the performance of the method.

The results are comparable with the Adams' method and considerably better than the Milne method. Program 6.9.1 implements Hamming's method.

Table 6.9.1. Using Hamming's method with $h = 0.05$ to solve $y' = -10y$
where $y = 1$ when $x = 0$.

x	y	Exact value
0	1	1
.1	.3681709	.3678795
.2	.1350355	.1353353
.3	4.958757E-02	4.978707E-02
.4	1.824622E-02	1.831563E-02
.5	6.719489E-03	6.737944E-03
.6	2.478912E-03	2.47875E-03
.7	9.13364E-04	9.118811E-04
.8	3.345062E-04	3.354623E-04
.9	1.225443E-04	1.234096E-04
1	4.532266E-05	4.539989E-05

```
100   REM -----------------------------------------------------------------
110   REM SOLUTION OF A DIFFERENTIAL EQUATION: HAMMING'S METHOD GL/JP (C)
120   REM -----------------------------------------------------------------
130   PRINT"SOLUTION OF A FIRST ORDER"
140   PRINT"DIFFERENTIAL EQUATION USING HAMMING'S METHOD"
150   PRINT
160   PRINT"ASSUMING YOUR DIFFERENTIAL EQUATION IS"
170   PRINT"OF THE FORM dy/dx=f(x,y)"
180   PRINT
190   PRINT"ENTER INITIAL AND FINAL VALUES OF INDEPENDENT VARIABLE x";
200   INPUT X0,XN
210   PRINT"ENTER STEP SIZE";
220   INPUT H
230   NM=(XN-X0)/H
240   PRINT"ENTER THE INITIAL VALUE FOR y";
250   INPUT Y0:X=X0:Y=Y0:Y(0)=Y:YN=Y
260   PRINT:PRINT" X";TAB(12);" Y"
270   FOR N=1 TO 3
280       XN=X:YN=Y
290       PRINT X;TAB(8);YN
300       GOSUB 660
305       REM CLASSICAL RUNGE KUTTA METHOD USED TO GIVE INITIAL VALUES
310       K1=H*FXY
320       XN=X+H/2:YN=Y+K1/2
330       GOSUB 660
340       K2=H*FXY:   YN=Y+K2/2
350       GOSUB 660
360       K3=H*FXY
370       YN=Y+K3:XN=X+H
380       GOSUB 660
390       K4=H*FXY
400       Y(N)=Y+(K1+2*K2+2*K3+K4)/6
410       X=X+H:Y=Y(N)
420   NEXT N
```

Program 6.9.1. (continues).

```
430   PRINT X;TAB(8);YN
440   FOR N=4 TO NM
445     REM NOW USE HAMMING PREDICTOR CORRECTOR
450     XN=X-2*H
460     YN=Y(1)
470     GOSUB 660
480     F1=FXY
490     XN=X-H:YN=Y(2)
500     GOSUB 660
510     F2=FXY
520     XN=X:YN=Y(3)
530     GOSUB 660
540     F3=FXY
550     Y4=Y(0)+4*H/3*(2*F3-F2+2*F1)
560     YM=Y4: IF N>4 THEN YM=Y4-112/121*(YP-YN)
570     YP=Y4:XN=X+H:YN=YM
580     GOSUB 660
590     F4=FXY
600     YN=(9*Y(3)-Y(1)+3*H*(F4+2*F3-F2))/8
610     YC=YN+9/121*(YM-YN)
620     X=X+H:Y(0)=Y(1):Y(1)=Y(2):Y(2)=Y(3):Y(3)=YC
630     PRINT X;TAB(8);YC
640   NEXT N
650   END
655   REM SUB TO CALC FUNCTION DEFINED IN STATEMENT 660 AS FXY=F(XN,YN)
660   FXY=YN
670   RETURN
```

Program 6.9.1

6.10 ERROR PROPAGATION IN DIFFERENTIAL EQUATIONS.

In the preceding sections as we have described various techniques for solving differential equations and the order, or a specific expression, for the truncation error at each step has been given. It is important to examine not only the magnitude of the error at each step but also how that error accumulates as the number of steps taken increases.

A crucial concept in examining the nature of the propagation of error in the numerical solution of differential equations is called stability. For the predictor-corrector method described above it can be shown that the predictor-corrector formula introduce additional spurious solutions. As the iterative process proceeds, for some problems, the effect these spurious solutions may be to overwhelm the true solution. This occurred with Milne's method in Table 6.8.2. In these circumstances the method is said to be unstable. Clearly we seek stable methods where the error does not develop in an unpredictable and unbounded way.

For some differential equations the value of the solution increases rapidly as the independent variable increases. In these circumstances the magnitude of the error may increase but as a proportion of the magnitude of the solution it may not be increasing. In this case the method is said to be relatively stable. For relative stability, the behaviour of the error parallels the behaviour of the solution.

Clearly it is important to examine each numerical method to see if it is stable. In addition, if it is not stable for all differential equations we should provide tests to determine when it can be used with confidence. The theoretical study of stability for differential equations is a major undertaking and it is not intended to include a detailed analysis here. The major results and their application to the numerical methods we have described will be briefly discussed in the next section.

6.11 THE STABILITY OF PARTICULAR NUMERICAL METHODS

A good discussion of the stability of many of the numerical methods for solving first order differential equations is given in Ralston and Rabinowitz (1978). Some of the more significant features are as follows.

(1)   The Euler Method: this method can be shown to be relatively stable.

(2)   The Milne Method: for some problems this method can be shown to be neither stable nor relatively stable.

(3)   The Adams Method: this method can be shown to be relatively stable and consequently is usually more reliable then Milne's method.

(4)   Hamming's Method: for certain forms of differential equation this method can be shown to both stable and relatively stable for specific ranges of the step size h.

(5)   Runge-Kutta methods: because Runge-Kutta methods do not introduce spurious solutions the stability analysis proceeds in a different way. Instability may arise for some values of h but sometimes this may be removed by reducing h to a sufficiently small value. We have already described how the Runge-Kutta methods are less efficient than the predictor-corrector methods because of the greater number of function evaluations that may be required at each step by the Runge-Kutta methods. If h is reduced too far the number of function evaluations required may make the method uneconomic. The restriction on the size of the interval required to maintain stability may be estimated from an inequality of the form

$$h|\partial f/\partial y| < K$$

where K is dependent on the particular Runge-Kutta method being used and may be estimated.

Tables 6.11.1, 6.11.2 and 6.11.3 illustrate the performance of the predictor-corrector methods for specified examples. The examples are:

1) $y' = xy^{(1/3)}$ where $y = 1$ when $x = 1$.

2) $y' = xy/(1 + x^2)$ where $y = 0$ when $x = 0$.

3) $y' = -2xy$ where $y = 1$ when $x = 1$.

Tables 6.11.1 and 6.11.2, which give the solution of problems 1 and 2, appear to show that there is little difference between the three methods considered and they are all fairly successful for the relatively large step size $h = 0.5$. However Table 6.11.3 deals with the solution of a difficult problem, problem 3, and demonstrates the instability of Milne's method.

Table 6.11.1. Comparison of the Milne, Adams and Hamming methods to solve $y' = xy/(1 + x^2)$ where $y = 1$ when $x = 1$ and $h = 0.5$.

x	Milne	Adams	Hamming	Exact value
0	1	1	1	1
5	.894233	.894233	.894233	.8944272
1	.7069448	.7069448	.7069448	.7071068
1.5	.5529183	.5529183	.5529183	.5547002
2	.4475529	.4489662	.4486109	.4472136
2.5	.3712456	.3730587	.3730595	.3713907
3	.316545	.3176451	.3181871	.3162278
3.5	.2745151	.2759177	.2765529	.2747211
4	.2428519	.2435685	.2442738	.2425356
4.5	.2166873	.2178404	.218511	.2169305
5	.1964409	.1969308	.1975756	.1961161

Table 6.11.2. Comparison of the Milne, Adams and Hamming methods to solve $y' = xy^{1/3}$ where $y = 1$ when $x = 1$ and $h = 0.5$.

x	Milne	Adam's	Hamming	Exact value
1	1	1	1	1
1.5	1.68609	1.68609	1.68609	1.686171
2	2.828245	2.828245	2.828245	2.828427
2.5	4.565929	4.565929	4.565929	4.560359
3	7.020848	7.020723	7.020921	7.021132
3.5	10.352	10.35189	10.35204	10.35238
4	14.69656	14.69637	14.69649	14.69694
4.5	20.19775	20.19759	20.19769	20.19822
5	26.99953	26.9993	26.99941	27

Table 6.11.3. Comparison of the Milne, Adams and Hamming methods to solve $y' = -2xy$ where $y = 1$ when $x = 1$ and $h = 0.1$.

x	Milne	Adams	Hamming	Exact value
1	1	1	1	1
2	.0497862	.0498011	4.980344E-02	4.978702E-02
3	3.170534E-04	3.354157E-04	3.342025E-04	3.354639E-04
4	-1.687471E-04	2.566005E-07	3.132709E-07	3.059064E-07
5	-2.796654E-03	9.946295E-10	-3.629791E-10	3.775221E-11

6.12. SYSTEMS OF SIMULTANEOUS DIFFERENTIAL EQUATIONS

Systems of differential equations arise naturally from mathematical models of the physical world. An interesting example of this is the system of differential equations which models the interaction of competing or predator-prey populations. This model is based on the Volterra equations and may be written in the form

$$dP/dt = K_1 P - CPQ$$
$$dQ/dt = -K_2 Q + DPQ \qquad\qquad (6.12.1)$$

together with the initial conditions

$$Q = Q_0 \text{ and } P = P_0 \text{ at time } t = 0$$

The variables P and Q give the size of the two interacting and competing populations at time t. Here K_1, K_2, C and D are positive constants. K_1 relates to the rate of growth of the prey population P and K_2 relates to the rate of decay of the predator population Q. It seems reasonable to assume that the number of encounters of predator and prey is proportional to P multiplied by Q and that a proportion C of these encounters will be fatal to members of the prey population. Thus the term CPQ gives a measure of the decrease in the prey population and the unrestricted growth in this population, which could occur assuming ample food, must be modified by the subtraction of this term. Similarly the decrease in the population of the predator must be modified by the addition of the term DPQ since the predator population gains food from its encounters with its prey and therefore more of the predators survive. The solution of the differential equation depends on the specific values of the constants and will usually result in the stable cyclic variation of the populations. This is because as the predators continue to eat the prey, the prey population will fall and become insufficient to support the predator population which itself then falls. However as the predator population falls more of the prey survive and consequently the prey population will now increase. This in turn leads to an increase in the predator population since it now has more food and the cycle begins again. This cycle maintains the predator-prey

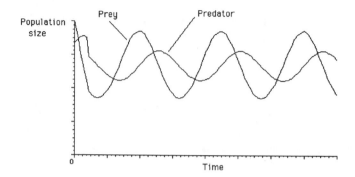

Fig. 6.12.1. Graph showing a stable relationship between a predator population and a prey population.

populations between certain upper and lower limits and this situation is illustrated graphically in Fig. 6.12.1 The Volterra differential equations can be solved directly but this solution does not provide a simple relation between the size of the predator and prey populations and therefore numerical methods of solution should be applied. An interesting description of this problem is given by Simmons (1972).

Systems of differential equations also arise as a means of solving higher order differential equations. To illustrate this consider the second order differential equation˙

$$2d^2x/dt^2 + 4(dx/dt)^2 - 2x = \cos x \qquad (6.12.2)$$

together with the initial conditions

$$x = 0 \text{ and } dx/dt = 10 \text{ when } t = 0$$

If we substitute $p = dx/dt$ then (6.12.2) becomes

$$2dp/dt + 4p^2 = \cos x + 2x$$
$$\qquad\qquad (6.12.3)$$
$$dx/dt = p$$

with initial conditions

$$p = 10 \text{ and } x = 0 \text{ when } t = 0.$$

The second order differential equations has been replaced by a system of first order differential equations. If we have an nth order differential equation of the form

$$a_n \, d^n y/dx^n + a_{n-1} d^{n-1} y/dx^{n-1} + \dots + a_0 y = f(x, y) \tag{6.12.4}$$

by making the substitutions

$$P_0 = y \text{ and } dP_{i-1}/dx = P_i \text{ for } i = 2, 3, \dots, n \tag{6.12.5}$$

then (6.12.4) becomes

$$a_n dP_{n-1}/dx = f(x,y) - a_{n-1} P_{n-1} - a_{n-2} P_{n-2} - \dots - a_0 P_0 \tag{6.12.6}$$

Now equations (6.12.5) and (6.12.6) together constitute a system of n first order differential equations. Initial values will be given for (6.12.4) in terms of the various order derivatives P_i for $i = 1, 2, 3, \dots, n$ at some initial value x_0 and these can easily be translated into initial conditions for the system of equations (6.12.5) and (6.12.6). In general the solution of the original nth order differential equation and the system of first order differential equations, (6.12.5) and (6.12.6), are the same. In particular the numerical solution will provide the values of y for a specified range of x. An excellent discussion of the equivalence of the solutions of the two problems is given in Simmons (1972). We can see from this description that any order differential equation of the form (6.12.4) with given initial values can be reduced to solving a system of first order differential equations. This argument is easily extended to the more general nth order differential equation by making exactly the same substitutions as above in

$$d^n y/dx^n = f(x, y, y', \dots, y^{(n-1)})$$

where $y^{(n-1)}$ denotes the $(n-1)$th order derivative of y.

The numerical techniques we have described for solving a single first order differential equations can be applied directly to solve systems of first order differential equations. Having chosen a suitable increment h for the independent variable x we may apply the Runge-Kutta method to each equation in turn to obtain values for each of the independent variables. The method may then be applied in the usual iterative way. A predictor-corrector method can equally well be used. To illustrate this procedure it is applied to a general system using a simple predictor-corrector method. Consider the system

$$dy_r/dx = f_r(x,y) \text{ for } r = 1, 2, \dots, m \tag{6.12.7}$$

In addition we have the initial value conditions

$$y_r(x_0) = a_r \text{ for } r = 1, 2, \dots, m$$

A simple predictor-corrector method based on the modified Euler technique takes the form

$$y_{r,n+1}^{(0)} = y_{r,n} + hf_{r,n} \quad \text{for } r = 1, 2, \ldots, m$$

Then for $k = 0, 1, 2, \ldots$

$$y_{r,n+1}^{(k+1)} = y_{r,n} + h(f_{r,n+1}^{(k)} + f_{r,n})/2 \quad \text{for } r = 1, 2, \ldots, m \qquad (6.12.8)$$

The second equation constitutes the corrector and is iterated until convergence is achieved by incrementing the index k. When convergence has been achieved n is incremented and the whole procedure is repeated until all the required values have been obtained. The Runge-Kutta methods have the advantage that they are self-starting and can give high accuracy. Consequently program 6.12.1 is based on a Runge-Kutta method. This program uses the classical method but any order method can be used by changing the data statements. *Care must be taken in using this program since the right-hand side of each equation must be defined in the appropriate section of the program and the initial conditions must be input in the correct order.*

```
100   REM ------------------------------------------------------------
110   REM SOLVING SIMULTANEOUS EQUNS USING RUNGE KUTTA METHODS GL/JP (C)
120   REM ------------------------------------------------------------
130   PRINT "SOLUTION OF DIFFERENTIAL EQUNS OF THE FORM dy/dx= f(x,y)."
140   PRINT "ENTER THE INITIAL AND FINAL VALUES OF x";
150   INPUT X0,XN
160   PRINT "ENTER THE STEP SIZE H";
170   INPUT H
180   NM=(XN-X0)/H
190   PRINT "ENTER THE NUMBER OF EQUATIONS";
200   INPUT NF
210   PRINT "ENTER INITIAL VALUES FOR DIFFERENTIAL EQUATION"
220   FOR S=1 TO NF
230       PRINT "Y(";S;")=";
240       INPUT Y(S)
250   NEXT S
260   X=X0
270   PRINT: PRINT "  X";TAB(12);"Y(1)";TAB(25);"Y(2)"
275   REM READ VALUES  FOR RUNGE KUTTA METHOD ORDER OD
280   OD=4
290   FOR I=1 TO OD
300       READ A(I)
310   NEXT I
320   FOR I=1 TO OD
330       FOR J=1 TO OD
340           READ B(I,J)
350       NEXT J
360   NEXT I
```

Program 6.12.1. (continues).

```
370    FOR I=1 TO OD
380        READ C(I)
390    NEXT I
395    REM LOOP FOR EACH STEP USING RUNGE KUTTA
400    FOR N=1 TO NM+1
410        PRINT X;TAB(8);Y(1);TAB(25);Y(2)
415        REM FOR EACH DEPENDENT VARIABLE CALCULATE NEXT VALUE USING RUNGE
KUTTA PROCEDURE
420        FOR S=1 TO NF
430            YO(S)=Y(S)
440        NEXT S
450        FOR I=1 TO OD
460            XN=X+A(I)*H
470            FOR S=1 TO NF
480                YN(S)=YO(S)
490                FOR J=1 TO I-1
500                    YN(S)=YN(S)+B(I,J)*K(J,S)
510                NEXT J
520            NEXT S
530            FOR S=1 TO NF
540                GOSUB 620
550                K(I,S)=H*F(S)
560                Y(S)=Y(S)+K(I,S)*C(I)
570            NEXT S
580        NEXT I
590        X=X+H
600    NEXT N
610    END
615    REM SUB FOR CALCULATING THE RIGHT-HAND SIDE VALUES OF EACH EQUATION
620    F(1)=6*YN(2)-YN(1)
630    F(2)=YN(1)
640    RETURN
650    REM DATA FOR RUNGE KUTTA COEFFICIENTS
660    DATA 0,.5,.5,1
670    DATA 0,0,0,0
680    DATA .5,0,0,0
690    DATA 0,.5,0,0
700    DATA 0,0,1,0
710    DATA .166666667,.333333333,.333333333,.166666667
```

Program 6.12.1.

Consider the solution of the second order differential equation

$$d^2y/dx^2 + dy/dx - 6y = 0$$

where $dy/dx = -3$ and $y = 1$ when $x = 0$

By substituting $Y(1) = dy/dx$ and $Y(2) = y$ we obtain

Table 6.12.1. Use of the Runga-Kutta method to solve
the simultaneous equations (6.12.9) with $h = 0.1$.

X	Y(1)	Y(2)	TRUE Y(1)	TRUE Y(2)
0	-3	1	-3	1
.1	-2.222512	.7408375	-2.222455	.7408182
.2	-1.64652	.5488403	-1.646435	.5488116
.3	-1.219804	.4066015	-1.219709	.4065697
.4	-.9036763	.3012257	-.9035826	.3011942
.5	-.6694771	.2231594	-.6693905	.2231302
.6	-.4959735	.165325	-.4958966	.1652989
.7	-.3674355	.122479	-.3673692	.1224564
.8	-.2722097	9.073722E-02	-.2721538	9.071794E-02
.9	-.2016628	6.722173E-02	-.2016165	.0672055
1	-.1493989	4.980061E-02	-.1493611	4.978704E-02

$$dY(1)/dx = -Y(1) + 6Y(2)$$

$$dY(2)/dx = Y(1)$$

(6.12.9)

where $Y(1) = -3$ and $Y(2) = 1$ when $x = 0$.

Program 6.12.1 is used to solve these equations and the results are given in Table 6.12.1. The program can be applied to any number of simultaneous differential equations of the appropriate type but convergence will depend on the step size and the nature of the right-hand sides of these equations.

It should be noted that methods exist based on Taylor's series expansions for solving systems of second order differential equations with initial value conditions; for example Newmark (1959).

6.13 SPECIAL TECHNIQUES

A further set of predictor-corrector equations may be generated by making use of an interpolation formula due to Hermite. An unusual feature of these equations is that they contain second order derivatives. It is usually the case that the calculation of second order derivatives is not particularly difficult and consequently this feature does not add a significant amount of work to the solution of the problem. However it should be noted that in using a computer program for this technique the user has to supply not only the function on the right-hand side of the differential equation but its derivative as well. To the general user this may be unacceptable. The equations for this method take the form

$$y_{n+1}^{(1)} = y_n + h(y'_n + 3y'_{n-1})/2 + h^2(17y''_n + 7y''_{n-1})/12$$

$$y^*_{n+1}^{(1)} = y_{n+1}^{(1)} + 31(y_n - y_n^{(1)})/30$$

$$y'_{n+1}^{(1)} = f(x_{n+1}, y^*_{n+1}^{(1)}) \tag{6.13.1}$$

For $k = 1, 2, 3, ...$

$$y_{n+1}^{(k+1)} = y_n + h(y'_{n+1}^{(k)} + y'_n)/2 + h^2(-y''_{n+1}^{(k)} + y''_n)/12$$

This method is stable and has a smaller truncation error at each step than either Hamming's method or Milne's method. Thus it may be worthwhile accepting the additional effort required from the user. The method is known as the Hermite method. We note that since we have

$$dy/dx = f(x,y)$$

then

$$d^2y/dx^2 = df(x,y)/dx$$

and thus y''_n etc. are easily calculated. Program 6.13.1 implements this method and Table 6.13.1 gives the results for a specific example.

Table 6.13.1. Use of Hermite's method with $h = 0.1$ to solve $y' = -10y$ where $y = 1$ when $x = 0$. A relatively good result for a difficult problem.

x	y	Exact y
0	1	1
.1	.375	.3678795
.2	.1381579	.1353353
.3	5.090025E-02	4.978707E-02
.4	1.875275E-02	1.831564E-02
.5	6.908933E-03	6.737947E-03
.6	2.545425E-03	2.478752E-03
.7	9.378185E-04	9.118815E-04
.8	3.455452E-04	3.354623E-04
.9	1.27342E-04	1.234097E-04
1	4.689282E-05	4.539989E-05

```
100  REM ----------------------------------------------------------------
110  REM HERMITE'S METHOD FOR THE SOLN OF DIFFERENTIAL EQUNS GL/JP (C)
120  REM ----------------------------------------------------------------
130  PRINT "SOLUTION OF A FIRST ORDER"
140  PRINT "DIFFERENTIAL EQUATION USING  HERMITE'S METHOD":PRINT
```

Program 6.13.1.(continues).

```
150   PRINT "ASSUMING YOUR DIFFERENTIAL EQUATION IS"
160   PRINT"OF THE FORM dy/dx=f(x,y)"
170   PRINT:PRINT"ENTER INITIAL & FINAL VALUES OF INDEPENDENT VARIABLE x";
180   INPUT X0,XN
190   PRINT"ENTER STEP SIZE";
200   INPUT H
210   NM=(XN-X0)/H
220   PRINT"ENTER THE INITIAL VALUE FOR y";
230   INPUT Y0:X=X0:Y=Y0:Y(0)=Y:YN=Y
240   PRINT:PRINT" X";TAB(12);" Y"
250   PRINT X;TAB(8);YN
260   XN=X:YN=Y
270   GOSUB 600
280   F1=FXY:D1=D
285   REM CLASSICAL RUNGE KUTTA METHOD
290   K1=H*FXY
300   XN=X+H/2:YN=Y+K1/2
310   GOSUB 600
320   K2=H*FXY
330   YN=Y+K2/2
340   GOSUB 600
350   K3=H*FXY
360   YN=Y+K3:XN=X+H
370   GOSUB 600
380   K4=H*FXY
390   Y(1)=Y+(K1+2*K2+2*K3+K4)/6
400   YN=Y(1)
410   H1=H/2:H2=H1*H/6
420   FOR N=2 TO NM+1
425      REM NOW USE HERMITE PREDICTOR CORRECTOR METHOD
430      XN=X+H
440      PRINT XN;TAB(8);YN
450      GOSUB 600
460      F2=FXY
470      D2=D
480      PP=Y4
490      Y4=Y(1)+H1*(3*F1+F2)+H2*(17*D2+7*D1)
500      YM=Y4:IF N>2 THEN YM=Y4+31/30*(Y(1)-PP)
510      XN=XN+H:YN=YM
520      GOSUB 600
530      F4=FXY:D4=D
540      YN=Y(1)+H1*(F4+F2)+H2*(D2-D4)
550      IF ABS(YN-YC)>.0000001 THEN YC=YN:GOTO 520
560      X=X+H:Y(1)=YN
570      F1=F2:D1=D2
580   NEXT N
590   END
595   REM SUB TO CALC FUNCN AND DERIV AT STATEMTS 600 AND 610 RESPECTIVELY
600   FXY=YN
610   D=YN
620   RETURN
```

Program 6.13.1.

Another method which uses second order derivatives is that of Numerov. This however is only used for equations which have the form

$$d^2y/dx^2 = f(x,y) \quad \text{with} \quad y(x_0) = y_0 \text{ and } y'(x_0) = y'_0 \qquad (6.13.2)$$

In this case the second derivative does not have to be evaluated since it is given directly by the original equation, (6.13.2). The method is based on the equations

$$y_{n+1} = 2y_n - y_{n-1} + h^2(10y''_n + y''_{n-1})/12$$

$$y''_{n+1} = f(x_{n+1}, y_{n+1}) \qquad (6.13.3)$$

for $k = 1, 2, 3, ...$

$$y_{n+1}^{(k+1)} = 2y_n - y_{n-1} + h^2(y''_{n+1}{}^{(k)} + 10y''_n + y''_{n-1})/12$$

These formula are used for each step $n = 1, 2, 3, ...$ until the interval is spanned. The value for y_1 required for the first step must be approximated by Taylor's series. If the step size is small this will be satisfactory but large step sizes will lead to significant errors throughout the iterations. Program 6.13.2 implements Numerov's method and results for a specified example are given in Table 6.13.2. In this case step size is sufficiently small to allow convergence.

One feature which may be used to improve many of the methods discussed above is step size adjustment. This means that we adjust the step size h according to the progress of the iteration. One criteria for adjusting h is to monitor the size of the truncation error. If the truncation error is smaller than the accuracy requirement we can increase h but if the truncation error is too large we can reduce h. Step size adjustment can lead to considerable additional work, for example if a predictor-corrector method is used new initial values must be calculated. The following method is an interesting alternative to this kind of procedure.

Table 6.13.2. Use of Numerov's method with $h = 0.1$ to solve $y'' = y$ where $y = 1$ and $y' = -1$ when $x = 0$.

x	y	Exact value
0	1	1
.1	.9048333	.9048374
.2	.8187225	.8187308
.3	.7408057	.7408182
.4	.6703031	.67032
.5	.6065091	.6065307
.6	.5487853	.5488116
.7	.4965539	.4965853
.8	.4492922	.4493289
.9	.4065272	.4065696
1	.3678308	.3678794

```
100   REM --------------------------------------------------------------------
110   REM SOLVING DIFFERENTIAL EQUATION'S USING NUMEROV'S METHOD GL/JP (C)
120   REM --------------------------------------------------------------------
130   PRINT"SOLUTION OF A FIRST ORDER"
140   PRINT"DIFFERENTIAL EQUATION USING  NUMEROV'S METHOD":PRINT
150   PRINT"ASSUMING YOUR DIFFERENTIAL EQUATION IS"
160   PRINT"OF THE FORM y''=f(x,y)"
170   PRINT:PRINT"ENTER INITIAL & FINAL VALUES OF INDEPENDENT VARIABLE x";
180   INPUT X0,XN
190   PRINT"ENTER STEP SIZE";
200   INPUT H
210   NM=(XN-X0)/H
220   PRINT"ENTER THE INITIAL VALUE FOR y";
230   INPUT Y0:X=X0:Y=Y0:Y(0)=Y
240   PRINT "ENTER INITIAL VALUE FOR y' ";
250   INPUT YD
255   REM FIND INITIAL APPROXIMATION FOR Y1 USING TAYLOR'S SERIES
260   XN=X:YN=Y
270   GOSUB 520
280   F0=FXY
290   Y(1)=Y0+H*YD+H*H/2*F0+H*H*H/6*YD
300   PRINT:PRINT" X";TAB(12);" Y"
310   PRINT X;TAB(8);YN
320   XN=X:YN=Y
330   GOSUB 520
340   F1=FXY
350   YN=Y(1)
360   H2=H*H/12
370   FOR N=2 TO NM+1
375       REM NOW USE NUMEROV PREDICTOR CORRECTOR
380       XN=X+H
390       PRINT XN;TAB(8);YN
400       GOSUB 520
410       F2=FXY
420       Y4=2*Y(1)-Y(0)+H2*(10*F2+F1)
430       XN=XN+H:YN=Y4
440       GOSUB 520
450       F4=FXY
460       YN=2*Y(1)-Y(0)+H2*(F4+10*F2+F1)
470       IF ABS(YN-YC)>.00001 THEN YC=YN:GOTO 440
480       X=X+H:Y(0)=Y(1):Y(1)=YN
490       F1=F2
500       NEXT N
510   END
515   REM SUB TO CALCULATE FUNCTION IN STATEMENT 520 AS FXY=F(XN,YN)
520   FXY=YN
530   RETURN
```

Program 6.13.2.

6.14 EXTRAPOLATION TECHNIQUES

This method is based on a procedure called Richardson's extrapolation, introduced in Chapter 1. This procedure can also be applied to the solution of differential equations and obtains successive initial approximations for y_{n+1} using a modified mid-point method. The interval sizes used for obtaining these approximations are calculated from

$$h_i = h_{i-1}/2 \text{ for } i = 1, 2, 3, ... \tag{6.14.1}$$

with the initial value h_0 given.

Once these initial approximations have been obtained we can use (6.14.2), the extrapolation formula, to obtain improved approximations.

$$T_{m,k} = (4^m T_{m-1,k+1} - T_{m-1,k})/(4^m - 1) \tag{6.14.2}$$

$$\text{for } m = 1, 2, ... , s \text{ and } k = 0, 1, 2, ... , s - m$$

The calculations are set out in an array in much the same way as the calculations for Romberg's method for integration described in Chapter 4. When $m = 0$ the values of $T_{0,k}$ for $k = 0, 1, 2, ... , s$ are taken as the successive approximations to the values of y_{n+1} using the h_i values obtained from (6.14.1).

The formula for calculating the approximations used for the initial values $T_{0,k}$ in the above array are given below.

$$y_1 = y_0 + hy'_0$$

$$y_{n+1} = y_{n-1} + 2hy'_n \text{ for } n = 1, 2, ... , N_k \tag{6.14.3}$$

Here $k = 1, 2, ...$ and N_k is the number of steps taken in the range of interest so that $N_k = 2^k$ since the size of the interval is halved each time. The $2h$ distance between y_{n+1} and y_{n-1} values may lead to significant variations in the magnitude of the error. Because of this, instead of using the final value of y_{n+1} given by (6.14.3), Gragg (1965) has suggested that at the final step these values are smoothed using the intermediate value y_n. This leads to the values for $T_{0,k}$

$$T_{0,k} = (y^k_{N-1} + 2y^k_N + y^k_{N+1})/4$$

where the superscript k denotes the value at the kth division of the interval.

Alternatives to the method of Gragg are available for finding the initial values in the table and various combinations of predictor-correctors may be used. It should be noted however that if the corrector is iterated until convergence is achieved this will improve the accuracy of the initial values but at considerable computational expense for smaller step sizes i.e. for larger N values. Program 6.14.1 implements this technique and the results for a specific problem are given in Table 6.14.1.

In Table 6.7.1 various forms of the Runge-Kutta method were used to solve the

```
100   REM -----------------------------------------------------------------
110   REM  SOLN OF DIFFERENTIAL EQUATIONS:EXTRAPOLATION METHOD. GL/JP (C)
120   REM -----------------------------------------------------------------
130   PRINT"SOLUTION OF A FIRST ORDER"
140   PRINT"DIFFERENTIAL EQUATION USING EXTRAPOLATION METHOD":PRINT
150   PRINT"ASSUMING YOUR DIFFERENTIAL EQUATION IS"
160   PRINT"OF THE FORM dy/dx=f(x,y)"
170   PRINT
180   PRINT"ENTER INITIAL AND FINAL VALUES OF INDEPENDENT VARIABLE x";
190   INPUT X0,XL
200   PRINT"MAXIMUM NUMBER OF INTERVAL DIVISIONS REQUIRED";
210   INPUT ND
220   PRINT
230   PRINT"ENTER THE INITIAL VALUE FOR y";
240   INPUT Y0
250   Y=Y0
255   REM CALCULATE INITIAL VALUES
260   FOR NI=1 TO ND
270      NM=2^NI: H=(XL-X0)/NM
280      X=X0: XN=X0:Y=Y0:YN=Y0
290      GOSUB 490:F1=FXY
300      Y1=Y+H*F1:X=X+H
310      FOR N=1 TO NM
300         XN=X : YN=Y1
320         GOSUB 490:F2=FXY
330         Y2=Y+2*H*F2
340         YF2=Y2:YF1=Y1:YF=Y
350         Y=Y1:Y1=Y2: X=X+H
360      NEXT N
365      REM SMOOTH INITIAL VALUES
370      A(NI)=(YF+2*YF1+YF2)/4
380   NEXT NI
385   REM IMPROVED VALUES CALCULATED BY EXTRAPOLATION METHOD
390   PRINT  : PRINT"APPROXIMATIONS:"
400   FOR M= 1 TO ND-1
410      PRINT "GROUP";M
420      FOR I=1 TO ND-M
430         B(I)=(A(I+1)*4^M-A(I))/(4^M-1) : A(I)=B(I)
440         PRINT B(I);
450      NEXT I
460   PRINT
470   NEXT M
480   END
485   REM SUB TO EVALUATE FUNCN DEFINED IN STATEMENT 490 AS FXY=F(XN,YN)
490   FXY=YN
500   RETURN
```

Program 6.14.1

Table 6.14.1 The use of the extrapolation method to solve $y' = y$ at $x = 10$ given $y = 1$ when $x = 0$. The approximation to this is the final entry in the table. The exact value is $y = 22026.466$.

```
THE EXTRAPOLATION METHOD FOR SOLVING DIFFERENTIAL EQUATIONS.
ENTER THE INITIAL AND FINAL VALUES OF x? 0,10

MAXIMUM NUMBER OF INTERVAL DIVISIONS REQUIRED?  7

ENTER THE INITIAL VALUE FOR y?  1

APPROXIMATIONS:
GROUP  1
 1165.167    5837.408    15392.71    20908.09    21925.97    22019.49
GROUP  2
 6148.891    16029.74    21275.78    21993.83    22025.73
GROUP  3
 16186.57    21359.05    22005.23    22026.23
GROUP  4
 21379.34    22007.77    22026.32
GROUP  5
 22008.38    22026.33
GROUP  6
 22026.34
```

same problem. However, the final value achieved in Table 6.14.1 is better than any of the results achieved by the Runge-Kutta method. It must be noted that only the final value is found, other values in a given interval can be obtained if intermediate ranges are considered.

6.15. STIFF EQUATIONS

When the solution of a differential equation contains components which change at significantly different rates for given changes in the independent variable the equation is said to be "stiff". A differential equation or a system of differential equations may be affected by this phenomena and when it is present a particularly careful choice of the step size must be made if stability is to be achieved.

As an illustration we consider the differential equation

$$dy/dx + 10y = x$$

with the initial condition $y = 0$ when $x = 0$. The exact solution of this differential equation is

$$y = 0.1x - 0.01 + 0.01e^{-10x} \qquad (6.15.1)$$

From (6.15.1) we see that the exponential term decays quickly for increasing x and therefore the solution is approximately linear. Consequently It would be reasonable to assume that a large value of the increment h coul be used. Table 6.15.1 demonstrates that this is not the case and we see that it is necessary to use a small value of h to ensure stability in the solution process and the value of h is very critical. Table 6.15.1 shows that good results are achieved for this problem for $h = 0.2$ over the whole range, $x = 0$ to $x = 5$. The method fails completely if h is taken as 0.3.

We will now consider how the stiffness phenomena arises in systems of differential equations.

Table 6.15.1. Use of Adams' method to solve $y' + 10y = x$ where $y = 0$ when $x = 0$ using step size $h = 0.2$. For this h the results are accurate.

```
ADAM'S METHOD FOR THE SOLUTION OF DIFFERENTIAL EQUATIONS.
ASSUMING YOUR DIFFERENTIAL EQUATION IS
D F THE FORM dy/dx=f(x,y)

ENTER INITIAL AND FINAL VALUES OF INDEPENDENT VARIABLE x?  0,5
ENTER STEP SIZE?  .2
ENTER THE INITIAL VALUE FOR y?  0

   x        y                    Exact value
   0        0                    0
   1        9.004341E-02         9.000046E-02
   2        .1899909             .19
   3        .2900039             .2900001
   4        .39                  .3900001
   5        .4899971             .49
```

Consider the system given below.

$$dy_1/dx = -by_1 - cy_2$$
$$dy_2/dx = y_1 \qquad (6.15.2)$$

This system may be written in matrix form as

$$\begin{bmatrix} dy_1/dx \\ dy_2/dx \end{bmatrix} = \begin{bmatrix} -b & -c \\ 1 & 0 \end{bmatrix} \begin{bmatrix} y_1 \\ y_2 \end{bmatrix} \qquad (6.15.3)$$

The solution of this system is

$$y_1 = A \exp(r_1x) + B \exp(r_2x)$$

and   (6.15.4)

$$y_2 = C \exp(r_1x) + D \exp(r_2x)$$

where it can be easily verified that r_1 and r_2 are the eigenvalues of the matrix.

$$\begin{bmatrix} -b & -c \\ 1 & 0 \end{bmatrix}$$ (6.15.5)

A definition of eigenvalues together with a description of methods for finding them is given in Chapter 10.

By taking various values of b and c we can generate many problems of the form (6.15.3) having solutions (6.15.4) where the eigenvalues r_1 and r_2 will of course change from problem to problem. If a numerical procedure is applied to solve these systems of differential equations the success of the method will depend crucially on the eigenvalues of the matrix (6.15.5). This is illustrated by Table 6.15.2. This table shows that the performance of the numerical method used depends on the ratio of the eigenvalues. If this ratio is small good results can be achieved; if the ratio is large to poor results are likely to be obtained unless a small step is taken. Thus the larger the ratio of the eigenvalues the smaller the step size required. Table 6.15.2 shows that failure occurs in one case because even the step 0.1 is too large.

Table 6.15.2. Values of y_1 at $x = 2$ for the simultaneous differential equations given by (6.15.2) with the initial conditions $y_1 = 0$ and $y_2 = 1$ at $x = 0$.

Values of b and c	Eigenvalue ratio	Step size			
		h=.5	h=.25	.1	Exact
3 and 2	2	.2513242	.2523186	.2523543	.2523545
10 and .1	1000	-33.808	.9811318	.9811623	.9811622
100 and 1	10000	Fail	Fail	Fail	.9803

The significance of the eigenvalue ratio in relation to the required step size can be generalised to systems of many equations. Consider the system of n equations

$$dy/dx = Ay + p(x)$$ (6.15.6)

Where y is an n component column vector, $p(x)$ is an n component column vector of functions of x and A an $n \times n$ matrix of constants. It can be shown that the solution of this system takes the form

$$\mathbf{y}(x) = \sum_{i=1}^{n} v_i \, \mathbf{d}_i \exp(r_i x) + \mathbf{s}(x) \tag{6.15.7}$$

Here $r_1, r_2, \ldots$ are the eigenvalues and $\mathbf{d}_1, \mathbf{d}_2, \ldots$ the eigenvectors of $\mathbf{A}$. The vector function $\mathbf{s}(x)$ is the particular integral of the system, sometimes called the steady state solution since for negative eigenvalues the exponential terms should die away with increasing x. If it is assumed that the $r_k < 0$ for $k = 1, 2, 3, \ldots$ and we require the steady state solution of system (6.15.6) then any numerical method applied to solve this problem may face significant difficulties as we have seen from the problems considered in Table 6.15.2. Problems will arise if there is a wide variation in the magnitude of the eigenvalues. We must continue the integration until the exponential components have been reduced to negligible levels and yet we must take sufficiently small steps to ensure stability, thus requiring many steps over a large interval. This is the most significant effect of stiffness.

The definition of stiffness can be extended to any system of the form (6.15.6). The stiffness ratio is defined as the ratio of the largest and smallest eigenvalues of $\mathbf{A}$ and gives a measure of the stiffness of the system. The larger its value the more difficult the system is to solve, a phenomena which is illustrated clearly by Table 6.15.2.

The methods used to solve stiff problems must be based on stable techniques. If we use a predictor-corrector method, not only must this method be stable but the corrector must be iterated to convergence. An interesting discussion of this topic is given in Ralston and Rabinowitz (1978). Specialised methods have been developed for solving stiff problems and Gear (1971) has provided a number of techniques which have been reported to be successful.

6.16. BOUNDARY VALUE PROBLEMS

The differential equations we have been dealing with up to this point are characterised by the initial conditions from where we begin the numerical solution procedure. These conditions are given in terms of some initial point. For example, consider a second order differential equation. We may be given that $y = 0$ and $y' = 5$ when $x = 0$ (i.e $y(0) = 0$ and $y'(0) = 5$) and be required to solve the differential equation over the range 0 to 2. The solution range need not, of course, be restricted to this range.

In the case of the boundary value problem with one independent variable the value of the dependent variable and sometimes its derivatives are given at two points which mark the boundaries of the problem domain over which the solution is required. The following example illustrates this type of problem. We wish to solve the differential equation

$$x^2y'' + xy' + y = x + 2$$

given the boundary conditions

$$y(0) = 1 \text{ and } y(2) = 2$$

For this differential equation the range of values of x for which we wish to obtain the solution lies between $x = 0$ and $x = 2$ and specified conditions must be satisfied at these points.

6.17. SHOOTING METHODS FOR BOUNDARY VALUE PROBLEMS

One approach to solving the boundary value problem is known as the shooting method. It will be seen that the name of the method is descriptive of the procedure. Consider the boundary value problem

$$p(x)y'' + q(x)y' + r(x)y = f(x) \tag{6.17.1}$$

subject to the boundary conditions

$$y(a) = a_0 \text{ and } y(b) = b_0$$

We may replace this problem by the initial value problem

$$p(x)y'' + q(x)y' + r(x)y = f(x) \tag{6.17.2}$$

$$y(a) = a_0 \text{ and } y'(a) = s$$

Consequently the boundary value problem defined by (6.17.1) is reduced to finding s, the shooting parameter such that when we numerically solve (6.17.2) starting at $x = a$ we will arrive at the value b_0 for $y(b)$ with an acceptable level of accuracy.

The principle of the shooting method is illustrated in Fig. 6.17.1 which shows solution trajectories, for various values of s, generated in the process of solving the equation

$$x^2y'' - 6y = 0 \text{ with boundary conditions } y = 1 \text{ when } x = 1 \text{ and } 2.$$

The process was initiated with $y = 1$ and $s = 0$ when $x = 1$ and the final value attained for y, when $x = 2$, is much larger than 1, the desired value. A further value of s was then chosen, $s = -1.9$, and the solution trajectory then gave too low a value for y when $x = 2$. Taking intermediate values for s between 0 and -1.9 gives better results. Proceeding in this way ultimately provides a suitable value for s equal to -1.516. The values of s correspond to the gradients of y at $x = 1$.

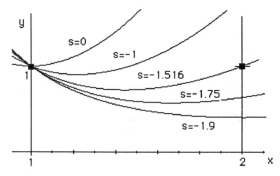

Fig. 6.17.1. Solution of $x^2y'' - 6y = 0$ where $y = 1$ when $x = 1$ using various shooting parameters. Only $s = -1.516$ satisfies the boundary condition $y = 1$ when $x = 2$.

The outline given above describes the general approach. Any method may be used to solve (6.17.2) such as a predictor-corrector method or the Runge-Kutta method. The major problem is to find an efficient procedure for choosing the best value of s. A significant improvement to the arbitary choice procedure described above is to use bisection or linear interpolation to find s once some solution trajectories have been found.

Applying the method to a specific problem provides the results given in Table 6.17.1. In this table the method of bisection is used to obtain improved values for the parameter s. At each stage it is important to check that the bisection is performed using s values where the corresponding y values enclose the required value, which in the problem of Table 6.17.1 is 2. Several bisections are required before a good value is obtained but this value provides reasonable approximations in the range of interest as can be seen from Table 6.17.2.

Table 6.17.1. Use of the shooting method to solve $y'' + y' - 6y = 0$ given the boundary conditions $y = 1$, $x = 0$ and $y = 2$, $x = 1$. This shows the final values of y attained for different values of the shooting parameter s. The last value of s (= −1.671875) gives a value of y close to 2 as required.

s	Final y
−2	1.517618
−1	2.985435
−1.5	2.251527
−1.75	1.884572
−1.625	2.06805
−1.6875	1.976312
−1.65625	2.022181
−1.671875	1.999246

Table 6.17.2. Use of the shooting method, with $s = -1.671875$, to solve
$y'' + y' - 6y = 0$ given the boundary conditions $y = 1$, $x = 0$ and $y = 2$, $x = 1$.

x	y	Exact y
0	1	1
.1	.8684869	.8685205
.2	.7993187	.7993917
.3	.782595	.7827134
.4	.8123665	.8125371
.5	.885918	.8861490
.6	1.003304	1.003606
.7	1.167091	1.167476
.8	1.38226	1.382746
.9	1.656271	1.656877
1	1.999246	2

6.18. FINITE DIFFERENCE APPROXIMATIONS

In this section a powerful alternative method is presented for the solution of some simple types of boundary value problems. This is the finite difference approximation method. We have seen that boundary value problems with one independent variable are described by ordinary differential equations. However, boundary value problems that arise in science and engineering are often in terms of several independent variables and these are described by partial differential equations.

Essentially, the finite difference method approximates the ordinary or partial differential equation by replacing it by a set of algebraic relationships. Ideally a computer program to solve boundary value problems would consist of two, inter-linked, stages as follows:

(1)   Automatic conversion of the differential equations to algebraic relationships.

(2)   Solution of the resulting algebraic relationships. These may be linear or non-linear equations, recurrence relations or an algebraic eigenvalue problem.

It is not easy to create general purpose programs, functioning on the lines described above, so that any given boundary value problem can be solved. This is because the variety of equation forms that can represent the boundary value problem is almost endless: furthermore there is an immense variety of boundary conditions and boundary shapes. Frequently the differential equation has to be reduced to some algebraic relationships by the user and the computer employed only to solve the resulting algebraic equations, recurrence relations or eigenvalue problem. Even then, considerable difficulty can remain if the algebraic relationships are non-linear.

6.19. SINGLE VARIABLE BOUNDARY VALUE PROBLEMS

The central difference approximation for the differential operators with equally spaced nodes are given in (5.2.3) and (5.3.1). For convenience these approximations are given diagrammatically in Fig. 6.19.1. If necessary, they can also be derived for unequally spaced nodes although the expressions are naturally more cumbersome. If required alternative approximations can be developed so that the errors are proportional to h^4, rather than h^2, but for most problems the use of equi-spaced nodes with an error of order h^2 is adequate.

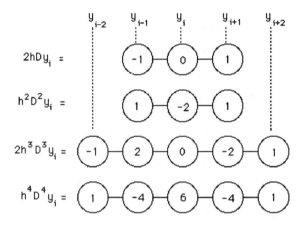

Fig. 6.19.1. Central difference operators with errors of order h^2. $D \equiv d/dx$.
The circled values are multipliers of the y values aligned directly above them.

Consider a second order linear differential equation with coefficients that are functions of the independent variable x. Thus

$$f(x)y'' + g(x)y' + p(x)y = q(x) \tag{6.19.1}$$

Without loss of generality we will consider the boundaries to be at $x = 0$ and $x = 1$. Whilst the two boundary conditions must be specified in this problem with one independent variable, there is no boundary shape to be considered. Multiplying (6.19.1) by h^2, where h is the distance between the nodes, and using the operator D ($\equiv d/dx$) we have

$$[f(x)]h^2D^2y + [hg(x)/2]2hDy + [h^2p(x)]y = h^2q(x)$$

Let the range of x be divided into n equal intervals of width h, so that $h = 1/n$. Using the central difference approximations at node i, gives

$$[f(x_i)](y_{i-1} - 2y_i + y_{i+1}) + [hg(x_i)/2](y_{i+1} - y_{i-1}) + [h^2p(x_i)]y_i = h^2q(x_i)$$

or $$[f(x_i) - hg(x_i)/2]y_{i-1} + [h^2p(x_i) - 2f(x_i)]y_i +$$

$$[f(x_i) + hg(x_i)/2]y_{i+1} = h^2q(x_i) \qquad (6.19.2)$$

Suppose the boundary conditions are at $x = 0$, $y = A$ and at $x = 1$, $y = B$. Equation (6.19.2) is applied at the $n - 1$ internal mesh points, $i = 1, 2, \ldots, n - 1$, and a set of $n - 1$ algebraic equations are obtained. For example, if $n = 4$ we have

$$[f(x_1) - hg(x_1)/2]A + [h^2p(x_1) - 2f(x_1)]y_1 + [f(x_1) + hg(x_1)/2]y_2 = h^2q(x_1)$$

$$[f(x_2) - hg(x_2)/2]y_1 + [h^2p(x_2) - 2f(x_2)]y_2 + [f(x_2) + hg(x_2)/2]y_3 = h^2q(x_2)$$

$$[f(x_3) - hg(x_3)/2]y_2 + [h^2p(x_3) - 2f(x_3)]y_3 + [f(x_3) + hg(x_3)/2]B = h^2q(x_3)$$

In matrix notation this becomes

$$\begin{bmatrix} h^2p(x_1) - 2f(x_1) & f(x_1) + hg(x_1)/2 & 0 \\ f(x_2) - hg(x_2)/2 & h^2p(x_2) - 2f(x_2) & f(x_2) + hg(x_2)/2 \\ 0 & f(x_3) - hg(x_3)/2 & h^2p(x_3) - 2f(x_3) \end{bmatrix} \begin{bmatrix} y_1 \\ y_2 \\ y_3 \end{bmatrix}$$

$$= \begin{bmatrix} h^2q(x_1) - [f(x_1) - hg(x_1)/2]A \\ h^2q(x_2) \\ h^2q(x_3) - [f(x_3) + hg(x_3)/2]B \end{bmatrix}$$

Because (6.19.1) is a second order differential equation the boundary conditions could be defined in terms of $y'(0)$ and $y'(1)$. To examine the effect of these particular boundary conditions let $y' = P$ at $x = 0$ and $y' = Q$ at $x = 1$. Equation (6.19.2) can be applied at the internal node points, $i = 1, 2, \ldots, n - 1$ without difficulty. However, applying (6.19.2) to node 0 would introduce a fictitious point outside the boundary at $x = -h$, denoted by subscript -1, and applying the equation to node n would introduce a second fictitious point outside the boundary at $x = 1 + h$, denoted by subscript $n + 1$. Now at $x = 0$, $y' = P$ and we can express y'_0 in terms of the finite difference approximation to give

$$y_1 - y_{-1} = P/(2h), \text{ thus } \quad y_{-1} = y_1 - P/(2h)$$

Similarly at $x = 1$ we can put y'_n in finite difference form to give

$$y_{n+1} = y_{n-1} + Q/(2h)$$

Using these two equations we can eliminate y_{-1} and y_{n+1} from the set of $n + 1$ linear equations that result from the application of (6.19.2).

In practice a mixture of boundary conditions can arise so that, for example, the boundary condition at $x = 0$ may be specified in terms of y whereas the condition at $x = 1$ may be expressed in terms of y'.

It will be observed that this second order differential equation gives rise to tri-diagonal matrices. Thus the equations can be solved rapidly by using the elimination technique, section 3.13. Program 6.19.1 solves boundary value problems described by a linear second order differential equation.

```
100   REM -------------------------------
110   REM 2ND ORDER B V PROBLEM JP/GL (C)
120   REM -------------------------------
125   REM DEFINE THE FUNCTIONS THAT ARE THE MULTIPLIERS OF Y'', Y' AND Y
126   REM AND THE RIGHT HAND SIDE IN STATEMENTS 130-160 RESPECTIVELY
130   DEF FNA(X)=1+X^2
140   DEF FNB(X)=X
150   DEF FNC(X)=-1
160   DEF FND(X)=X^2
170   PRINT "SOLUTION OF 2ND ORDER"
180   PRINT "BOUNDARY VALUE PROBLEM IN ONE DIMENSION"
190   PRINT:PRINT "AT X = 0"
200   PRINT "ENTER 1 IF Y IS PRESCRIBED OR 0 IF Y' IS PRESCRIBED";
210   INPUT C0
220   PRINT "AT X = 1"
230   PRINT "ENTER 1 IF Y IS PRESCRIBED OR 0 IF Y' IS PRESCRIBED";
240   INPUT C1
250   IF C0=1 THEN 280
260   PRINT "ENTER Y'(0)";
270   GOTO 290
280   PRINT "ENTER Y(0)";
290   INPUT P0
300   IF C1=1 THEN 330
310   PRINT "ENTER Y'(1)";
320   GOTO 340
330   PRINT "ENTER Y(1)";
340   INPUT P1
350   PRINT "ENTER NUMBER OF INTERVALS";
360   INPUT N
370   H=1/N:M=N-C1
380   DIM A(M),D(M),B(M),C(M)
385   REM CALCULATE COEFFICIENTS OF THE TRIDIAGONAL SYSTEM
386   REM D(I) ARE DIAG COEFFICIENTS
390   FOR I=C0 TO M
400      D(I)=FNC(I*H)*H^2-2*FNA(I*H)
410   NEXT I
415   REM A(I) ARE THE COEFFICIENTS ABOVE MAIN DIAGONAL
420   FOR I=C0 TO M-1
```

Program 6.19.1.(continues)

```
430      A(I)=FNA(I*H)+H*FNB(I*H)/2
440   NEXT I
450   IF C0=0 THEN A(0)=2*FNA(0)
455   REM B(I) ARE THE COEFFICIENTS BELOW MAIN DIAGONAL
460   FOR I=C0+1 TO M
470      B(I)=FNA(I*H)-H*FNB(I*H)/2
480   NEXT I
490   IF C1=0 THEN B(N)=2*FNA(1)
495   REM C(I) ARE THE COEFFICIENTS OF THE RIGHT-HAND SIDE
500   FOR I=C0 TO M
510      C(I)=FND(I*H)*H^2
520   NEXT I
530   IF C0=1 THEN C(1)=C(1)-P0*(FNA(H)-H*FNB(H)/2):GOTO 550
540   C(0)=C(0)+2*H*P0*(FNA(0)-H*FNB(0)/2)
550   W=(N-1)*H:IF C1=1 THEN C(N-1)=C(N-1)-P1*(FNA(W)+H*FNB(W)/2):GOTO 570
560   C(N)=C(N)-2*H*P1*(FNA(1)+H*FNB(1)/2)
565   REM SOLUTION OF TRIDIAGONAL SYSTEM BY ELIMINATION
570   FOR I=C0+1 TO M
580      D(I)=D(I)-B(I)*A(I-1)/D(I-1)
590      C(I)=C(I)-B(I)*C(I-1)/D(I-1)
600   NEXT I
610   C(M)=C(M)/D(M)
620   FOR I=M-1 TO C0 STEP -1
630      C(I)=(C(I)-A(I)*C(I+1))/D(I)
640   NEXT I
650   PRINT "SOLUTION:":PRINT
660   PRINT TAB(2);"X";TAB(12);"Y(X)"
670   FOR I=C0 TO M
680      PRINT TAB(2);I*H;TAB(12);C(I)
690   NEXT I
700   END
```

Program 6.19.1

If the boundary values of x are not 0 and 1 then the equation must be reformulated by using an appropriate substitution. For example, if the boundary conditions are specified at a and b then they can be transformed to 0 and 1 by introducing a new variable v as follows.

$$x = (b-a)v + a = rv + a \text{ where } r = b - a$$

Thus $dv/dx = 1/r$

This results in

$$dy/dx = (dy/dv)(dv/dx) = (1/r)(dy/dv)$$

$$d^2y/dx^2 = (d/dx)(dy/dx) = (d/dx)[(1/r)(dy/dv)] = (1/r^2)(d^2y/dv^2)$$

Table 6.19.1. Solution of $(1 + x^2)y'' + xy' - y = x^2$ with
boundary conditions $y(0) = 1$ and $y'(1) = 1$.

```
SOLUTION OF 2ND ORDER
BOUNDARY VALUE PROBLEM IN ONE DIMENSION

AT X = 0
ENTER 1 IF Y IS PRESCRIBED OR 0 IF Y' IS PRESCRIBED?  1
AT X = 1
ENTER 1 IF Y IS PRESCRIBED OR 0 IF Y' IS PRESCRIBED?  0
ENTER Y (0)?  1
ENTER Y'(1)?  1
```

ENTER NUMBER OF INTERVALS? 4 SOLUTION:		ENTER NUMBER OF INTERVALS? 8 SOLUTION:		EXACT SOLUTION
X	Y(X)	X	Y(X)	Y(X)
.25	1.055714	.125	1.020024	1.020006
.5	1.172143	.25	1.055553	1.0555
.75	1.345205	.375	1.106256	1.106153
1	1.570663	.5	1.171654	1.171494
		.625	1.251199	1.250977
		.75	1.344341	1.344057
		.875	1.45057	1.450225
		1	1.569439	1.569036

Consider the solution of

$$(1 + x^2)y'' + xy' - y = x^2$$

subject to the boundary conditions that when $x = 0$, $y = 1$ and when $x = 1$, $y' = 1$. This equation is solved using program 6.19.1 and Table 6.19.1 shows the output from the program using 4 and 8 mesh points. This differential equation has the analytical solution given by

$$y = Ax + B \sqrt{(1 + x^2)} + (2 + x^2)/3,$$

where $A = (1 - \sqrt{2}/2)/3$ and $B = 1/3$. It should be noted that since the errors are of order h^2 we can refine the accuracy of the solution by using Richardson's extrapolation technique.

Two examples of boundary value problems are now briefly discussed. These have been chosen to illustrate situations where program 6.19.1 could not be used.

Example 1

$$y'' + yy' = p(x) \text{ where } y = y(x)$$

Using the finite difference approximations we have

$$(y_{i-1} - 12y_i + y_{i+1}) + [hy_i/2](y_{i+1} - y_{i-1}) = h^2 p(x_i) \quad i = 1, 2, ...$$

This equation gives rise to terms in $y_i y_{i-1}$ and $y_i y_{i+1}$. Consequently, instead a set of linear equations, we obtain a set of non-linear algebraic equations which are much more difficult to solve. This particular example can be more easily solved using the shooting method.

Example 2

$$y^{(iv)} + k^4 y = 0 \text{ where } y = y(x)$$

where k is an unknown quantity to be determined together with values of y at specified values of x. Using central difference approximations at i gives

$$(y_{i-2} - 4y_{i-1} + 6y_i - 4y_{i+1} + y_{i+2}) + (kh)^4 y_i = 0 \quad i = 1, 2, ...$$

This set of equations represents an algebraic eigenvalue problem and these can only be solved using the specialised techniques described in Chapter 10. In addition, the variety of the boundary conditions that may arise with a 4th order differential equation makes it difficult to provide a systematic technique for programming such problems. For example, suppose the boundary conditions at $x = 0$ take the form $y'' = 0$ and $y' = Ay'''$ where A is a known constant. Using the finite difference approximations we have

$$y_{-1} - 2y_0 + y_1 = 0$$

and $$(-y_{-1} + y_1) = h^2 A(-y_{-2} + 2y_{-1} - 2y_1 + y_2)$$

Solving these equations allows the fictitious nodes y_{-2} and y_{-1} to be eliminated.

It should be clear from the above that there is a major problem in writing general purpose software that can cater for all possible boundary conditions. For most situations it is generally easier for the user to develop the finite difference equations and to use the computer to solve the resulting algebraic relationships.

6.20. TWO VARIABLE BOUNDARY VALUE PROBLEMS

Boundary value problems with two independent variables are described by partial differential equations. The following discussion will be restricted to elliptical partial differential equations. Consider the partial differential equation

$$x^2 \partial^2 z/\partial x^2 + \partial^2 z/\partial y^2 = xy^3 \qquad\qquad (6.20.1)$$

subject to boundary condition $z = 0$ on the rectangular domain given by

$$x = 0 \text{ and } 1, y = 0 \text{ and } 1$$

This is an example of a two dimensional boundary value problem and it should be noted that the shape of the boundary, as well as the boundary conditions, must be specified. When using a finite difference approximation the x-y plane is divided up so as to form a mesh and the nodes or mesh points are the points at the intersection of the lines of the mesh. Many shapes of mesh can be used but here only a square mesh finite difference approximation will be considered.

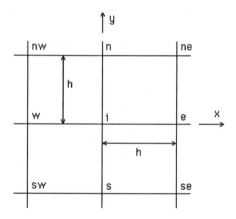

Fig. 6.20.1. Identification of nodes in the region of node i. For example node n lies to the north and node nw lies to the north-west of node i.

The partial derivatives $\partial z/\partial x$ and $\partial z/\partial y$, where $z = f(x,y)$, are denoted by $D_x z$ and $D_y z$ and we require finite difference approximations to them. The finite difference approximations for one dimension are given in Fig. 6.19.1. Here we have two dimensions to consider and instead of using the subscript labelling system of Fig. 6.19.1 we will now use that of Fig. 6.20.1. For example, in the x direction the sequence of nodes $i - 1, i, i + 1$ will be replaced by w, i, e and in the y direction by s, i, n. The letters n, e, s and w are obviously chosen because they identify the nodes lying to the north, east, south and west of node i. Thus, interpreting the definition of $2hD$ of Fig. 6.19.1 in terms of the notation of Fig. 6.20.1 we have

$$2hD_x z_i = z_e - z_w$$

$$2hD_y z_i = z_n - z_s$$

where z_e for example is the value of z evaluated at the point $x_i + h, y_i$.

The expression for the second order mixed derivative of z with respect to x and y, D_{xy} can be obtained by applying the operator D_x to the operator D_y. Thus

$$2hD_x(2hD_y z_i) = (z_n - z_s)_e - (z_n - z_s)_w$$

$$= z_{ne} - z_{se} - z_{nw} + z_{sw}$$

where, for example z_{nw} is the value of z evaluated at $x_i - h, y_i + h$. This approximation is shown diagrammatically in Fig. 6.20.2.

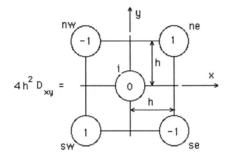

Fig.6.20.2. Central difference operator D_{xy} with errors of order h^2.

Similarly the second partial derivatives of z with respect to x, denoted by D_{xx} and with respect to y, denoted by D_{yy} are obtained from Fig. 6.19.1 as follow

$$h^2 D_{xx} z_i = z_w - 2z_i + z_e$$

$$h^2 D_{yy} z_i = z_n - 2z_i + z_s$$

From these approximations we can obtain an approximation for $\nabla^2 z$ thus

$$h^2 \nabla^2 z = h^2 D_{xx} z_i + h^2 D_{yy} z_i = z_w + z_e + z_n + z_s - 4z_i$$

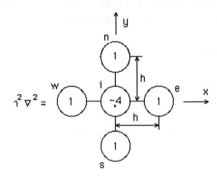

Fig. 6.20.3. Central difference operator ∇^2 with errors of order h^2

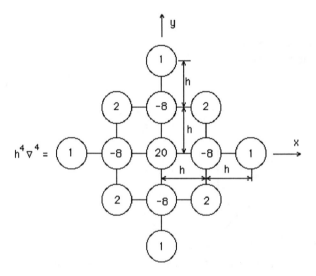

Fig. 6.20.4. Central difference operator ∇^4 with errors of order h^2

This approximation is shown diagrammatically in Fig. 6.20.3. Using a similar approach an approximation for $h^4\nabla^4 z$ can be determined. For completeness it is shown in diagram form in Fig. 6.20.4.

The two dimensional finite difference approximations are now used to solve an equation of the form

$$\nabla^2 z + f(x,y) = 0 \tag{6.20.2}$$

This is Poisson's Equation and if $f(x,y)$ vanishes it becomes Laplace's Equation. These two equations are of great importance and govern many phenomena in electromagnetism, heat conduction and elasticity.

Considering only solutions enclosed by rectangular boundaries the finite difference approximation of Fig. 6.20.3 can be used thus

$$z_w + z_e + z_n + z_s - 4z_i + h^2 f_i = 0$$

where f_i is the value of $f(x,y)$ at the node i and the subscripts n, e, s and w are defined in Fig. 6.20.1. Assuming that z is specified at all points along the boundary then repeated application of this equation to all the internal mesh points gives a set of linear algebraic equation. Whilst any appropriate technique can be used to solve these equations iterative methods are generally preferred. If Jacobi iteration is used, then writing the nth iteration of z_i by $z_i^{(n)}$ gives

$$z_i^{(n+1)} = (z_w^{(n)} + z_e^{(n)} + z_n^{(n)} + z_s^{(n)} + h^2 f_i)/4 \text{ for interior points}$$

The internal points can, of course be taken in any order. In practice this method is not used because convergence is very slow. To improve convergence, we may use the Gauss-Seidel iteration, which, when applied to Poisson's equation it is sometimes called the unextrapolated Liebmann method. In this procedure, the most up to date iterative values are used as soon as they become available. Assuming that the rectangular mesh has been scanned row by row from left to right then at a given stage in this process

$$z_i^{(n+1)} = (z_w^{(n+1)} + z_e^{(n)} + z_n^{(n+1)} + z_s^{(n)} + h^2 f_i)/4$$

The above equation can be rearranged thus

$$z_i^{(n+1)} = z_i^{(n)} + (z_w^{(n+1)} + z_e^{(n)} + z_n^{(n+1)} + z_s^{(n)} - 4z_i^{(n)} + h^2 f_i)/4$$

$$= z_i^{(n)} + R/4$$

R is the current residual and the iteration scheme is one of adding $R/4$ to the current estimate of z_i to obtain a new estimate of z_i.

It has been shown that adding $\omega R/4$, where $1 < \omega < 2$, to the current estimate of z_i increases the rate of convergence. This is the basis of the extrapolated Liebmann or successive-over-relaxation (SOR) scheme. Thus

$$z_i^{(n+1)} = z_i^{(n)} + \omega R/4 \text{ where } 1 < \omega < 2$$

Rearranging this equation gives

$$z_i^{(n+1)} = (1 - \omega)z_i^{(n)} + (z_w^{(n+1)} + z_e^{(n)} + z_n^{(n+1)} + z_s^{(n)} + h^2 f_i)\omega/4$$

Clearly we must now consider how to obtain an optimum value for ω. Experiments have shown that the rate of convergence is extremely sensitive to the value chosen for ω. For example, it has been reported (Carre, 1961) that for a particular problem $\omega = 1.9$ gave a forty fold increase in the rate of convergence when compared with the normal Gauss-Siedel scheme but $\omega = 1.875$ gave a convergence which was only twice as fast as the Gauss-Seidel scheme. Generally the optimum value of ω cannot be easily calculated but for Poisson's equation over a rectangular region it has been shown (Frankel,1950) that the optimum value of ω is given by the smallest root of

$$[\cos(\pi /p) + \cos(\pi /q)]^2\omega^2/16 - \omega + 1 = 0$$

where p and q are the number of mesh intervals in the x and y directions.

Program 6.20.1 solves Laplace's and Poisson's equations using a rectangular mesh. Instead of using the finite difference operator for $\nabla^2 z$ given in Fig. 6.20.3 the one shown in Fig. 6.20.5 is used. This operator allows the mesh interval in all four directions to be different. This feature allows the user to create a rectangular rather than a square mesh and to adjust the position of nodes conform to curved or awkwardly placed boundaries as shown in Fig. 6.20.6.

```
100   REM --------------------------------------------------------
110   REM FINITE DIFFERENCE SOLN OF POISSON'S EQUATION JP/GL (C)
120   REM --------------------------------------------------------
130   PRINT"SOLUTION OF POISSON'S EQUATION"
140   PRINT"Uxx+Uyy+f(x,y)=0"
150   PRINT"DETERMINES U(x,y) AT DISCRETE POINTS"
160   PRINT
170   PI=3.141593
180   READ N,M,H,K
190   PRINT "NUMBER OF INTERNAL NODES = ";N
200   PRINT "NUMBER OF BOUNDARY NODES = ";M
210   DIM A$(N,4),B(M),F(N),V(N),R(N,4)
215   REM READ VALUES OF f(x,y)
220   FOR I=1 TO N
230      READ F(I)
240   NEXT I
245   REM READ VALUES OF BOUNDARY NODES
250   FOR I=1 TO M
260      READ B(I)
270   NEXT I
280   FOR J=1 TO N
290      READ I
310      FOR C=1 TO 4
312         R(I,C)=1
320         READ P$
330         A$(I,C)=P$
340         IF LEFT$(P$,1)<>"B" THEN 380
350         READ R
```

Program 6.20.1.(continues).

```
360         R(I,C)=R
380      NEXT C
400   NEXT J
410   PRINT"APPROXIMATE NUMBER OF INTERVALS"
420   PRINT"      IN x DIRECTION";
430   INPUT NX
440   PRINT"      IN y DIRECTION";
450   INPUT NY
455   REM ESTIMATE OPTIMUM OMEGA FOR EXTRAPOLATED LIEBMANN ITERATION
460   A=(COS(PI/NX)+COS(PI/NY))^2
470   W=(4-SQR(16-4*A))*2/A
480   PRINT"ESTIMATED OPTIMUM VALUE OF OMEGA  IS ";W
490   PRINT"ENTER REQUIRED VALUE OF OMEGA (>=1)";
500   INPUT W
510   PRINT"REQUIRED CONVERGENCE ACCURACY";
520   INPUT Ep
525   REM BEGIN EXTRAPOLATED LIEBMANN ITERATION
530   It=0
540   Flg=0
550   FOR I=1 TO N
560      S=0
562      ZA=H^2*R(I,2)*R(I,4)
564      ZB=K^2*R(I,1)*R(I,3)
570      Z=1/ZA+1/ZB
580      FOR C=1 TO 4
585         REM DETERMINE IF NODE IS A BOUNDARY NODE
590         IF LEFT$(A$(I,C),1)<>"B" THEN 630
600         LE=LEN(A$(I,C)): B$=RIGHT$(A$(I,C),LE-1)
610         BN=VAL(B$): U(C)=B(BN)
620         GOTO 640
630         NN=VAL(A$(I,C)): U(C)=V(NN)
640      NEXT C
650      S=(U(1)/R(I,1)+U(3)/R(I,3))/(K^2*(R(I,1)+R(I,3)))
660      S=S+(U(2)/R(I,2)+U(4)/R(I,4))/(H^2*(R(I,2)+R(I,4)))
670      VT=(S+F(I)/2)*W/Z+(1-W)*V(I)
680      IF ABS(VT-V(I))>Ep THEN Flg=1
690      V(I)=VT
700   NEXT I
710   It=It+1
720   IF Flg=1 THEN 540
730   PRINT"NUMBER OF ITERATIONS = ";It
740   PRINT"NODE";TAB(20);"VALUE"
750   FOR I=1 TO N
760      PRINT I;TAB(20);V(I)
770   NEXT I
780   END
815   REM No of internal nodes,of boundary nodes and the h and k values.
820   DATA 32,6,1,1
825   REM n values of f(x,y) at the internal nodes, 1......n
830   DATA 0,0,0,0,0,0,0,0,0,0,0,0,0,0,0,0,0,0,0,0,0,0,0,0,0,0,0,0,0,0,0,0
831   REM m boundary values at the boundary nodes 1....m
832   DATA 150,0,25,50,75,100
```

Program 6.20.1.(continues).

```
835  REM Next n data statements begin with the node reference number,
836  REM followed by the N,E,S,W internal node nos or the the boundary
837  REM node number prefixed by B eg B4. These are followed by an extra
838  REM value,the dist. to the boundary(=<1) as a proportion of h or k.
840  DATA 1,B6,.236068,2,6,B6,.236068
850  DATA 2,B6,.8284271,3,7,1
860  DATA 3,B6,1,4,8,2
870  DATA 4,B6,1,5,9,3
880  DATA 5,B6,1,4,10,4
890  DATA 6,1,7,11,B6,.8284271
900  DATA 7,2,8,12,6
910  DATA 8,3,9,13,7
920  DATA 9,4,10,B1,1,8
930  DATA 10,5,9,B1,1,9
940  DATA 11,6,12,14,B5,1
950  DATA 12,7,13,15,11
960  DATA 13,8,B1,1,B1,1,12
970  DATA 14,11,15,16,B4,1
980  DATA 15,12,B1,1,17,14
990  DATA 16,14,17,18,B3,1
1000 DATA 17,15,B1,1,19,16
1010 DATA 18,16,19,20,B2,1
1020 DATA 19,17,B1,1,21,18
1030 DATA 20,18,21,23,B2,1
1040 DATA 21,19,22,24,20
1050 DATA 22,B1,1,B1,1,25,21
1060 DATA 23,20,24,28,B2,.8284271
1070 DATA 24,21,25,29,23
1080 DATA 25,22,26,30,24
1090 DATA 26,B1,1,27,31,25
1100 DATA 27,B1,1,26,32,26
1110 DATA 28,23,29,B2,.236068,B2,.236068
1120 DATA 29,24,30,B2,.8284271,28
1130 DATA 30,25,31,B2,1,29
1140 DATA 31,26,32,B2,1,30
1150 DATA 32,27,31,B2,1,31
```

Program 6.20.1

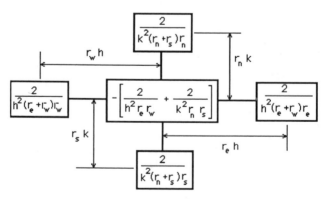

Fig 6.20.5. The operator ∇^2 for unequispaced nodes. If $k = h$ and $r_n = r_e = r_s = r_w = 1$ this operator simplifies to that shown in Fig.6.20.3. The values of $r_n, r_e, r_s, r_w \lesseqgtr 1$.

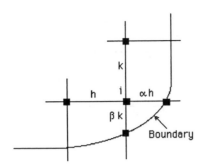

Fig. 6.20.6. Use of unevenly spaced nodes at a boundary.
$r_n = 1, r_e = \alpha, r_s = \beta$ and $r_w = 1$. Note that α and β are less than 1.

Consider the solution of the Laplace equation over a symmetric region with symmetric boundary conditions. By virtue of this symmetry the solution will be symmetric and thus need only be determined over half the region. Fig. 6.20.7 shows such a problem symmetric about the axis A-A. In order to solve this problem using program 6.20.1 the user must define a finite difference mesh and number the internal

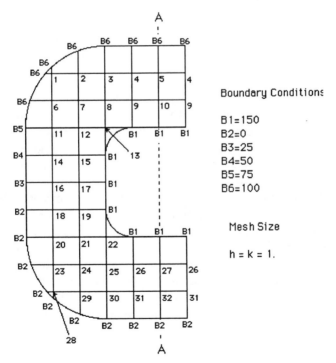

Fig.6.20.7. Boundary and internal node numbering scheme for the temperature distribution problem described in the text. The problem is symmetrical about the line A-A.

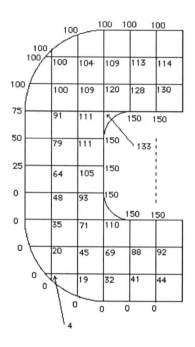

Fig.6.20.8. Solution of the temperature distribution problem described in the text.
The data are the temperatures at the nodes, computed using program 6.20.1.

nodes, in this case the total number of internal nodes is 32. It is necessary to give
corresponding nodes adjacent to the axis of symmetry identical reference numbers
and this requirement is applied to nodes 4, 9, 26, 31 in Fig. 6.20.7: symmetry demands
that these points will have identical values in the solution. The boundary nodes must
also be numbered and points along the boundary that have identical boundary
conditions may as in this case be given identical reference numbers ie B1 to B6.
Fig. 6.20.8 shows the internal nodal values obtained from using program 6.20.1 on
this problem, determined to the nearest integer value. It is not worth while seeking
any improvement in the accuracy of these values by further iteration because the
obtainable accuracy is limited by the finite difference approximation.

Many features of finite difference approximations for partial differential equations have not been discussed in detail. For example, finite difference approximations can be developed with errors of order h^4 rather than of order h^2 and mesh shapes
other than square may be used. Finite difference approximations for skew, polar and
triangular meshes have been published (Salvadori & Baron, 1961).

6.21 CONCLUSIONS AND SUMMARY

In this chapter we have described numerical methods for solving first order ordinary differential equations, systems of first order ordinary differential equations, boundary value problems and partial differential equations. The performance of these methods is problem dependent and consequently they should be used with care. Table 6.21.1 provides the reader with a summary of the methods considered together with advice on how they should be used.

Table 6.21.1. Summary of methods for the solution of differential equations.

Method	Program	Problem type	Order	Self-starting
Euler	6.4.1	Initial value	1st	Yes
Mod Euler	6.5.1	Initial value	1st	Yes
Runge-Kutta	6.7.1	Initial value	1st	Yes
Milne	6.8.1	Initial value	1st	No
Adams	6.8.2	Initial value	1st	No
Hamming	6.9.1	Initial value	1st	No
Runge-Kutta	6.12.1	System of initial value	1st	Yes§
Hermite	6.13.1	Initial value	1st	No¶
Numerov	6.13.2	Initial value	2nd*	Yes
Extrapolation	6.14.1	Initial value	1st	Yes
Shooting	None	Boundary value	2nd	Yes
Finite difference	6.19.1	Boundary value	2nd	Yes
Finite difference	6.20.1	Partial(elliptic)‡	2nd	Yes

§ Algorithms other than Runge-Kutta could be used.
¶ Requires the derivative of $f(x,y)$ to be available.
* Only differential equations of the form $y'' = f(x,y)$.
‡ Program limited to Poisson and Laplace equations.

PROBLEMS

6.1   A radioactive material decays at a rate that is proportional to the amount that remains. The differential equation which models this process is

$$dy/dt = -ky \text{ where } y = y_0 \text{ when } t = t_0$$

Here y_0 represents the mass at the time t_0. Solve this equation for $t = 0$ to 10 and where $y_0 = 50$ and $k = 0.05$. Using

(i) Euler's method, program 6.4.1, with $h = 1, 0.1, 0.01$.
(ii) Euler's modified method, program 6.5.1, with $h = 1, 0.1$.
(iii) Classical Runge-Kutta method, program 6.7.1, with $h = 1$.

Compare your results with the exact solution, $y = 50 \exp(-0.05t)$

6.2 Solve $y' = 2xy$ with initial conditions $y_0 = 2$ when $x_0 = 0$ in the range $x = 0$ to 2 using the following Runge-Kutta methods, all implemented in program 6.7.1, with $h = 0.2$.

      (i) Classic   (iv) Butcher
      (ii) Gill   (v) Ralston.
      (iii) Merson

Note the exact solution is $y = 2\exp(x^2)$.

6.3 Repeat problem 6.1 using the following predictor-corrector methods using $h = 2$ for $t = 0$ to 50:

      (i) Milne's method, program 6.8.1.
      (ii) Adam's method, program 6.8.2
      (iii) Hamming's method, program 6.9.1.

6.4 Solve $(1 + t^2)y' + 2ty = \cos t$ given $y_0 = 0$ at $t_0 = 0$ for $t = 0$ to 5 by the predictor-corrector methods listed in problem 6.3. Use a step $h = 0.5$. The solution is $y = \sin t/(1 + t^2)$

6.5 Express the following second order differential equation as a pair of first order equations

$$xy'' - y' - 8x^3y^3 = 0$$

with the initial conditions $y = 1/2$ and $y' = -1/2$ at $x = 1$. Solve the pair of first order equations using program 6.12.1 in the range 1 to 2 in steps of 0.1. The solution is $y = 1/(1 + x^2)$.

6.6 Use Hermite's method, program 6.13.1, to solve problem 6.1 with $h = 1$, problem 6.2 with $h = 2$ and problem 6.3 with $h = 0.2$.

6.7 Use the extrapolation method, program 6.14.1, to solve the following problems. In each case use 8 divisions.

      (i) $y' = 3y/x$ with initial conditions $x = 1, y = 1$. Determine y when $x = 20$.
      (ii) $y' = 2xy$ with initial conditions $x = 0, y = 2$. Determine y when $x = 2$.

6.8 Use Numerov's method, program 6.13.2, to solve

$$\ddot{y} + \omega^2 y + ky^3 = 0$$

with the initial values $t = 0$, $y = 1, \dot{y} = 0$. Determine y when $t = 5$ for the cases (i) $\omega = 2$ and $k = 0$, and (ii) $\omega = 2$ and $k = 1$. Compare your solution with the following solution which is exact for $k = 0$ and approximate for $k \neq 0$.

$$y(t) = y_0[\cos \omega_1 t + \varepsilon(\cos 3\omega_1 t - \cos \omega_1 t) + \varepsilon^2 (\cos 5\omega_1 t - \cos \omega_1 t)$$

$$+ \varepsilon^3 (\cos 7\omega_1 t - 6 \cos 3\omega_1 t + 5 \cos \omega_1 t) \dots]$$

where $\omega_1^2 = \omega^2 [1 + \frac{3}{4}\sigma - \frac{3}{128}\sigma^2 + \frac{9}{512}\sigma^3 \dots]$

$$\varepsilon = ky_0^2/(32\omega_1^2) \text{ and } \sigma = ky_0^2/\omega^2$$

6.9 Consider the boundary value problem

$$vy'' + 2y' - vy = e^v \text{ given } y(0) = 1/2 \text{ and } y(2) = 3.694528.$$

By using the substitution $x = v/2$ transform the problem to one with the range of the independent variable from 0 to 1. Then solve the equation using the finite difference method implemented in program 6.19.1 with 64 intervals. The exact solution is $y = e^v/2$.

6.10 A semi-circular lamina of radius R and uniform conductivity has its diameter kept at constant temperature 0°C and its circumference of 100°C. The steady state temperature distribution within the lamina is governed by Laplace's equation, i.e., $\nabla^2 \phi = 0$. Use program 6.20.1 to determine a finite difference solution for the steady state temperatures at the nodal points of a square mesh of side $R/4$. Use the node numbering scheme shown in the diagram and note

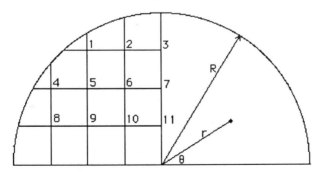

the symmetry of the problem. The exact solution is given by

$$\phi(r,\theta) = (400/\pi) \sum_{n=1}^{\infty} (r/R)^{2n-1}\sin\{(2n-1)\theta\}/(2n-1)$$

where r and θ are defined in the diagram. Evaluate $\phi(r,\theta)$ for the nodal points used in the finite difference analysis and compare the results. You will find it helpful to write a simple program to compute the summations required in the evaluation $\phi(r,\theta)$. Neglect all terms beyond the twentieth in the summations.

6.11 For a twisted solid, prismatic bar of constant cross section the torsion function $\psi(x,y)$ satisfies the equation

$$\nabla^2\psi + 2 = 0$$

subject to the boundary condition $\psi = 0$. Use program 6.20.1 to determine a finite difference approximation for the stress function ψ for a bar of semi-circular cross section. Use the same node numbering system as shown in the diagram for problem 6.10 and observe the symmetry of the problem. Note that the data for this problem is very similar to problem 6.10.

From the theory of elasticity the following relationship can be derived:

$$\frac{\theta G}{TL} = \frac{1}{2\iint \psi\, dx\, dy}$$

where T is the applied torque, θ is the twist, L is the length of the section, G is a constant of the material. The repeated integral is over the surface of the section. Using the values of the stress function obtained from the finite difference solution obtain an estimate for the integral of this function over the surface of the section by using the trapezoidal rule over the surface and hence obtain an estimate for $\theta G/TL$.

7

Approximating functions efficiently

7.1 INTRODUCTION

In this chapter we consider various methods for the approximate evaluation of mathematical functions in one variable. The aim is to show that accurate and efficient approximations can be obtained for any well behaved function for specific values of the independent variable. The term "well behaved" can lead to a lengthy theoretical discussion; here we simply state that in general such a function will be continuous and that its derivative can be evaluated throughout the interval in which it is defined. We will show how the approximations which are used on microcomputers for the most commonly used functions such as $\cos x$, $\sin x$ and $\log x$ are derived. In addition we will indicate how approximations can be obtained for the wide range of functions which, although not so frequently used as the standard functions mentioned above, have practical application in science and engineering.

Any approximation must be such that it provides a specified level of accuracy for a given range of the function's independent variable. The discussion will be initiated by considering the simplest type of approximation to functions and the problems which arise from using it.

7.2 SIMPLE POLYNOMIAL APPROXIMATIONS

We discussed in Chapter 1 how simple polynomial approximations can be obtained for a wide range of functions using Maclaurin's or Taylor's series. If, for example, we apply Maclaurin's series to expand the function e^x we obtain

$$e^x = 1 + x + x^2/2! + x^3/3! + \dots + x^{r+1}/(r + 1)! + \dots \qquad (7.2.1)$$

We may truncate this series by removing all terms after the fifth to obtain the approximation

$$e^x \approx 1 + x + x^2/2! + x^3/3! + x^4/4! \qquad (7.2.2)$$

Any number of terms may be taken and in general the more terms we take the more accurate the approximation.

The basis for any form of polynomial approximation is given by the Weierstrass approximation theorem which states that if $f(x)$ is a continuous function defined on the finite closed interval $[a,b]$ then for any ε there exists an nth degree polynomial $p_n(x)$ such that

$$|p_n(x) - f(x)| < \varepsilon$$

for all x in $[a,b]$. This theorem indicates that there will always exist a polynomial approximation to the function $f(x)$ which will satisfy the required error-bound, i.e. everywhere in the interval $[a,b]$ the error will be less than ε. It is important to note that the polynomial approximation is not necessarily derived from a Taylor's series expansion. These interesting theoretical points are discussed in detail by Simmons (1963).

We will now consider the difficulties that arise in obtaining approximations to specified functions over various intervals and how to deal with them. Fig. 7.2.1 shows the variations in the error of the approximation (7.2.2) in the range $-1 < x < 1$ are not constant over the interval. For small values of x the results obtained by using (7.2.2)

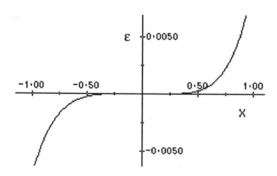

Fig. 7.2.1. Error in the simple polynomial approximation for e^x.

are very accurate and in fact fewer terms would suffice to give the required accuracy. However, towards the end-points of the interval, the error increases. Since our approximation must provide the required accuracy over the whole interval, all terms in the approximation must be included to ensure this. This observation indicates that if we obtained an approximation which provided a more even distribution of the error in the given interval then we should require fewer terms to obtain the same overall accuracy.

The next section describes the theoretical basis for polynomial approximations which minimise the overall error in the approximation.

7.3 THE MINIMAX POLYNOMIAL APPROXIMATION

The minimax polynomial of degree n is that polynomial which minimises the size of the maximum error in the approximation to a given function for all polynomials of degree n. Clearly the minimax polynomial provides not only an approximation satisfying a specified accuracy requirement but in addition provides the best approximation in terms of overall error.

We now establish the key characteristics of minimax polynomials and consequently this will enable us to find specific minimax polynomial approximations. The equi-oscillation or Chebyshev theorem provides a means of indentifying minimax polynomials. This theorem was proved by the Russian mathematician Chebyshev and states that if $p_n(x)$ is a minimax polynomial approximation of degree n to $f(x)$ on the interval $[a,b]$ with error polynomial $\varepsilon_n(x)$, where

$$\varepsilon_n(x) = p_n(x) - f(x)$$

then there are at least $n + 2$ points $x_1, x_2, \dots, x_{n+2}$ in $[a,b]$ with $a < x_1 < x_2 < x_3 < \dots < x_{n+2} < b$ such that

$$(1) \ |\varepsilon_n(x_i)| = E_n \quad i = 1, 2, \dots, n + 2$$

where E_n is the error bound for the approximation.

$$(2) \ \varepsilon_n(x_i) = -\varepsilon_n(x_{i+1}) \quad i = 1, 2, \dots, n + 1$$

In addition there exists only one polynomial of degree less than or equal to n having this property.

To illustrate the meaning of this theorem we will consider a specific example. We shall find the minimax polynomial approximation of the form

$$a_0x + a_1$$

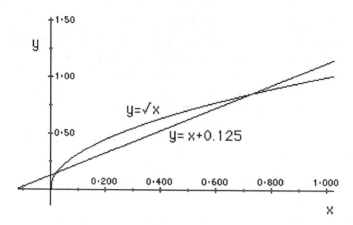

Fig. 7.3.1. This shows that in the range 0 to 1 the error of the
approximating function is a maximum at $x = 0$, $x = 0.25$ and $x = 1$.

for the function $\sqrt{x}$. The error function in this case can be written

$$\varepsilon(x) = a_0 x + a_1 - \sqrt{x} \tag{7.3.1}$$

The function $\sqrt{x}$ and the approximating line (which we are now to determine) can be
drawn on the same graph for the range of the independent variable values $0 \le x \le 1$.
This is illustrated in Fig. 7.3.1. The errors between the $\sqrt{x}$ function and an approximating line

$$y = a_0 x + a_1$$

are easily found by differencing the value of $\sqrt{x}$ and the value of $a_0 x + a_1$ for specific
values of x. It can be seen that the error attains its maximum value in magnitude for
three values of x which are the points where $x = 0$, $x = 1$ and a point where x lies
between these values. Applying the theorem gives

$$\varepsilon(0) = a_1 = E$$
$$\tag{7.3.2}$$
$$\varepsilon(1) = a_0 + a_1 - 1 = E$$

To find the intermediate value of x for which the maximum error is found we must
differentiate $\varepsilon(x)$. This gives

$$\varepsilon'(x) = a_0 - 0.5x^{-0.5} = 0 \tag{7.3.3}$$

Using (7.3.2) gives

$$a_1 = E \text{ and } a_0 = 1$$

Consequently from (7.3.3) the value of x which gives the maximum is

$$x = 0.25$$

Now by the minimax polynomial theorem the sign of the maximum error changes at each successive point. Consequently the sign at the point between 0 and 1 will be negative. Thus we have, when $x = 0.25$

$$a_0 x + a_1 - \sqrt{x} = -E$$

On substituting the known values of a_0, a_1 and $x = 0.25$ we have

$$0.25 + E - 0.5 = -E$$

and therefore
$$E = 0.125$$

This gives us the minimax polynomial approximation

$$\sqrt{x} \approx x + 0.125 \quad \text{in the range } 0 \leq x \leq 1$$

We can apply this method to more complex functions and use higher degree polynomial approximation but the systems of equations we obtain are generally non-linear and become increasingly complex. Thus the procedure we have described is not, in general, suitable for finding the minimax polynomial approximation. The method of Remez, which we describe in Chapter 8, provides a systematic and efficient procedure for determining minimax polynomials.

An alternative method, described in the following section, provides good minimax polynomial approximations and is based on a special type of polynomial known as a Chebyshev polynomial.

7.4 CHEBYSHEV POLYNOMIALS

Consider the function

$$y = \cos nt$$

for $n = 1, 2, 3, \ldots$ and for t in the range $-\pi$ to π. The graphs of these functions are given in Fig. 7.4.1 for $n = 1, 2$ and 3.

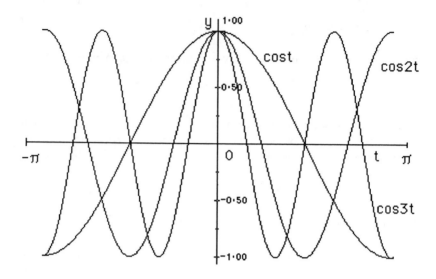

Fig. 7.4.1. Graphs of $y = \cos t$, $y = \cos 2t$ and $y = \cos 3t$ in the range $-\pi \le t \le \pi$.

These functions have alternate maxima and minima of opposite signs and equal magnitude. Consequently this type of function has the characteristics of a minimax polynomial. However the functions as they stand are clearly not polynomials but using the substitution $x = \cos t$ they can be reformulated as polynomials in x. As an example we will consider the function $\cos 2t$. Using a standard identity we have

$$\cos 2t = 2 \cos^2 t - 1$$

On letting $x = \cos t$ we have

$$\cos 2t = 2x^2 - 1$$

To obtain the general form of this modification we apply DeMoivre's theorem. This takes the form

$$\cos nt + i \sin nt = (\cos t + i \sin t)^n \qquad (7.4.1)$$

for any n and where $i = \sqrt{-1}$. To illustrate the use of this theorem we apply (7.4.1) with $n = 4$ so we have

$$\cos 4t + i \sin 4t = (\cos t + i \sin t)^4 \tag{7.4.2}$$

Expanding the right-hand side of (7.4.2) gives

$$\cos 4t + i \sin 4t =$$
$$\cos^4 t + i4 \cos^3 t \sin t - 6 \cos^2 t \sin^2 t - i4 \cos t \sin^3 t + \sin^4 t$$

On equating the real parts on both sides of this equation we obtain

$$\cos 4t = \cos^4 t - 6 \cos^2 t \sin^2 t + \sin^4 t$$
$$= \cos^4 t - 6 \cos^2 t (1 - \cos^2 t) + (1 - \cos^2 t)^2 \tag{7.4.3}$$

If we substitute $x = \cos t$ in (7.4.3) it becomes

$$\cos 4t = 8x^4 - 8x^2 + 1$$

This may be written

$$\cos(4 \cos^{-1} x) = 8x^4 - 8x^2 + 1 \tag{7.4.4}$$

where x lies in the interval 1 to -1. The graph of this function is given in Fig. 7.4.2.

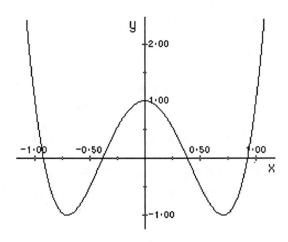

Fig. 7.4.2. The graph of $y = 8x^4 - 8x^2 + 1$ in the range -1 to 1, showing
the oscillatory nature of the function.

We may write Chebyshev functions as

$$T_n(x) = \cos nt \text{ where } x = \cos t \tag{7.4.5}$$

or

$$T_n(x) = \cos(n \cos^{-1}x)$$

There is a much simpler way of generating Chebyshev polynomials which is based on a recurrence formula. This formula is derived by considering the trigonometric identity

$$\cos(n + 1)t + \cos(n - 1)t = 2 \cos nt \cos t$$

In view of (7.4.5) this gives

$$T_{n+1}(x) + T_{n-1}(x) = 2xT_n(x)$$

or $$T_{n+1}(x) = 2xT_n(x) - T_{n-1}(x) \tag{7.4.6}$$

Since $T_0(x) = 1$ and $T_1(x) = x$, this formula allows us to generate any Chebyshev polynomial.

7.5 CHEBYSHEV POLYNOMIAL APPROXIMATIONS

To examine how Chebyshev approximations to functions can be obtained we consider the specific function $f(x)$ where

$$f(x) = x^{n+1}$$

in the interval -1 to 1. We can express $f(x)$ in terms of Chebyshev functions as follows

$$x^{n+1} = a_0T_0(x) + a_1T_1(x) + a_2T_2(x) + \ldots + a_{n+1}T_{n+1}(x) \tag{7.5.1}$$

Taking $p(x)$ as our approximation, where $p(x)$ consists of all the terms on the right-hand side of (7.5.1) up to and including $a_nT_n(x)$, we obtain an error function $\varepsilon(x)$ having the form

$$\varepsilon(x) = x^{n+1} - p(x) = a_{n+1}T_{n+1}(x)$$

Since this error term is a Chebyshev polynomial it has alternate extrema of the form $\pm a_{n+1}$ at $n + 2$ points and consequently $p(x)$ is a minimax polynomial of degree n or less. In general we can extend this to any function $f(x)$ by writing

$$f(x) = \sum_{i=0}^{n} a_i T_i(x) \tag{7.5.2}$$

which expresses $f(x)$ as a sum of minimax polynomials. If we consider the difference between $f(x)$ and $p(x)$ it will consist of the sum of the remaining terms after the $(n+1)$th term. Consequently, if the terms decrease quickly, the term $a_{n+1} T_{n+1}(x)$ will be good approximation to the value of these remaining terms. Hence the error function

$$\varepsilon(x) = f(x) - p(x)$$

will have the equi-oscillation or Chebyshev property and $p(x)$ will be a good approximation to the minimax polynomial. To use such an approximation in practice we must calculate the coefficients a_i. The procedure for doing this is now described.

We will begin by writing (7.5.2) in the more convenient form

$$f(x) = a_0 T_0/2 + \sum_{i=1}^{n} a_i T_i(x) \tag{7.5.3}$$

On multiplying (7.5.3) by $T_k(x)/\sqrt{(1-x^2)}$ and integrating between -1 and 1 we obtain

$$\int_{-1}^{1} \frac{f(x) T_k(x) dx}{\sqrt{(1-x^2)}} = \frac{a_0}{2} \int_{-1}^{1} \frac{T_0(x) T_k(x) dx}{\sqrt{(1-x^2)}} + \sum_{i=1}^{n} a_i \int_{-1}^{1} \frac{T_i(x) T_k(x) dx}{\sqrt{(1-x^2)}} \tag{7.5.4}$$

Now consider the integral

$$\int_{-1}^{1} \frac{T_i(x) T_k(x) dx}{\sqrt{(1-x^2)}}$$

Let $x = \cos t$. Since by definition $T_k(x) = \cos(k \cos^{-1} t)$ we have

$$T_k(\cos t) = \cos kt \text{ and } T_i(\cos t) = \cos it$$

In addition

$$dx = -\sin t \, dt \text{ and } \sqrt{(1-x^2)} = \sin t$$

The range of integration is transformed to 0 to π so the integral becomes

$$\int_{0}^{\pi} \cos kt \cos it \, dt$$

The values of this integral are given by

$$\int_0^\pi \cos kt \cos it \, dt = \begin{cases} 0 & \text{if } k \neq i \\ \pi/2 & \text{if } k = i \neq 0 \\ \pi & \text{if } k = i = 0 \end{cases} \tag{7.5.5}$$

Substituting from (7.5.5) in (7.5.4) gives

$$a_k = \frac{2}{\pi} \int_{-1}^1 \frac{f(x)T_k(x)dx}{\sqrt{(1-x^2)}} \qquad \text{for } k = 0, 1, 2, \dots, n \tag{7.5.6}$$

By putting $x = \cos t$ we have

$$a_k = \frac{2}{\pi} \int_0^\pi \cos kt \cos it \, dt \quad \text{for } k = 0, 1, 2, \dots, n$$

This integral can be evaluated either by using numerical integration methods or by approximating it by a Chebyshev Summation. The latter method will be described in section 7.12.

In the next section we shall show how Chebyshev polynomials can be used to obtain practical and efficient approximations to mathematical functions.

7.6 ECONOMISATION OF POWER SERIES APPROXIMATIONS

Starting with a given power series approximation to a specified function, Chebyshev functions may be used to convert the given series to one expressed in terms of Chebyshev functions valid in the range –1 to 1. The higher order Chebyshev terms will be small in magnitude and consequently a number of these terms can be removed. This leaves a Chebyshev approximation which can then be converted to a new polynomial approximation having fewer terms but giving enhanced accuracy. This procedure is called the economisation of power series approximations and is illustrated by the following example involving the Bessel function. The Bessel function is one of the most important functions of mathematical physics and plays a important role in models of the physical world. The Bessel function of order n is written $J_n(x)$ and has the standard expansion in powers of x thus

$$J_n(x) = (n/2)^n \sum_{k=0}^\infty (-x^2/4)^k \Big/ \{k! \, \Gamma(n+k+1)\}$$

where Γ is the Gamma function and for integer values of z, $\Gamma(z) = (z-1)!$ For non-integer values of z, $\Gamma(z)$ is defined differently and this definition is given in Abramowitz and Stegun (1964). Using the first three terms of this series for $J_n(x)$ with $n = 0$ we obtain the following approximation for $J_0(x)$.

$$J_0(x) \approx 1 - x^2/4 + x^4/64 \tag{7.6.1}$$

We require to find the Chebyshev approximation in the range -1 to 1. To transform (7.6.1) into the form of a Chebyshev function we use the following equations

$$
\begin{aligned}
1 &= T_0 \\
x &= T_1 \\
x^2 &= (T_0 + T_1)/2 \\
x^3 &= (3T_1 + T_3)/2 \\
x^4 &= (3T_0 + 4T_2 + T_4)/8
\end{aligned}
\tag{7.6.2}
$$

....

To simplify the notation we use T_i to denote $T_i(x)$ in (7.6.2) and subsequent equations. Substituting for x^2 and x^4 from (7.6.2) into the series (7.6.1) gives

$$J_0(x) \approx 0.8808593T_0 - 0.1171875T_2 + 0.00195312T_4 \tag{7.6.3}$$

The third term of (7.6.3) can be removed and the approximation to $J_0(x)$ can be converted back into a series in terms of x by using the following table of Chebyshev polynomials.

$$
\begin{aligned}
T_0 &= 1 \\
T_1 &= x \\
T_2 &= 2x^2 - 1 \\
T_3 &= 4x^3 - 3x \\
T_4 &= 8x^4 - x^2 + 1
\end{aligned}
\tag{7.6.4}
$$

This gives

$$J_0(x) \approx 0.8808593 - 0.1171875(2x^2 - 1)$$

$$J_0(x) \approx 0.998047 - 0.234374x^2 \tag{7.6.5}$$

The two term approximation for the Bessel function using the standard definition with $n = 0$ is

$$J_0(x) \approx 1 - x^2/4 \tag{7.6.6}$$

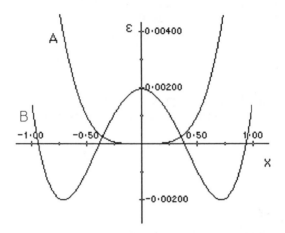

Fig. 7.6.1. Error in the simple polynomial approximation (A) and error in the
economised approximation (B) for the Bessel function. The economised
approximation provides a better overall error performance and
the curve has the characteristic minimax polynomial form.

Fig. 7.6.1 shows how the economised approximation (7.6.5) compares with the
approximation (7.6.6).

7.7 CHANGING THE RANGE OF AN APPROXIMATION

The Chebyshev approximations described are valid for functions in the range −1 to
1. For functions where we require the approximation for a range of the independent
variable different from −1 to 1 we may use a simple transformation to obtain
the standard form. Consider the general interval $[a,b]$. To transform this interval to
$[-1,1]$ so that the ordinary Chebyshev functions can be used we apply the transformation

$$x = (b-a)t/2 + (b+a)/2 \qquad\qquad (7.7.1)$$

or in terms of t

$$t = (2x - b - a)/(b - a) \qquad\qquad (7.7.2)$$

From these we can see that when $x = b$ then $t = -1$ and when $x = a$ then $t = 1$ so the
transformation has achieved the required result.

As an example of the use of the above transform consider the function $\sin x$
defined in the range 0 to 1. Applying (7.7.2) with $a = 0$ and $b = 1$ we find

$$t = 2x - 1 \text{ or } x = (t + 1)/2$$

Hence the function is transformed to

$$\sin\{(t + 1)/2\} \text{ for } t \text{ in the interval } [0,1]$$

After obtaining the required approximation we may return to the original variable by using the relation $t = 2x - 1$.

7.8 COMPUTING ECONOMISED APPROXIMATIONS

In this section we derive a systematic approach to the economisation procedure described in section 7.6. This procedure involves two main steps, firstly transforming the original polynomial approximation to a Chebyshev series and then truncating it and converting back to a polynomial approximation form. Both these processes can be defined precisely and combined to produce a program for the economisation procedure. The first step is achieved by writing (7.4.6) in the form

$$xT_n = 0.5(T_{n+1} + T_{n-1}) \tag{7.8.1}$$

where $xT_0 = T_1$. Writing the polynomial in nested form and applying (7.8.1) allows the polynomial to be converted to the form

$$\Sigma b_k T_k \tag{7.8.2}$$

Having expressed the polynomial approximation in the form (7.8.2) the Chebyshev form is truncated and this must now be expressed as a polynomial in x. This is achieved by using (7.4.6)

$$T_{n+1} = 2xT_n + T_{n-1}$$

and $T_0 = 1$ and $T_1 = x$. It is a simple matter to express (7.8.2) as a polynomial using these equations. Program 7.8.1 provides an efficient implementation of this procedure.

To demonstrate the effectiveness of program 7.8.1 the economised form of a number of functions has been determined and the results are shown in Table 7.8.1.

```
100  REM ----------------------------------------------------------------
110  REM CONVERSION OF POLYNOMIAL TO ECONOMISED CHEBYSHEV FORM GL/JP (C)
120  REM ----------------------------------------------------------------
130  PRINT "ECONOMISATION OF POLYNOMIALS"
140  PRINT"ENTER THE DEGREE OF THE POLYNOMIAL";
150  INPUT N:N1=N+1
160  FOR I =1 TO N1
170      PRINT "ENTER COEFFICIENT OF X^";I-1;
180      INPUT A(I)
190  NEXT I
195  REM CHEBYSHEV COEFFICIENTS ARE CALCULATED AND PLACED IN ARRAY T(I)
200  FOR K=1 TO N1
210      Q(1)=A(N1+1-K)  :Q(2)=T(1)
220      IF K<=2 THEN 270
230      FOR I=2 TO K-1
240          Q(I-1)=Q(I-1)+T(I)/2
250          Q(I+1)=Q(I+1)+T(I)/2
260      NEXT I
270      FOR I =1 TO N1
280          T(I)=Q(I):Q(I)=0
290      NEXT I
300  NEXT K
310  PRINT "CHEBYSHEV COEFFICIENTS ARE:"
320  FOR K=1 TO N1
330      PRINT "T(";K-1;")=";T(K)
340  NEXT K
345  REM  POLYNOMIAL COEFFICIENTS CALCULATED AND PLACED IN ARRAY S1(I)
350  FOR K=1 TO N-1
360      S1(1)=S1(1)+T(N1-K)
370      FOR I=1 TO K
380          IF N1-K<=2 THEN 420
390          S2(I+1)=S2(I+1)+2*S1(I)
400          S3(I)=S3(I)-S1(I)
410          GOTO 440
420          IF I=1 THEN S2(1)=S2(1)+T(1)
430          S2(I+1)=S2(I+1)+S1(I)
440      NEXT I
450      FOR R=1 TO N
460      S1(R)=S2(R):S2(R)=S3(R):S3(R)=0
470      NEXT R
480  NEXT K
490  PRINT "POLYNOMIAL COEFFICIENTS LOWEST DEGREE FIRST ARE:"
500  FOR K=1 TO N
510      PRINT "A(";K-1;")=";S1(K)
520  NEXT K
530  END
```

Program 7.8.1.

Table 7.8.1. Coefficients of the Chebyshev approximation for the functions indicated. The maximum error of these approximations is less than the maximum error in the simple polynomial approximation of degree four in the interval $-1 \le x \le 1$.

Function	Coefficient of :			
	1	x	x squared	x cubed
exp(x)	.9947916	1	.5416667	.1666667
J(x)	.9980469	0	-.234375	0
log(1+x)	.03125	1	-.75	.3333333

7.9 EVALUATING APPROXIMATIONS IN CHEBYSHEV FORM

If the function $f(x)$ is approximated by a Chebyshev series we may write

$$f(x) \approx a_0 T_0/2 + \sum_{i=1}^{n} a_i T_i \qquad (7.9.1)$$

Clearly we require to evaluate the function $f(x)$ using (7.9.1) for any specified value of x in the appropriate range. The obvious way of doing this is to use the polynomial expressions for T_i and evaluate these for the given x and since the a_i are known we can obtain the approximate value. This is a cumbersome process and a more efficient method will now be described.

The calculation of $f(x)$ using (7.9.1) may be achieved using the recurrence relationships which define the $u_i(x)$ as follows

$$u_i(x) = a_i + xu_{i+1}(x) - u_{i+2}(x)$$

and

$$u_{n-1}(x) = a_{n-1} + 2xu_n(x) \qquad (7.9.2)$$

$$u_n(x) = a_n$$

where $i = n-2, n-3, \dots, 1$. Finally the value of the function is given by $u_0(x)$ where

$$u_0(x) = 0.5a_0 + xu_1(x) + u_2(x)$$

This result is easily proved by multiplying the general term of (7.9.2) by T_i. This gives

$$u_i T_i = a_i T_i + 2x u_{i+1} T_i - u_{i+2} T_i$$

or

$$a_i T_i = u_i T_i - 2x u_{i+1} T_i + u_{i+2} T_i \qquad (7.9.3)$$

for $i = n - 2, n - 3, \dots, 1$

In addition, putting $i = n - 1$ in (7.9.3) we have

$$a_{n-1} T_{n-1} = u_{n-1} T_{n-1} - 2x u_n T_{n-1} + u_{n+1} T_{n-1} \qquad (7.9.4)$$

and from the last equation of (7.9.2) we have

$$a_n T_n = u_n T_n \qquad (7.9.5)$$

Adding (7.9.3), (7.9.4) and (7.9.5) gives, on collecting terms in $u_n, u_{n-1}, \dots$ etc. on the right-hand side,

$$a_0 T_0/2 + \sum_{i=1}^{n} a_i T_i = u_n (T_n - 2x T_{n-1} + T_{n-2})$$

$$+ u_{n-1}(T_{n-1} - 2x T_{n-2} + T_{n-3})$$

$$+ u_{n-2}(T_{n-2} - 2x T_{n-3} + T_{n-4})$$

$$+ \dots + u_1(T_1 - x T_0) + u_0 T_0$$

All terms on the right-hand side except the last are zero by virtue of (7.4.6) and since $T_0 = 1$ we have

$$u_0 = a_0 T_0/2 + \sum_{i=1}^{n} a_i T_i$$

So we conclude the recurrence relations (7.9.2) may be used to evaluate the Chebyshev approximation. Program 7.9.1 uses this procedure to evaluate the Chebyshev approximation given the constants $a_0, a_1, a_2, \dots, a_{n-1}$.

We have described above how the Chebyshev polynomial is a good approximation to the minimax polynomial. However polynomial approximations are not the only type of approximations to specified functions. An alternative approximation is often used which is called a rational approximation. This type of approximation is expressed as the ratio of two polynomials. In the next two sections we describe methods for finding rational approximations to specified functions.

```
100    REM------------------------------------------
110    REM CALCULATION OF CHEBYSHEV SERIES GL/JP (C)
120    REM------------------------------------------
130    PRINT "CALCULATION OF CHEBYSHEV SERIES"
140    PRINT "ENTER VALUE OF INDEPENDENT VARIABLE "
150    PRINT "FOR WHICH CHEBYSHEV SERIES IS TO BE CALCULATED"
160    INPUT X
170    PRINT "ENTER ORDER OF CHEBYSHEV SERIES";
180    INPUT N
190    FOR I=0 TO N
200        PRINT "ENTER COEFFICIENT A(";I;")";
210        INPUT A(I)
220    NEXT I
225    REM USE RECURRENCE FORMULA TO FIND VALUE OF CHEBYSHEV SERIES
230    U2=A(N)
240    U1= A(N-1)+2*X*U2
250    FOR I= N-2 TO 1 STEP -1
260        U3=A(I)+2*X*U1-U2
270        U2=U1:U1=U3
280    NEXT I
290    U0=A(0)+U3*X-U2
300    PRINT "VALUE OF CHEBYSHEV SERIES FOR X= ";X;" IS "; U0
310    END
```

Program 7.9.1.

7.10 PADÉ APPROXIMATIONS

Padé approximations to a given function $f(x)$ take the form

$$f(x) = P_m(x)/Q_n(x) \tag{7.10.1}$$

where

$$P_m(x) = a_0 + a_1 x + a_2 x^2 + \ldots + a_m x^m \tag{7.10.2}$$

$$Q_n(x) = b_0 + b_1 x + b_2 x^2 + \ldots + b_n x^n \tag{7.10.3}$$

If $f(x)$ has a Maclaurin expansion then we may obtain the series for $f(x)$

$$f(x) = c_0 + c_1 x + c_2 x^2 + \ldots + c_r x^r + \ldots$$

which may contain an infinity of terms. Now one criteria for choosing our rational approximation to $f(x)$, which we can denote by R_{mn}, is to require that the power series expansion of

$$f(x) - R_{mn} = f(x) - P_m(x)/Q_n(x)$$

$$= \{Q_n(x)f(x) - P_m(x)\}/Q_n(x) \tag{7.10.4}$$

has as many as possible of the leading coefficients of x equal to zero. This will ensure that $f(x) - R_{mn}$ contains only high degree terms in x and hence the difference between $f(x)$ and R_{mn} is small for small x.

From (7.10.2) and (7.10.3) we see that we have $m + n + 2$ constants to determine. These may be obtained by setting the first $m + n + 2$ terms of (7.10.4) equal to zero. In this way we obtain $m + n + 2$ linear equations in $m + n + 2$ unknowns. It is customary to reduce the size of this system by setting $b_0 = 1$.

To illustrate how this method proceeds we find the Padé approximation R_{32} for the function $\tan^{-1}x$ in the range -1 to 1. The Maclaurin expansion for this function is

$$\tan^{-1}x \approx x - x^3/3 + x^5/5$$

Thus we have from (7.10.4)

$$(1 + b_1x + b_2x^2)(x - x^3/3 + x^5/5) \approx a_0 + a_1x + a_2x^2 + a_3x^3 \qquad (7.10.5)$$

If the first six terms in this expansion are to be zero then the coefficients of the first six powers of x up to x^5 on both sides of (7.10.5) must be equal. This gives the equations

For the constant term: $a_0 = 0$
For the x term: $a_1 = 1$
For the x^2 term: $b_1 = a_2$
For the x^3 term: $b_2 - 1/3 = a_3$
For the x^4 term: $-(1/3)b_1 = 0$
For the x^5 term: $-1/5 + b_2/3 = 0$

This is a linear system of six equations with six variables. Solving the system gives

$$b_1 = 0; \ b_2 = 3/5; \ a_0 = 0; \ a_1 = 1; \ a_2 = 0; \ a_3 = 4/15$$

Hence the rational approximation is

$$\tan^{-1}x \approx (x + 0.26667x^3)/(1 + 0.6x^2) \qquad (7.10.6)$$

Fig. 7.10.1 gives the results of computing $\tan^{-1}x$ from (7.10.6) compared with the result using four terms of the standard polynomial expansion for $\tan^{-1}x$. It can be seen from Fig. 7.10.1 that for larger values of x the results become less accurate. Unlike the Chebyshev method there is no redistribution of the error throughout the interval and an improvement to this method based on the use of Chebyshev functions is now described.

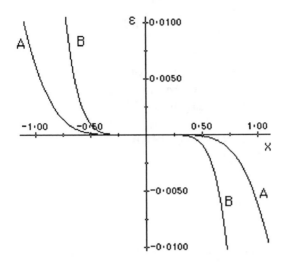

Fig.7.10.1. Errors in the Padé approximation (A) and in the simple
polynomial approximation (B) for the inverse tan function. Clearly
the Padé approximation gives significantly improved accuracy
at the ends of the interval.

7.11 MAEHLY'S METHOD OF RATIONAL APPROXIMATION

Maehly's method uses a Chebyshev approximation to the function $f(x)$ for which
we require a rational approximation and finds this rational approximation in terms
of Chebyshev functions. This provides significantly improved results when
compared with the Padé method and is widely used to obtain efficient function
approximations.

From (7.9.1) we may write the rational approximation in the form

$$f(x) = P_m(x)/Q_n(x)$$

Consequently using Chebyshev approximations for each of the functions $f(x)$,
$P_m(x)$ and $Q_n(x)$ we have

$$c_0/2 + \sum_{k=1}^{p} c_k T_k = (\sum_{s=0}^{m} a_s T_s)/ (\sum_{r=0}^{n} b_r T_r)$$

Consequently

$$P_m(x) - f(x)Q_n(x) = \sum_{s=0}^{m} a_s T_s - \sum_{r=0}^{n} b_r T_r \, (c_0/2 + \sum_{k=1}^{p} c_k T_k) = 0 \qquad (7.11.1)$$

where p is set to provide sufficient equations to determine the rational approximation we require. The coefficients c_0, c_k may be calculated from (7.5.6); this itself may be a significant task but sometimes values are available in tables. On simplification and rearrangement of (7.11.1) we obtain the following system of equations for the a_s and b_r values. These equations arise from the requirement that the first $m + n + 1$ terms in the Chebyshev expansion for (7.11.1) are zero.

$$a_0 = 0.5 \sum_{i=0}^{n} b_i c_i \qquad (7.11.2)$$

$$a_k = 0.5 \sum_{i=0}^{n} b_i (c_{|k-i|} + c_{k+i}) \quad \text{for } k = 1, 2, \dots, m+n \qquad (7.11.3)$$

where $a_k = 0$ for $k > m$. We can write this set of equations for which $a_k = 0$ separately as

$$\sum_{i=0}^{n} b_i (c_{|k-i|} + c_{k+i}) = 0 \quad \text{for } k = m+1, 2, \dots, m+n$$

The details of this derivation are given in Ralston and Rabinowitz (1978). Setting $b_0 = 1$ gives a system of $m + n + 1$ equations in $m + n + 1$ unknowns.

We illustrate the use of this method by obtaining the rational Chebyshev approximation for e^x in the range -1 to 1. For $m = 2$ and $n = 1$ we may represent this approximation by

$$e^x = (a_0 T_0 + a_1 T_1 + a_2 T_2)/(b_0 T_0 + b_1 T_1) \qquad (7.11.4)$$

We set $b_0 = 1$ and we must find the c_j coefficients from (7.5.6). In this case these coefficients can be obtained directly from mathematical tables on page 428 of Abramowitz and Stegun (1964). It can be shown the c_j coefficients for the function e^x are twice the modified Bessel functions of the first kind i.e. $I_n(1)$ for $n = 0, 1, 2, \dots, 6$. This is an unusual situation and the c_j coefficients normally have to be evaluated using the integral expressions (7.5.6) or approximations to them. The values of c_j for e^x are as follows:

$$c_0 = 2.53213; \quad c_1 = 1.130318; c_2 = 0.271495; c_3 = 0.044337$$

$$c_4 = 0.005474; \quad c_5 = 0.000543; c_6 = 0.000045$$

Now $m = 2$ and $n = 1$; therefore we have four equations for the four unknown values a_0, a_1, a_2, b_1. From (7.11.2) we have

$$a_0 = (b_0 c_0 + b_1 c_1)/2$$

and from (7.11.3) we have

$$
\begin{aligned}
\text{for } k = 1 \quad & a_1 = (2b_0 c_1 + b_1(c_0 + c_2))/2 \\
\text{for } k = 2 \quad & a_2 = (2b_0 c_2 + b_1(c_1 + c_3))/2 \\
\text{for } k = 3 \quad & 0 = (2b_0 c_3 + b_1(c_2 + c_4))
\end{aligned}
\qquad (7.11.5)
$$

Noting that $b_0 = 1$, the above system can be solved to give the values:

$$b_1 = -0.320159; \quad a_0 = 1.08512; \; a_1 = 0.681515; \; a_2 = 0.0834568.$$

This gives the approximation in terms of Chebyshev functions

$$e^x \approx (1.08512 T_0 + 0.681515 T_1 + 0.0834568 T_2)/(1 - 0.320159 T_1) \tag{7.11.6}$$

We may rewrite this in terms of x as

$$e^x \approx (1.00166 + 0.681515x + 0.166914x^2)/(1 - 0.320159x) \tag{7.11.7}$$

Fig. 7.11.1 compares the nature of the approximation given by (7.11.7) with the Padé approximation of the same order. Note the error reduction at the end of the interval given by Maehly's method.

The methods considered so far for obtaining the coefficients of the Chebyshev approximations involve the evaluation of difficult integrals to high accuracy. An alternative approach for finding the Chebyshev coefficients is provided by the method of Chebyshev summation. This method provides relatively simple formulae for evaluating the Chebyshev coefficients that do not involve the evaluation of integrals. The formulae provide approximations to the Chebyshev coefficients and the complexity of the formulae depends on the accuracy required.

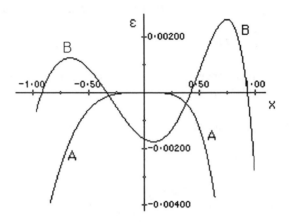

Fig 7.11.1. Error in the Padé approximation (A) and the Chebyshev
rational approximation (B) for e^x

7.12 CHEBYSHEV SUMMATION

This method seeks to provide an approximation to a specified function $f(x)$ in the
range -1 to 1 based on the summation

$$Q(x) = a_0/2 + \sum_{k=1}^{n} a_k T_k(x) \qquad\qquad (7.12.1)$$

The coefficients a_i are obtained by choosing n specific values for x in the range -1
to 1 and solving the system of equations

$$f(x_i) = Q(x_i) \qquad\qquad (7.12.2)$$

where x_i for $i = 0, 1, 2, \ldots, n-1$ are the chosen values.

The choice of the values x_i for $i = 0, 1, 2, \ldots, n-1$ must be made carefully. If we
assume the function $f(x)$ is a nth degree polynomial then a good choice of the x_i would
be those points which make the error zero in the best polynomial approximation to
$f(x)$. Since this is given by the summation (7.12.1) the error is $a_n T_n(x)$. By virtue of
the definition of the Chebyshev coefficients this error is zero at the points

$$x_i = \cos\{(2i + 1)\pi/(2n)\} \qquad\qquad (7.12.3)$$

The system of equations (7.12.2) can now be solved for the a_i. A better method is available to solve the equations which depends on the following orthogonality property. For the points given by (7.12.3) the following result holds

$$T_m(x_i)T_s(x_i) = \begin{cases} n & \text{if } m = s = 0 \\ 0 & \text{if } m \neq s \\ n/2 & \text{if } m = s \neq 0 \end{cases} \tag{7.12.4}$$

We can now use this property to obtain expressions for the coefficients a_i. Multiplying both sides of (7.12.2) by $T_k(x_i)$ and using (7.12.4) gives

$$a_k = (2/n)\sum_{i=0}^{n-1} f(x_i)T_k(x_i) \quad \text{for } k = 0, 1, 2, \dots, n-1. \tag{7.12.5}$$

Values of $T_k(x_i)$ can be calculated using (7.4.6). For $k = 0$ since $T_0(x) = 1$ we have

$$a_0 = (1/n)\sum_{i=0}^{n-1} f(x_i) \tag{7.12.6}$$

These formulae can now be used to determine the coefficients of the Chebyshev summation. The accuracy depends on our choice of n, the number of points that we take. Program 7.12.1 implements the Chebyshev summation method. Table 7.12.1 compares the exact values of the Chebyshev coefficients with those obtained using the Chebyshev summation method.

Table 7.12.1. Comparison of the exact Chebyshev coefficients with those determined by the Chebyshev summation method for the Chebyshev approximation for e^x. The results show very good agreement except when only 3 points are used.

	Exact	Chebychev Summation Approx's.		
k	Coeff's	6 point	4 point	3 point
0	1.266066	1.266066	1.266066	1.266021
1	1.130318	1.130318	1.30315	1.129772
2	0.271495	0.271495	0.271450	0.266021
3	0.044337	0.044337	0.043794	0.5E-10

```
100   REM -------------------------------------------------------------------
110   REM CHEBYSHEV SUMMATION METHOD FOR FUNCTION APPROXIMATION GL/JP (C)
120   REM -------------------------------------------------------------------
125   REM DEFINE FUNCT'N IN 130.MUST BE REAL AND FINITE IN RANGE -1<=X<=1
130   DEF FNA(X)=EXP(X)
140   PRINT "CHEBYSHEV SUMMATION METHOD FOR FUNCTION APPROXIMATION"
150   PRINT "ENTER NUMBER OF POINTS FOR APPROXIMATION";
160   INPUT N
170   DIM A(N),X(N)
175   REM CALCULATE CHEBYSHEV COEFFICIENTS A(0)AND A(1)
180   FOR I=0 TO N-1
190       X(I)=COS((2*I+1)*3.14159265/(2*N))
200       P=X(I)
210       A(0)=A(0)+FNA(P)
220       A(1)=A(1)+P*FNA(P)
230   NEXT I
240   A(0)=A(0)/N
250   A(1)=2*A(1)/N
255   REM CALCULATE REST OF CHEBYSHEV COEFFICIENTS
260   FOR K=2 TO N-1
270       A(K)=0:ORDER =K
280       FOR I=0 TO N-1
290           P=X(I):GOSUB 505
300           A(K)=A(K)+FNA(P)*T
310       NEXT I
320       A(K)=2*A(K)/N
330   NEXT K
340   PRINT "APPROXIMATE CHEBYSHEV COEFFICIENTS ARE:"
350   FOR I =0 TO N-1
360       PRINT "A(";I;")=";A(I)
370   NEXT I
375   REM CALCULATE VALUE OF CHEBYSHEV APPROXIMATION
380   PRINT "DO YOU WISH TO CALCULATE VALUES OF FUNCTION (Y/N)";
390   INPUT A$
400   IF A$="N" THEN STOP
410   PRINT "ENTER VALUE FOR INDEPENDENT VARIABLE";:INPUT X
420   U2=A(N)
430   U1= A(N-1)+2*X*U2
440   FOR I= N-2 TO 1 STEP -1
450       U3=A(I)+2*X*U1-U2
460       U2=U1:U1=U3
470   NEXT I
480   U0=A(0)+U3*X-U2
490   PRINT "VALUE OF CHEBYSHEV SERIES FOR X= ";X;" IS "; U0
500   END
505   REM SUBROUTINE TO CALCULATE CHEBYSHEV COEFFICIENTS
510   B=P:C=1
520   FOR M= 2 TO ORDER
530       T=2*P*B-C
540       C=B:B=T
550   NEXT M
560   RETURN                    Program 7.12.1.
```

7.13 THE METHOD OF HASTINGS

The method of Hastings provides an iterative procedure for finding Chebyshev approximations. The main steps may be described as follows.

Step 1. Solve the following system of equations for the values b_k for $k = 0, 1, 2, \dots, n - 1$

$$b_0 + b_1 x_i + b_2 x_i^2 + \dots + b_{n-1} x_i^{n-1} = f(x) \qquad (7.13.1)$$

where the values x_i, the minimum error points, are obtained from (7.12.3).

Step 2. Obtain the extreme error values by using the approximation provided by step 1. If these do not meet the minimax criteria of having equal values but of alternating sign, obtain improved approximations to the extremal points.

Step 3. Calculate improved approximations to the extremal points using quadratic interpolation.

Step 4. Estimate the deviation from the function $f(x)$ of the approximation using the values provided by step 3. Using weights 1 for the end extremal points and 2 for the others. Let this value be d.

Step 5. Solve the equation system

$$b_0 + b_1 x_i + b_2 x_i^2 + \dots + b_{n-1} x_i^{n-1} = f(x_i) \pm d \qquad (7.13.2)$$

Here the value d is alternately added or subtracted from each equation. Note that since there are $n - 1$ extreme points and n unknowns we add a further equation by taking a further arbitrary value of x. The process may now be repeated from step 2 until the minimax criteria is satisfied to the required level of accuracy.

This technique has some similarity to the method of section 7.12 and consequently no program is given for it.

7.14 CONCLUSIONS AND SUMMARY

Simple polynomial approximations give poor error distributions but the use of Chebyshev functions enables us to provide much improved polynomial apprioximations. The alternative, rational approximation, generally provides further improvements and we have discussed the methods of Padé and Maehly. Furthermore, the utilisation of the Chebyshev approximation in the Maehly method ensures the approximation has a smoother error distribution. The Chebyshev summation method provides an easily programmable technique giving automatically Chebyshev approximations for any well behaved function we desire.

PROBLEMS

7.1 Use program 7.8.1. to find the truncated Chebyshev approximations for the following functions for which the simple Taylor's series approximations are given.

(a) $\sin x = x - x^3/3! + x^5/5! - x^7/7!$

(b) $\sin^{-1}x = x + x^3/2.3 + 1.3x^5/2.4.5 + 1.3.5x^7/2.4.6.7$
$$\text{providing } x^2 < 1 \text{ and } -\pi/2 < \sin^{-1}x < \pi/2$$

(c) $\exp(\sin x) = 1 + x + x^2/2! - 3x^4/4! - 8x^5/5!$

(d) $\log_e(1 + x) = x - x^2/2 + x^3/3 - x^4/4$ providing $-1 < x \le 1$

(e) $\tanh^{-1}x = x + x^3/3 + x^5/5 + x^7/7$ providing $x^2 < 1$

Check your solution for (a) by using it to evaluated sin 1. Compare this value with the "exact" value of sin 1 (obtained from your calculator) and also with the value obtained by using three terms of the given series approximation.

7.2 Use program 7.12.1 to obtain the 8 point Chebyshev approximation for the following functions and calculate the values of $\exp(\sin x)$ when $x = 0.25$ and $x = 1$:

(a) $\exp(2x)$ (b) $\cos x$

(c) $\log(1 + x)$ (d) $\exp(\sin x)$

7.3 The Jacobi elliptic function $sn(u,m)$ can be approximated thus

$$sn(u,m) \approx \sin u - m(u - \sin u \cos u)\cos u/4 \quad \text{if } m \text{ is small}$$

$$sn(u,m) \approx \tanh u + (1 - m)(\sinh u \cosh u - u)/(4 \cosh^2 u)$$
$$\text{if } (1 - m) \text{ is small.}$$

Use program 7.12.1 to obtain an 8 point Chebyshev approximation for these functions and evaluate $sn(0.5,0.01)$ and $sn(0.5,0.99)$. *Hint.* When defining the second expression in the program it is helpful to define the hyperbolic functions in separate definitions. For example, for sinh x,

DEF FNS(X) = (EXP(X)–EXP(–X))/2

and then use these functions in the definition of FNA(X).

8

Fitting functions to data

8.1 INTRODUCTION

In this chapter we are concerned with the process of fitting particular classes of functions to discrete data in an exact or approximate manner. To achieve this we must develop procedures to find the required coefficients to determine the specific function.

We consider first the case of exact data. In this situation one of the following methods can be used to generate the required function:

(1)   Collocation; meaning that a function is chosen and its coefficients adjusted so that it passes through all the data points.

(2)   Osculation; meaning that a function is chosen and its coefficients adjusted so that it both passes through all the data points *and* has given values for some of its derivatives.

(3)   Piece-wise curve fitting; meaning that a function is chosen and its coefficients are adjusted so that different forms of the function fit sub-groups of the data. Thus the data is represented by a series of different forms of the function in a piece-wise manner rather than by a single function. An example of this is a cubic spline.

Considering now the case of data that is known to be inaccurate, then it is more important that our function follows the trend of the data, rather than trying to pass through each and every data point. Thus we wish to adjust the coefficients of a selected function so that it fits the data in an approximate way according to some specified criterion. The least squares criterion is most frequently used but others such as the minimax criterion may also be employed. Each of these procedures and the circumstances under which they may be used will now be examined in some detail.

8.2 COLLOCATION AND INTERPOLATION

Interpolation is the process of estimating an intermediate value from a set of tabulated values. These tabulated values may, for example, be a series of precise empirical results or perhaps values of function which is laborious to calculate.

A collocating function is a suitable basis from which to carry out interpolation because it fits all the given data points. The particular type of collocating function most frequently used for interpolation is the polynomial and some procedures to find a specific polynomial to fit given data are now described.

Many of the procedures used for interpolation evaluate the polynomial, for a particular argument, without explicitly determining its coefficients. This is advantageous because it avoids unnecessary calculation. Some procedures, for example the Newton, Gauss, Stirling and Everett interpolation formulae, make use of difference tables, a device which is an invaluable aid to hand calculation but which can only be used when the data is tabulated with equally spaced arguments. These methods are adequately covered in many standard text books on numerical analysis, for example see Hildebrand (1974) and will not be discussed further; here we are concerned with techniques suited for implementation on a microcomputer.

8.3 LINEAR INTERPOLATION

This is the simplest form of interpolation. The interpolated value is obtained from the equation of the straight line that passes through two tabulated values, one each side of the required value. This straight line is a first degree collocating polynomial. Because information from only the two neighbouring data values is used the accuracy of the interpolation may not be very high and methods which give greater accuracy by making use of more data values will now be considered.

8.4 THE METHOD OF UNDETERMINED COEFFICIENTS

This is a simple method for explicitly obtaining the coefficients of a collocating polynomial without placing any restriction on the spacing of the data. Consider a set of $n + 1$ data points, (x_i, y_i) where $i = 0, 1, \ldots, n$. By substituting these values into the nth degree polynomial

$$y = a_0 + a_1 x + a_2 x^2 + \ldots + a_n x^n \qquad (8.4.1)$$

we can generate $n + 1$ equations

$$y_i = a_0 + a_1 x_i + a_2 x_i^2 + \ldots + a_n x_i^n \qquad i = 0, 1, \ldots, n.$$

These $n + 1$ equations can be expressed in matrix form

$$\mathbf{y} = \mathbf{Za} \tag{8.4.2}$$

where $\mathbf{y}$ and $\mathbf{a}$ are column vectors with components y_i and a_i respectively and $i = 0, 1, \ldots, n$. $\mathbf{Z}$ is given by

$$\mathbf{Z} = \begin{bmatrix} 1 & x_0 & x_0^2 & x_0^3 \ldots x_0^n \\ 1 & x_1 & x_1^2 & x_1^3 \ldots x_1^n \\ \multicolumn{4}{c}{\cdots\cdots\cdots\cdots\cdots\cdots} \\ \multicolumn{4}{c}{\cdots\cdots\cdots\cdots\cdots\cdots} \\ 1 & x_n & x_n^2 & x_n^3 \ldots x_n^n \end{bmatrix}$$

Solving these equations we can determine the coefficients a_i.

Although this procedure for finding the unknown coefficients is straightforward, (8.4.2) may be ill-conditioned and present some of the computational difficulties described in Chapter 3. As an example of the use of this procedure we will fit a fourth degree polynomial to the five data points given in Table 8.4.1.

Table 8.4.1. Data based on the function $y = 1 - 2x - 3x^2 + x^3$.

x	-1	0	1	3	5
y	-1	1	-3	-5	41

Substituting values of x and y into a fourth degree polynomial, (8.4.1), gives the matrix equation

$$\begin{bmatrix} 1 & -1 & 1 & -1 & 1 \\ 1 & 0 & 0 & 0 & 0 \\ 1 & 1 & 1 & 1 & 1 \\ 1 & 3 & 9 & 27 & 81 \\ 1 & 5 & 25 & 125 & 625 \end{bmatrix} \begin{bmatrix} a_0 \\ a_1 \\ a_2 \\ a_3 \\ a_4 \end{bmatrix} = \begin{bmatrix} -1 \\ 1 \\ -3 \\ -5 \\ 41 \end{bmatrix}$$

Solving this set of equation gives the following values for the undetermined coefficients

$$a_0 = 1, a_1 = -2, a_2 = -3, a_3 = 1, a_4 = 0.$$

The data of Table 8.4.1 was generated from a third degree polynomial. Consequently the result that $a_4 = 0$ is to be expected since the fourth degree term is absent. In addition, discarding any one pair of data values and fitting a third degree polynomial to the remaining data will give the same values for the coefficients a_0, a_1, a_2 and a_3. If one further point is discarded only a second degree polynomial can be fitted to the three remaining points and this polynomial will not pass through the discarded point.

Table 8.4.2. Data generated from $y = \cosh \beta x - \cos \beta x - a(\sinh \beta x - \sin \beta x)$.

x	y
0.0	0.0
0.1	0.45614
0.2	1.20901
0.3	1.51248
0.4	1.05185
0.5	0.03937
0.6	-0.94753
0.7	-1.31485
0.8	-0.78975
0.9	0.45702
1.0	2.0

As a further example consider the eleven data points of Table 8.4.2. This data has not been generated from a polynomial but from the following function

where
$$y = \cosh \beta x - \cos \beta x - a(\sinh \beta x - \sin \beta x)$$

$$a = 0.9992245 \text{ and } \beta = 7.85475743.$$

Table 8.4.2 contains 11 items of data but for our illustration we will only choose five data points with x values 0, 0.3, 0.5, 0.7, 1 and generate a fourth degree polynomial that only passes through these values. This polynomial is shown in Fig. 8.4.1.

There are 11 items of data in table 8.4.2, consequently a tenth degree polynomial could be fitted to the data and would pass through every point. However, this would require the solution of eleven linear simultaneous equations and we now consider the practicality of solving this system. The determinant of the coefficient matrix for these equations is approximately 10^{-28} and the condition number $V(A)$, defined in Chapter 3, is equal to about 10^{-29} implying these equations are very ill-conditioned. To appreciate the significance of this value it must be recalled that $V(A) = 1$ means

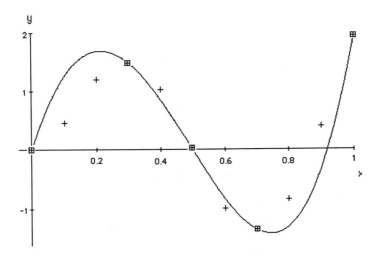

Fig. 8.4.1. Fourth degree polynomial that fits selected points of Table 8.4.2
denoted by □.   The remaining points in this table are denoted by +.

that the equations are perfectly conditioned and $V(\mathbf{A}) = 0$ means that the equations
are singular. In spite of this difficulty the coefficients of the polynomial have been
evaluated using a 9 digit precision microcomputer which provides sufficient
accuracy to give satisfactory values of the polynomial for values of x in the range
[0,1]. However, although the coefficients have been evaluated satisfactorily in this
case we are, none the less, working very close to the limits of numerical accuracy.
Clearly this procedure should be avoided or used with extreme care.

   We will now turn to two procedures for interpolation that do not explicitly
determine the coefficients of the collocating polynomial and thereby avoid the
need to solve a set of linear equations.

8.5 LAGRANGE'S FORMULA

Consider a set of unequally spaced points $(x_i, y_i), i = 0, 1, \ldots, n$. Lagrange's formula
generates a polynomial of degree n from

$$p(x) = \sum_{i=0}^{n} L_i(x) y_i \qquad\qquad (8.5.1)$$

where $L_i(x)$, the Lagrange multiplier function is given by

$$L_i(x) = \frac{(x - x_0)(x - x_1) \dots (x - x_{i-1})(x - x_{i+1}) \dots (x - x_n)}{(x_i - x_0)(x_i - x_1) \dots (x_i - x_{i-1})(x_i - x_{i+1}) \dots (x_i - x_n)}$$

$$L_i(x) = \prod_{\substack{j=0 \\ j \neq i}}^{n} (x - x_j) \Big/ \prod_{\substack{j=0 \\ j \neq i}}^{n} (x_i - x_j) \qquad (8.5.2)$$

From the definition of $L_i(x)$ it is seen that when $x = x_k$ then

$$L_i(x_k) = 0 \text{ for } k \neq i$$
$$L_i(x_k) = 1 \text{ for } k = i$$

Thus $p(x_k) = \sum_{i=0}^{n} L_i(x_k) y_i = y_k.$

Hence the polynomial passes through all points (x_k, y_k) where $k = 0, 1, \dots, n$.

From these formula the coefficients of the polynomial can be obtained explicitly. However, the labour involved is considerable and for this reason the Lagrange polynomial is not normally formulated explicitly, it is evaluated only for specific values of x. Computer programs to evaluate $L_i(x_k)$ and hence $p(x)$ are not presented since Aitken's procedure, which is described in the following section, is computationally more efficient.

8.6 AITKEN'S PROCEDURE

This elegant procedure may be used to interpolate between both even and uneven spaced data points. The procedure is equivalent to generating a sequence of Lagrange polynomials but it is a very much more efficient formulation. The procedure is to compute an interpolated value using successively higher degree polynomials until further increases in the degree of the polynomial give a negligible improvement on the interpolated value.

Suppose we wish, for the purposes of interpolation, to fit a polynomial function to the data (x_i, y_i) where $i = 0, 1, \dots, n$. Consider the following expression

$$p_{0m}(x) = M_{m0} \begin{vmatrix} y_0 & v_0 \\ y_m & v_m \end{vmatrix} \qquad (8.6.1)$$

where m can take any value from 1 to n, $v_m = x_m - x$ and $M_{m0} = 1/(x_m - x_0)$. Equation (8.6.1) represents a first degree polynomial and is equivalent to a linear interpolation using the points (x_0, y_0) and (x_m, y_m). This is easily verified by substituting

$x = x_0$ and $x = x_m$ in (8.6.1) to give $p_{0m}(x_0) = y_0$ and $p_{0m}(x_m) = y_m$. Now consider

$$p_{01m}(x) = M_{m1} \begin{vmatrix} p_{01}(x) & v_1 \\ p_{0m}(x) & v_m \end{vmatrix} \qquad (8.6.2)$$

where m can now take any value from 2 to n and $M_{m1} = 1/(x_m - x_1)$. Equation (8.6.2) represents a second degree polynomial; a fact that can easily be verified since $p_{01}(x)$ and $p_{0m}(x)$ are first degree polynomials and when they are multiplied by v_m and v_1 respectively in (8.6.2), second degree polynomials are generated. Now

$$p_{01m}(x_0) = M_{m1} \begin{vmatrix} y_0 & x_1 - x_0 \\ y_0 & x_m - x_0 \end{vmatrix} = y_0$$

$$p_{01m}(x_1) = M_{m1} \begin{vmatrix} y_1 & x_1 - x_1 \\ p_{0m}(x_1) & x_m - x_1 \end{vmatrix} = y_1$$

$$p_{01m}(x_m) = M_{m1} \begin{vmatrix} p_{0m}(x_m) & x_1 - x_m \\ y_m & x_m - x_m \end{vmatrix} = y_m \quad m = 2, \dots, n$$

Thus the polynomial fits the data at three points (x_i, y_i) where $i = 0$, 1 and m. By repeated use of this procedure higher degree polynomials can be generated and in general

$$p_{012\dots km}(x) = M_{mk} \begin{vmatrix} p_{01\dots(k-1)k}(x) & v_k \\ p_{01\dots(k-1)m}(x) & v_m \end{vmatrix} \qquad (8.6.3)$$

where m can take any value in the range $k + 1$ to n. This is a polynomial of degree $k + 1$ and it can be shown that it fits the data at (x_i, y_i) where $i = 0, 1, \dots, k$ and m. The highest degree polynomial is given by

$$p_{012\dots n}(x) = M_{n(n-1)} \begin{vmatrix} p_{01\dots(n-2)(n-1)}(x) & v_{n-1} \\ p_{01\dots(n-2)n}(x) & v_n \end{vmatrix} \qquad (8.6.4)$$

This polynomial is of degree n and fits all the data. Thus the procedure is to generate a set of n first degree polynomials and, from these, $n - 1$ second degree polynomials are generated and so on until the highest degree polynomial is determined by using all the data. This is illustrated for three data points in Fig. 8.6.1.

When using Aitken's method in practice only the values of the polynomials for specified values of x are computed; the coefficients of the polynomials are not determined explicitly. Furthermore if, for a specified x, the stage is reached when the difference in value between successive degree polynomials is negligible then the

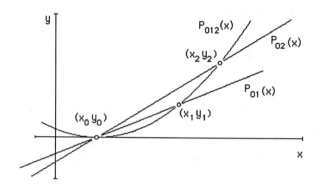

Fig. 8.6.1. Interpolating polynomials generated by Aitken's method for three data points.

procedure can be terminated. This is a particular advantage of this procedure when compared with Lagrange's method; it is not necessary to decide in advance the degree of the polynomial to be used.

```
100   REM ------------------------------
110   REM AITKEN INTERPOLATION JP/GL (C)
120   REM ------------------------------
130   PRINT "INTERPOLATION USING AITKENS PROCEDURE"
140   READ N
150   DIM X(N),Y(N)
160   FOR I=0 TO N-1
170       READ X(I)
180   NEXT I
19    FOR I=0 TO N-1
200       READ Y(I)
210   NEXT I
220   PRINT "VALUE OF X FOR ESTIMATE OF f(X)";
230   INPUT X
240   PRINT "ESTIMATE OF f(";X;"):-"
250   FOR J=0 TO N-2
260       PRINT "USING POLYNOMIAL OF DEGREE ";J+1
265       REM GENERATE INTERPOLATING POLYNOMIALS
270       FOR I=J+1 TO N-1
280           Y(I)=((X-X(J))*Y(I)-(X-X(I))*Y(J))/(X(I)-X(J))
290           PRINT Y(I);" ";
300       NEXT I
310       PRINT
320   NEXT J
330   END
340   DATA 6: REM NO OF DATA POINTS
350   DATA -1,1,2,5,6,7: REM VALUES OF X
360   DATA 12.5,-8.5,-151,-9746.5,-23231,-48743.5:REM VALUES OF Y
```

Program 8.6.1.

Program 8.6.1 generates a sequence of estimates for the interpolated value of the function for a specified value of x. To show the gradual improvement in the estimate the program continues until the highest degree polynomial has been generated. It would be a simple matter to modify the program so that when the difference in value between successive degree polynomials is below a preset value then the procedure can be terminated.

To illustrate the use of Aitken's procedure the data of Table 8.4.1 is used to obtain a sequence of estimates for the function when $x = 2$. The values for the polynomials are:

$$P_{01} = \underline{5}$$
$$P_{02} = -4 \qquad P_{012} = \underline{-13}$$
$$P_{03} = -4 \qquad P_{013} = -1 \qquad P_{0123} = \underline{-7}$$
$$P_{04} = 20 \qquad P_{014} = 11 \qquad P_{0124} = -7 \qquad P_{01234} = \underline{-7}$$

It is seen that there is no need to generate the last polynomial, which is fourth degree, because there is no change from the previously calculated values using a third degree polynomial. This accords with the fact already noted; the data of Table 8.4.1 was generated from a third degree polynomial. In the above layout the values along the main diagonal, which are underlined, converge to the required result.

Table 8.6.1. Data for problem described in text.

x	y
-1	12.5
1	-8.5
2	-151
5	-9746.5
6	-23231
7	-48743.5

As a further example of the use of Aitken's procedure, the data of Table 8.6.1 was used to estimate the value of the unknown function when $x = 3$. The output from the program 8.6.1 is shown in Table 8.6.2. The data of Table 8.6.1 was generated from the function

$$y = 1 - 2x + 3x^2 - 6x^3 - 2x^4 - 2.5x^5$$

Table 8.6.2 shows that Aitken's interpolation for $x = 3$ gives a value of $y = -909.5$. Substituting $x = 3$ in the above equation confirms the exact value of y is -909.5.

Table 8.6.2. Aitken's method converges to the required value,
giving high accuracy.

```
INTERPOLATION USING AITKENS PROCEDURE
VALUE OF X FOR ESTIMATE OF f(X)?  3

ESTIMATE OF f( 3 ):-
USING POLYNOMIAL OF DEGREE  1
-29.5  -205.5  -6493.5  -13269.5  -24365.5
USING POLYNOMIAL OF DEGREE  2
-381.5  -3261.5  -5325.5  -8141.5
USING POLYNOMIAL OF DEGREE  3
-1341.5  -1617.5  -1933.5
USING POLYNOMIAL OF DEGREE  4
-789.5  -749.5
USING POLYNOMIAL OF DEGREE  5
-909.5
```

8.7 OSCULATING FUNCTIONS

The osculating function, like the collocating function, fits all the prescribed data points. In addition, the derivatives of the osculating function must equal given values at specified points. This method of curve fitting has advantages when one or more values of the derivatives are known at specific points. This situation might arise in science and engineering where the circumstances of the problem demand that a derivative is zero. For example a beam will have a known deflection and slope at any point where it is rigidly clamped.

Table 8.7.1. Data for problem described in text.

x	y	dy/dx
0	2	
1	1	2
2	8	

The simplest way of determining the coefficients of an osculating polynomial is by using the method of undetermined coefficients. Suppose we wish to fit a polynomial to three points from the set of data given in Table 8.7.1. Let

$$y = a_0 + a_1 x + a_2 x^2 + a_3 x^3 \tag{8.7.1}$$

Thus, differentiating with respect to x gives

$$dy/dx = a_1 + 2a_2x + 3a_3x^2 \tag{8.7.2}$$

The three pairs of values of x and y and the value $dy/dx = 2$ at $x = 1$ given in Table 8.7.1 are substituted into (8.7.1) and (8.7.2) respectively to give the following system of equations.

$$\begin{bmatrix} 1 & 0 & 0 & 0 \\ 1 & 1 & 1 & 1 \\ 1 & 2 & 4 & 8 \\ 0 & 1 & 2 & 3 \end{bmatrix} \begin{bmatrix} a_0 \\ a_1 \\ a_2 \\ a_3 \end{bmatrix} = \begin{bmatrix} 2 \\ 1 \\ 8 \\ 2 \end{bmatrix}$$

Solving these equations gives

$$a_0 = 2, \ a_1 = -3, \ a_2 = 1, \ a_3 = 1$$

The additional information given by the derivative has enabled us to fit a cubic rather than a quadratic polynomial to this data. This provides an improved fit.

8.8 PIECE-WISE CURVE FITTING: THE CUBIC SPLINE

Only one example of a piecewise curve fitting technique will be discussed; the use of the 'cubic spline'. The cubic spline is a smooth curve that passes through a set of data points and thus it is an alternative to a collocating function. The term 'spline' is derived from the name of a device traditionally used by draftsmen to join points on a drawing by a smooth curve. It consisted of a steel strip, held in position by weights. Since a steel strip is subject to the laws of elastic deflexion its shape, between adjacent weights, is a cubic polynomial function. The polynomials are different in each interval but at each fixed point there is no discontinuity in slope or curvature. The cubic spline is mathematically generated to mimic this behaviour.

We will now outline the development of a procedure to generate a series of cubic polynomials that pass through a set of n data points and have continuity of slope and rate of change of slope (curvature) at the $n-2$ internal points. Between (x_i, y_i) and (x_{i+1}, y_{i+1}) let the third degree polynomial be

$$y = a_i + b_i(x - x_i) + c_i(x - x_i)^2 + d_i(x - x_i)^3 \tag{8.8.1}$$

At one end of the interval, $x = x_i$, $y = y_i$ and at the other end, $x = x_{i+1}$, $y = y_{i+1}$. Substituting these conditions in (8.8.1) we obtain

$$y_i = a_i \tag{8.8.2}$$

$$y_{i+1} = a_i + b_ih_i + c_ih_i^2 + d_ih_i^3 \tag{8.8.3}$$

where $h_i = x_{i+1} - x_i$. Since we require the slope and the rate of change of slope or curvature at the end of each polynomial to match that of its neighbour we must differentiate (8.8.1) with respect to x twice to obtain expressions for y' and y''. If r_i is the rate of change of slope at i then from the expression for y'' at $x = x_i$ we can obtain

$$c_i = r_i/2 \tag{8.8.4}$$

and at $x = x_{i+1}$ we obtain

$$d_i = (r_{i+1} - r_i)/6h_i \tag{8.8.5}$$

Substituting in the expressions for a_i, c_i and d_i in (8.8.3) leads to

$$b_i = (y_{i+1} - y_i)/h_i - h_i(2r_i + r_{i+1})/6 \tag{8.8.6}$$

We now have expressions for the coefficients b_i, c_i and d_i in terms of the curvature at i and $i + 1$. However, as yet r_i and r_{i+1} are unknown but they may be determined as follows. We have, when $x = x_i$ is substituted in the derivative of (8.8.1)

$$y'_i = b_i \tag{8.8.7}$$

Considering now the previous interval, the subscript i in (8.8.1) is replaced by $i - 1$. Differentiating this modified form of (8.8.1) and putting $x = x_i$ we have

$$y'_i = b_{i-1} + 2c_{i-1}h_{i-1} + 3d_{i-1}h_{i-1}^2 \tag{8.8.8}$$

where

$$h_{i-1} = x_i - x_{i-1}$$

Substituting for b_i, b_{i-1}, c_{i-1} and d_{i-1}, in (8.8.7) and (8.8.8) and equating the two expressions for y'_i because of the requirement that the slopes of the two adjoining cubics at the point (x_i, y_i) are identical leads to

$$h_{i-1}r_{i-1} + 2(h_i + h_{i-1})r_i + h_i r_{i+1}$$
$$= 6[(y_{i+1} - y_i)/h_i - (y_i - y_{i-1})/h_{i-1}] \tag{8.8.9}$$

where $i = 2, 3, \ldots, n - 1$. This equation can be applied to each internal data point where two cubic polynomials join, giving a set of $n - 2$ equations. However, two further equations are required, in order that we can solve for the n unknown curvatures, and assumptions must be made as to the curvature of the polynomials at each of the extreme ends of the complete set of data where there is no neighbouring polynomial. This assumption will effect the shape of the polynomial near the end points but the effects will diminish rapidly towards the middle of the data.

Let the extreme points be labelled 1 and n. One possible assumption is that the curvature at these extreme points is zero; i.e. $r_1 = 0$ and $r_n = 0$. It is then only necessary to solve the $n - 2$ equations of (8.8.9) to obtain values for r_i where $i = 2, 3, ... , n - 1$. An alternative assumption, which we shall use here, is to make r_1 a linear extrapolation from r_3 and r_2, and r_n a linear extrapolation from r_{n-2} and r_{n-1}. This latter assumption gives

$$h_2 r_1 - (h_1 + h_2) r_2 + h_1 r_3 = 0 \qquad (8.8.10)$$

$$h_{n-1} r_{n-2} - (h_{n-2} + h_{n-1}) r_{n-1} + h_{n-2} r_n = 0 \qquad (8.8.11)$$

Using (8.8.9), (8.8.10) and (8.8.11) we obtain a set of n simultaneous equations which must be solved to obtain the n unknown curvatures. Once these curvature coefficients are obtained the coefficients of the polynomials can be deduced from (8.8.2), (8.8.4), (8.8.5) and (8.8.6).

Irrespective of the assumption made regarding the curvature at the ends of the data set, it is necessary to solve a set of simultaneous equations which may be large. However, the matrix of coefficients is diagonally dominant and banded, (but not tri-diagonal) so that an iterative method can be used to advantage since convergence is guaranteed for diagonally dominant systems.

Program 8.8.1 generates the coefficients of the cubic spline fit for given data and then offers the user the facility to estimate a value for y for a chosen value of x.

```
100   REM -----------------------
110   REM CUBIC SPLINE JP/GL (C)
120   REM -----------------------
130   PRINT "COEFFICIENTS OF CUBIC SPLINE":PRINT
140   DEF FNA(X)=((C(I0,3)*X+C(I0,2))*X+C(I0,1))*X+C(I0,0)
150   READ N
160   DIM X(N),Y(N),W(N),H(N),A(N,N+1),C(N,3)
170   FOR I=1 TO N
180       FOR J=1 TO N+1
190           A(I,J)=0
200       NEXT J
210   NEXT I
220   FOR I=1 TO N
230       READ X(I)
240   NEXT I
250   FOR I=1 TO N
260       READ Y(I)
270   NEXT I
```

Program 8.8.1. (continues).

```
280   FOR I=1 TO N-1
290      H(I)=X(I+1)-X(I)
300   NEXT I
305   REM FILL ARRAY A. PLACE RHS VECTOR IN  COLUMN N+1 OF A
310   A(1,1)=H(2):A(1,2)=-(H(1)+H(2)):A(1,3)=H(1)
320   FOR I=2 TO N-1
330      A(I,I-1)=H(I-1):A(I,I)=2*(H(I-1)+H(I)):A(I,I+1)=H(I)
340   NEXT I
350   A(N,N-2)=H(N-1):A(N,N-1)=-(H(N-2)+H(N-1)):A(N,N)=H(N-2)
370   FOR I=2 TO N-1
380      A(I,N+1)=(Y(I+1)-Y(I))/H(I)-(Y(I)-Y(I-1))/H(I-1)
390      A(I,N+1)=A(I,N+1)*6
400   NEXT I
405   REM SUB AT 780 SOLVES EQU'NS BY GAUSS-SEIDEL ITER'N BASED ON
406   REM PROGRAM 3.16.2.(NOT LISTED). TAKES ADVANTAGE OF BANDED
407   REM FORM OF A TO SIMPLIFY ITER'N. RETURNS SOL'N IN W
410   GOSUB 780
415   REM COMPUTE COEFFICIENTS OF CUBIC SPLINE
420   FOR I=1 TO N-2
430      IF I=2 THEN 460
440      C(I,3)=(W(I+1)-W(I))/(6*H(I)):C(I,2)=W(I)/2
450      C(I,1)=(Y(I+1)-Y(I))/H(I)-(2*H(I)*W(I)+H(I)*W(I+1))/6:C(I,0)=Y(I)
460   NEXT I
470   FOR I=1 TO N-2
480      IF I=2 THEN 570
490      IF I=1 THEN K=I+2:GOTO 520
500      IF I=N-2 THEN K=N:GOTO 520
510      K=I+1
520      PRINT "FOR THE RANGE X=";X(I);"TO";X(K)
530      FOR J=3 TO 0 STEP -1
540      PRINT "COEFF OF  (X-";X(I);")^";J;"= ";C(I,J)
550      NEXT J
560      PRINT
570   NEXT I
580   PRINT "EVALUATION OF f(X) FOR GIVEN X"
590   PRINT "VALUE OF X";
600   INPUT X
610   I=3
620   IF X<=X(I) THEN 670
630   I=I+1
640   IF I<N+1 THEN 620
650   PRINT "OUT OF RANGE"
660   GOTO 710
670   I0=I-1
680   IF I0=2 THEN I0=1
690   IF I0=N-1 THEN I0=N-2
700   PRINT " X= "; X;" Y= ";FNA(X-X(I0))
710   PRINT "ANOTHER EVALUATION (Y/N)";
720   INPUT A$
730   IF A$="Y" THEN 590
740   END
750   DATA 7: REM NUMBER OF DATA POINTS
760   DATA 1,2,3,4,6,8,10: REM VALUES OF X
770   DATA 10,5,3.333,5,1.667,1.25,1:REM VALUES OF Y
```

Program 8.8.1.

8.9 USING THE CUBIC SPLINE

The collocating function performs satisfactorily when interpolating a well-behaved function, or where the number of data points available greatly exceeds the expected number of inflection in the original function. However, the smoothness of the cubic spline is of considerable advantage when the these conditions are not met. Figs 8.9.1 and 8.9.2 show examples of collocating polynomials and cubic splines fitting the same data. Fig. 8.9.1 illustrates a spline and a tenth degree polynomial fitted to the data of Table 8.4.2: it is seen that on this scale the two curves are coincident so that using a spline appears to offer no particular advantage. However, determining the coefficients of the tenth degree polynomial requires the solution of a set of ill-conditioned equations whereas in the case of the spline fit we are solving a banded matrix. The second example, Fig. 8.9.2, shows the result of fitting a spline and a sixth degree polynomial to the data of Table 8.9.1; here the spline appears to give a much better fit. It must be stressed that we can never be certain that one fit is better than another unless the original function is known. However, with this data, it does appear the spline fit is more realistic.

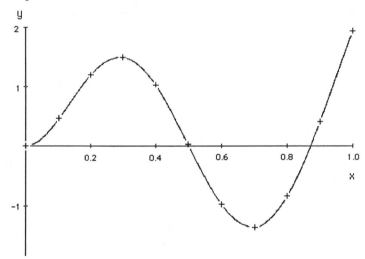

Fig. 8.9.1. Cubic spline and 10th degree polynomial fit to the 11 data points of Table 8.4.2.
Note that the two curves cannot be separated at this resolution.

Table 8.9.1. Data for Fig. 8.9.2.

x	1	2	3	4	6	8	10
f(x)	10	5	3.333	5	1.667	1.25	1

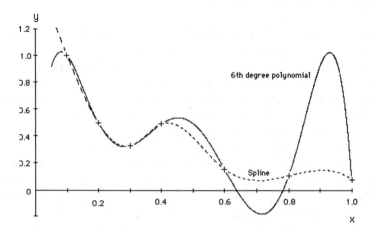

Fig.8.9.2. Cubic spline and 6th degree polynomial passing through the data points of Table 8.9.1.

A further advantage of the spline arises when derivatives or integrals of the function from which the data is taken are required. Using the collocating polynomial for this purpose is unsatisfactory because

(1)   Algorithms such as Aitken's do not compute the coefficients of the collocating polynomials explicitly so that the function cannot be differentiated or integrated. The method of undetermined coefficients would have to be used with all the potential numerical problems.

(2)   If the collocating polynomial is oscillating spuriously, as we might suspect in the example shown in Fig. 8.9.2, the derivatives of the polynomial will be quite meaningless.

To illustrate the use of splines for differentiation and integration, the following estimates were obtained from a cubic spline fit to the data of Table 8.4.2. Since this data was generated from the function defined in (8.4.3) the exact values were obtained by direct differentiation and integration.

When $x = 0.14$, $y' = 7.9516$ (Exact value $= 7.9562$)
When $x = 0.46$, $y' = -10.657$ (Exact value $= -10.659$)
When $x = 0.14$, $y'' = -3.5473$ (Exact value $= -6.3209$)
When $x = 0.46$, $y'' = -23.360$ (Exact value $= -24.404$)

Integral from 0 to 1 $= 0.2548$ (Exact value $= 0.2544$)

With the exception of the second derivative when $x = 0.14$ the accuracy obtained is quite satisfactory for many applications. Furthermore, widely spaced intervals were used involving only eleven data points. Using the Trapezoidal rule with $h = 0.1$ gives the value 0.2674 for the integral and using the central difference formula for differentiation with $h = 0.1$ gives the value 6.6269 for the derivative when $x = 0.14$. Neither of these estimates are as good as those using splines and in general splines should be used for differentiation and integration when the data is discrete and widely spaced. The methods of Chapters 4 and 5 are more appropriate when the data is either closely spaced or in the form of a function. For a more detailed discussion of splines and their application see Greville (1969).

8.10 APPROXIMATING FUNCTIONS

It is often necessary to construct the best smooth curve through a set of data points which are known to contain random errors. The data itself might have been obtained experimentally, it might be historical data from several different sources or it may be data that has been taken from an existing graph of an unknown function and thus subject to human or machine sampling errors. In such a situation it would not be helpful to try and fit a collocation function or spline to the data since these functions would pass through every data point with the implicit assumption that all points were equally correct. Indeed it may be impossible to fit this type of function to the data since it is not uncommon for experimental data to have more than one value of the dependent variable (each subjected to differing amounts of error) for a single value of the independent variable.

The reader may ask what is meant by the best smooth curve. How smooth is smooth!? No definitive answer can be given to that particular question, the user must make such a decision for himself based upon his requirements and on a knowledge of a likely mathematical model which will describe the physical, economic or sociological system, etc, from which the data has been taken. Thus we must proceed with extreme caution when fitting approximate functions to data if we are not to make deductions from the approximating function that are wholly inappropriate.

The reader might also ask what is meant by "best" in the context of a best smooth curve. Again there is not a unique answer to this question. However criteria can be proposed in terms of which the curve can be judged. It may then be said that the curve which is generated is best according to the particular objectives of these criteria. In the following sections two of these criteria, the 'Least Squares' and the 'Minimax' will be examined in detail.

8.11 WEIGHTED LEAST SQUARES CRITERION

The most widely used criterion for finding the best curve to fit a set of data is the Least Squares. This criterion requires that sum of the squares of the differences between the approximating function and the data should be a minimum.

Assuming we have $n + 1$ sets of data (x_i, y_i) where $i = 0, 1, \ldots, n$ and we wish to generate a function $y = f(x)$ that approximately fits this data using a least squares criterion, then the difference between the function and a particular data point i is

$$d_i = y_i - f(x_i) \tag{8.11.1}$$

This difference is also called the residual. The sum of all these differences squared, S, is

$$S = \sum_{i=0}^{n} d_i^2 = \sum_{i=0}^{n} [y_i - f(x_i)]^2 \tag{8.11.2}$$

It is this sum that we wish to minimise. Before proceeding with the mathematical detail of the process, a further factor must be considered. Is all the data known to be equally reliable; that is, is it subject to the same level of random error? If the data is experimental or historic this may not be the case. For example, historic data could be taken from more than one source and each may have collected the data with differing degrees of precision. Thus we may wish to introduce an allowance, called a weighting factor, to compensate for this variation in the quality of the data. Data considered to have a high level of reliability is given a high weighting factor, less reliable data is given a lower weighting factor. Only the user, with his or her knowledge of the nature and likely reliability of the data can decide the actual numerical values to ascribe to the weightings and, by the nature of the problem, the values used will only be reasonable estimates. For this reason the weightings should be kept to simple integer quantities such as 1, 2, 5 and 10 for example. Letting the weighting factor associated with (x_i, y_i) be w_i then (8.11.2) is modified to become

$$S = \sum_{i=0}^{n} w_i [y_i - f(x_i)]^2 \tag{8.11.3}$$

The function $f(x)$ will be a specific function of the coefficients $a_0, a_1, \ldots, a_m$, where the number of coefficients, $m + 1$, is less than the number of data points, $n + 1$. Thus, (8.11.3) can be written

$$S = \sum_{i=0}^{n} w_i [y_i - f(x_i, a_0, a_1, \ldots, a_m)]^2 \tag{8.11.4}$$

These coefficients, which at this stage are unknown, must be adjusted to make S a minimum. The weights w_i for $i = 0, 1, 2, \ldots, n$ are all positive and if the weight associated with a particular point is zero then it is ignored. A necessary condition for the minimisation of (8.11.4) is that the partial derivatives of the expression with respect to each coefficient in turn must be zero. Thus the values of the coefficients to make S a minimum may be determined by solving the set of simultaneous equations

$$\partial S/\partial a_j = 0 \quad j = 0, 1, 2, \ldots ,m \tag{8.11.5}$$

This condition may also be sufficient for minimisation but this will depend on the specific function $f(x, a_0, a_1, \ldots ,a_m)$ used this function may be a linear or non-linear function of the coefficients a_j. For example the polynomial function

$$y = a_0 + a_1 x + a_2 x^2 + \ldots + a_m x^m$$

is a linear function of the coefficients a_j although it is a non-linear function of x. The solution of (8.11.5) will now be investigated for both linear and non-linear approximating functions.

8.12 LEAST SQUARES USING A LINEAR COMBINATION OF FUNCTIONS

Let $f(x)$ be a linear combination of the functions $g_j(x)$ where $j = 0, 1, \ldots ,m$. Thus $f(x)$ may be written

$$f(x, a_0, a_1, \ldots ,a_m) = a_0 g_0(x) + a_1 g_1(x) + \ldots + a_m g_m(x) \tag{8.12.1}$$

Substituting this expression into (8.11.4) gives

$$S = \sum_{i=0}^{n} w_i [y_i - a_0 g_0(x_i) - a_1 g_1(x_i) - \ldots - a_m g_m(x_i)]^2$$

For this fit, employing a linear combination of functions, the solution obtained by equating the first order partial derivatives to zero is also a minimum. Thus we have for a minimum

$$\partial S/\partial a_j = -2\sum_{i=0}^{n} w_i [y_i - a_0 g_0(x_i) - a_1 g_1(x_i) - \ldots - a_m g_m(x_i)]g_j(x_i) = 0$$

where $j = 0, 1, 2, \ldots ,m$

$$\sum_{i=0}^{n} w_i g_0(x_i)g_j(x_i)a_0 + \sum_{i=0}^{n} w_i g_1(x_i)g_j(x_i)a_1 + \ldots$$

$$+ \sum_{i=0}^{n} w_i g_m(x_i)g_j(x_i)a_m = \sum_{i=0}^{n} w_i y_i g_j(x_i) \tag{8.12.2}$$

where $j = 0, 1, 2, \ldots ,m$

Writing equation (8.12.2) in matrix form, we have

$$Ga = b \tag{8.12.3}$$

where a is a vector of the unknown coefficients and

$$g_{pq} = \sum_{i=0}^{n} w_i g_p(x_i) g_q(x_i) \quad p, q = 0, 1, 2, \dots, m$$

$$b_p = \sum_{i=0}^{n} w_i y_i g_p(x_i) \quad\quad p = 0, 1, 2, \dots, m$$

If the system is non singular a solution exists and these equations can be solved using the techniques of Chapter 3.

Before the simultaneous equations of (8.12.3) can be solved it is necessary to decide on the form and number of the functions $g_j(x)$ that are to be used. Here a knowledge of the nature of the system from which the data has been taken is of value; the curve fitting is more likely to be successful if the functions chosen can readily model the true relationship between the parameters from which the data was sampled. To clarify this, suppose the true relationship between x and y is $y = 3 + 1.5 \log x$ and let us assume that the data follows this relationship with the addition of random measurement errors. Although it is possible to fit other functions to this data, the best and most economical fit will be $y = a + b \log x$ where a and b are to be determined and should be found to be close to 3 and 1.5 respectively.

The form of the relationship between x and y is often unknown and under these circumstances the usual practice is to try fitting a low degree polynomial to the data. The polynomial function is the first choice because of its simplicity. However if this is unsatisfactory other common functions that can be tried include linear combinations of trigonometric, logarithmic and exponential functions. The use of trigonometric functions with equi-spaced data has a special importance and an examination of the use and properties of these functions is given in section 8.18.

A general purpose computer program can be written to allow the user to specify a wide range of function types for the least squares process. Such a program is 8.12.1. However, the program is cumbersome to use since the functions to be used must be defined within the program.

```
100   REM ------------------------------------------
110   REM GENERAL WEIGHTED LEAST SQUARES JP/GL (C)
120   REM ------------------------------------------
130   PRINT "WEIGHTED LEAST SQUARE FIT"
140   PRINT "USING PRESCRIBED FUNCTIONS"
150   PRINT
160   READ N,N0
170   DIM A(N,N+1),T(N,N0),D(N),X(N0),Y(N0),W(N0),R(N)
180   FOR I=1 TO N0
190      READ X(I)
200   NEXT I
210   FOR I=1 TO N0
220      READ Y(I)
230   NEXT I
240   FOR I=1 TO N0
250      READ W(I)
260   NEXT I
270   GOSUB 620
280   FOR I=1 TO N
290      FOR J=I TO N
300         A(I,J)=0
310      NEXT J
320      A(I,N+1)=0
330   NEXT I
335   REM PERFORMS SUMMATION TO GENERATE ARRAY OF COEFFICIENTS
340   FOR K=1 TO N0
350      FOR I=1 TO N
360         FOR J=I TO N
370            A(I,J)=A(I,J)+T(I,K)*T(J,K)*W(K)
380         NEXT J
390         A(I,N+1)=A(I,N+1)+Y(K)*T(I,K)*W(K)
400      NEXT I
410   NEXT K
420   FOR I=1 TO N
430      FOR J=I TO N
440         A(J,I)=A(I,J)
450      NEXT J
460   NEXT I
465   REM SUB 720 SOLVES SIMULTANEOUS EQUATIONS (NOT LISTED)
466   REM BASED ON PROG 3.6.1.  RETURNS SOLUTION IN VECTOR D
470   GOSUB 720
480   FOR I=1 TO N
490      PRINT "COEFFICIENT ";I;" = ";D(I)
500   NEXT I
510   PRINT
520   PRINT "TABLE OF DATA AND APPROXIMATING FUNCTION"
530   PRINT " X";TAB(15);" Y";TAB(30);" f(X)"
535   REM GENERATE FUNCTION APPROXIMATION
540   FOR I=1 TO N0
550      F=0
560      FOR J=1 TO N
570         F=F+D(J)*T(J,I)
580      NEXT J
590      PRINT X(I);TAB(15);Y(I);TAB(30);F
600   NEXT I
```

Program 8.12.1. (continues)

```
610   END
620   FOR K=1 TO N0
630      T(1,K)=1
640      T(2,K)=LOG(X(K))
645      REM CONTINUE AS REQUIRED DEFINING FUNCTIONS IN TERMS OF X(K)
650   NEXT K
660   RETURN
670   DATA 2: REM NUMBER OF FUNCTIONS
680   DATA 10 :REM NUMBER OF DATA POINTS
690   DATA .2,.6,1.2,1.2,3,5,7,7,7.5,9.6: REM VALUES OF X
700   DATA .615,2.33,3.10,3.27,4.88,5.25,6.21,5.62,5.76,6.40:REM Y VALUES
710   DATA 1,1,1,1,1,1,1,1,1,1:REM WEIGHTING VALUES FOR DATA
```

Program 8.12.1.

8.13 LEAST SQUARES USING POLYNOMIAL FUNCTIONS

Equations (8.12.2) and (8.12.3) will now be developed using a polynomial in place of the summation of arbitrary functions $g_j(x)$ shown in (8.12.1). If $f(x)$ is a polynomial function then $g_j(x)$ will be replaced by x^j in (8.12.1) to give

$$f(x, a_0, a_1, \dots, a_m) = a_0 + a_1x + a_2x^2 + \dots + a_mx^m \tag{8.13.1}$$

Using this relationship it can be shown that g_{pq} and b_p of (8.12.3) will become

$$g_{pq} = \sum_{i=0}^{n} w_i x_i^{p+q} \quad p, q = 0, 1, 2, \dots, m \tag{8.13.2}$$

$$b_p = \sum_{i=0}^{n} w_i y_i x_i^p \quad p = 0, 1, 2, \dots, m \tag{8.13.3}$$

Substituting these coefficients in (8.12.3) results in equations that are simple to program. In particular the coefficient matrix of (8.12.3) is symmetric and can be shown to be positive definite, (Powell, 1981). It follows from this that the system of equations have a unique solution. Program 8.13.1 generates the terms given in (8.13.2) and (8.13.3) and solves the resulting simultaneous equations by Gaussian elimination (see section 3.4). In this way the coefficients of (8.13.1) can be found. The program is simpler to use than program 8.12.1 but it is less versatile. Whilst program 8.13.1 works quite satisfactorily for problems of modest size, the simultaneous equations that must be solved can be ill-conditioned. In these circumstances accurate solutions are difficult to obtain for the reasons given in section 3.10. It can be shown that if the data is equi-spaced in the range $x = 0$ to 1 then for large values of m and n the matrix approximates to a Hilbert matrix see Hamming, (1973). In section 3.19 the ill-conditioned nature of this matrix is illustrated. Usually the degree of the polynomial chosen to approximate the data is very much smaller than the number of data values. However, if the degree of the polynomial is one less than the number of data values, (i.e $m = n$) then the polynomial will fit the data exactly

```
100    REM ----------------------------------
110    REM POLYNOMIAL LEAST SQUARES JP/GL (C)
120    REM ----------------------------------
130    PRINT "WEIGHTED LEAST SQUARE FIT"
140    PRINT "USING POLYNOMIAL"
150    PRINT
160    READ N0,M
170    N=N0+1
180    DIM A(N,N+1),S(2*N0),X(M),Y(M),W(M),R(N),D(N)
190    FOR I=1 TO M
200       READ X(I)
210    NEXT I
220    FOR I=1 TO M
230       READ Y(I)
240    NEXT I
250    FOR I=1 TO M
260       READ W(I)
270    NEXT I
280    FOR P=0 TO 2*N0
290       S(P)=0
300    NEXT P
310    FOR P=1 TO N
320       A(P,N+1)=0
330    NEXT P
335    REM FOR SUMMATIONS FOR ARRAY ELEMENTS
340    FOR I=1 TO M
350       G=W(I)
360       FOR P=0 TO 2*N0
370          S(P)=S(P)+G:G=G*X(I)
380       NEXT P
390       G=W(I)
400       FOR P=1 TO N
410          A(P,N+1)=A(P,N+1)+Y(I)*G:G=G*X(I)
420       NEXT P
430    NEXT I
435    REM FILL ARRAY A WITH SUMMATIONS
440    FOR I=1 TO N
450       FOR J=1 TO N
460          A(I,J)=S(I+J-2)
470       NEXT J
480    NEXT I
485    REM SUB 620 SOLVES SIMULTANEOUS EQUATIONS (NOT LISTED)
486    REM BASED ON PROG 3.6.1.  RETURNS SOLUTION IN VECTOR D
490    GOSUB 620
500    FOR I=1 TO N
510       PRINT "COEFF OF X^";I-1;"=    ";D(I)
520    NEXT I
530    END
540    DATA 3: REM DEGREE  OF POLYNOMIAL
550    DATA 11: REM NUMBER OF DATA VALUES
560    DATA 0,.5,1,1.5,2,2.5,3,3.5,4,4.5,5:REM VALUES OF X
570    DATA 9.6,7.5,4.9,3.3,3.8,8.7,18.5,37.6,60.7,99.9,150.6:REM Y VALUES
580    DATA 1,1,1,1,1,1,1,1,1,1,1:REM VALUES OF WEIGHTS
```

Program 8.13.1.

and the sum of the differences squared will be zero. Under these circumstances the polynomial determined will be the collocating polynomial but the amount of computation required to obtain it will be very much greater than necessary.

8.14 LEAST SQUARES USING NON-LINEAR FUNCTIONS

We will now examine how we may fit an approximating function to data where the function chosen is a non-linear function of the unknown coefficients. One simple approach is to linearise the function. This can sometimes be done by taking logarithms. For example, if we wish to fit $y = ax^b$ to some data, then taking logarithms we have $\log y = \log a + b \log x$. Thus $\log y$ is a linear function of the coefficients $\log a$ and b. Note that in applying a conventional least squares procedure to this function we minimise the sum of the squares of the differences in $\log y$, not of y itself. No general procedure can be given for replacing non-linear functions by a linear equivalent; each function must be treated individually. Indeed, it may not even be possible to linearise the non-linear function selected. However, it is always possible to obtain a linear approximation to the non-linear function using Taylor's expansion. Suppose we wish to fit the following function to data.

$$y = f(x, a_0, a_1, \dots , a_m)$$

Let us assume some approximate values of the coefficients $a_0, a_1, \dots , a_m$ and these will be denoted by $a_0^{(0)}, a_1^{(0)}, \dots , a_m^{(0)}$. Writing

$$f^{(0)} = f(x, a_0^{(0)}, a_1^{(0)}, \dots , a_m^{(0)})$$

then Taylor's expansion, neglecting higher order terms, gives the approximation for y thus

$$y = f^{(0)} + \Delta a_0 [\partial f / \partial a_0]^{(0)} + \Delta a_1 [\partial f / \partial a_1]^{(0)} +$$

$$\dots + \Delta a_m [\partial f / \partial a_m]^{(0)} \qquad (8.14.1)$$

In (8.14.1) Δa_i is a small change in a_i from $a_i^{(0)}$ and $[\partial f / \partial a_j]^{(0)}$ is evaluated using the approximate values of the coefficient $a_0^{(0)}, a_1^{(0)}$ etc. Using the approximation provided by (8.14.1) in (8.11.3), the weighted sum of the differences squared, S, is

$$S = \sum_{i=0}^{n} w_i [y_i - f_i^{(0)} - \Delta a_0 [\partial f / \partial a_0]^{(0)} -$$

$$\Delta a_1 [\partial f_i / \partial a_1]^{(0)} - \dots - \Delta a_m [\partial f_i / \partial a_m]^{(0)}]^2 \qquad (8.14.2)$$

where $[\partial f_i/\partial a_j]^{(0)}$ is the partial derivative of the function f with respect to a_j evaluated at $x = x_i$ and using the approximate values of the coefficient $a_0^{(0)}, a_1^{(0)}$ etc. We will

now minimise the S by taking partial derivatives with respect to the coefficients Δa_j and setting them equal to zero. Thus

$$\partial S/\partial\{\Delta a_j\} = -2\sum_{i=0}^{n}w_i[y_i - f_i^{(0)} - \Delta a_0[\partial f_i/\partial a_0]^{(0)} -$$

$$\Delta a_1[\partial f_i/\partial a_1]^{(0)} - \dots - \Delta a_m[\partial f_i/\partial a_m]^{(0)}[\partial f_i/\partial a_j]^{(0)} = 0$$

$$j = 0, 1, 2, \dots, m$$

Rearranging the above equations we obtain

$$\sum_{i=0}^{n}w_i[\partial f_i/\partial a_0]^{(0)}[\partial f_i/\partial a_j]^{(0)}\,\Delta a_0 + \sum_{i=0}^{n}w_i[\partial f_i/\partial a_1]^{(0)}[\partial f_i/\partial a_j]^{(0)}\Delta a_1 +$$

$$\dots + \sum_{i=0}^{n}w_i[\partial f_i/\partial a_m]^{(0)}[\partial f_i/\partial a_j]^{(0)}\,\Delta a_m = \sum_{i=0}^{n}w_i[y_i - f_i^{(0)}][\partial f_i/\partial a_j]^{(0)}$$

$$j = 0,1, 2, \dots, m \qquad (8.14.3)$$

Expressing (8.14.3) in matrix form gives

$$\mathbf{Kv = b} \qquad (8.14.4)$$

where

$$k_{pq} = \sum_{i=0}^{n}w_i[\partial f_i/\partial a_p]^{(0)}[\partial f_i/\partial a_q]^{(0)}$$

$$b_p = \sum_{i=0}^{n}w_i[y_i - f_i^{(0)}][\partial f_i/\partial a_p]^{(0)}$$

$$v_p = \Delta a_p$$

Once the function $f(x)$ is chosen, its derivatives with respect to the coefficients a_j can be found and $[\partial f_i/\partial a_j]^{(0)}$ can be evaluated for the initial set of estimates for the coefficients a_j, together with the values of x_i. Solving (8.14.4) gives a set of corrections for the coefficients, Δa_j. These corrections can then be used to obtain a new set of coefficients $a_j^{(1)}$ thus

$$a_j^{(1)} = a_j^{(0)} + \Delta a_j \quad j = 0, 1, 2, \dots, m$$

This process is repeated in an iterative manner until the corrections, Δa_j, approach zero. Convergence of this procedure is not guaranteed and for this reason the initial values of the coefficients should be as close as possible to the correct values. Assuming that the user has a reasonable understanding of the problem or experiment from which the data arises the need to begin with good estimates for the unknown coefficients should not present too great a problem.

It is not easy to write a general purpose program to fit non-linear functions to data. In addition to the function itself, it is necessary to define all its derivatives with respect to the coefficients a_i. In principle, values of the derivatives could be determined numerically but this process could introduce further errors. Program 8.14.1 to fit a non-linear function to data is listed below. As well as providing the data the user must define the function to be used and its derivatives with respect to all the coefficients.

```
100   REM ----------------------------------
110   REM NON LINEAR LEAST SQUARES JP/GL (C)
120   REM ----------------------------------
130   PRINT "NON-LINEAR WEIGHTED LEAST SQUARE FIT"
140   READ N,NO
150   DIM A(N,N+1),T(N,NO),R(N),D(N),X(NO),Y(NO),W(NO),P(N)
170   FOR I=1 TO NO
180       READ X(I)
190   NEXT I
200   FOR I=1 TO NO
210       READ Y(I)
220   NEXT I
230   FOR I=1 TO NO
240       READ W(I)
250   NEXT I
260   PRINT
270   FOR I=1 TO N
280       PRINT "ENTER TRIAL VALUE, COEFF ";I;
290       INPUT P(I)
300   NEXT I
310   PRINT:PRINT "ITERATION OF COEFFICIENTS"
320   FOR I=1 TO N
330       PRINT P(I);TAB(15*I)
340   NEXT I
350   PRINT
355   REM SUB 760 EVALUATES FUNCTION AND DERIVATIVES
360   GOSUB 760
370   FOR I=1 TO N
380       FOR J=I TO N+1
390           A(I,J)=0
400       NEXT J
420   NEXT I
```

Program 8.14.1 (continues).

```
425   REM COMPUTE COEFFICIENTS OF A MATRIX AND RHS VECTOR
430   FOR K=1 TO N0
440      FOR I=1 TO N
450         FOR  J=I TO N
460            A(I,J)=A(I,J)+T(I,K)*T(J,K)*W(K)
470         NEXT J
480         A(I,N+1)=A(I,N+1)+(Y(K)-T(0,K))*T(I,K)*W(K)
490      NEXT I
500   NEXT K
510   FOR I=1 TO N
520      FOR J=I TO N
530         A(J,I)=A(I,J)
540      NEXT J
550   NEXT I
555   REM SUB 880 SOLVES SIMULTANEOUS EQUATIONS (NOT LISTED)
556   REM BASED ON PROG 3.6.1.  RETURNS SOLUTION IN VECTOR D
560   GOSUB 880
570   FOR I=1 TO N
580      P(I)=P(I)+D(I)
590      PRINT P(I);TAB(15*I);
600   NEXT I
610   PRINT
620   FLAG=0
630   FOR I=1 TO N
640      IF ABS(D(I))>.001 THEN FLAG=1
650   NEXT I
660   IF FLAG=1 THEN 360
670   PRINT
680   PRINT " X";TAB(15);" Y";TAB(30);" f(X)"
690   FOR I=1 TO N0
700      PRINT X(I);TAB(15);Y(I);TAB(30);T(0,I)
710   NEXT I
720   PRINT:PRINT "REPEAT WITH SAME DATA ";
730   INPUT Z$
740   IF Z$="Y" THEN 300
750   END
755   REM DEFINE FUNCTION TO BE FITTED IN TERMS OF VARIABLES X(K)
756   REM P(1) ... P(N) ARE UNKNOWN COEFFICIENTS
757   REM T(0,K) IS THE DEFINITION OF THE FUNCTION
758   REM T(1,K) IS THE DERIVATIVE OF THE FUNCTION WRT P(1) ETC
760   FOR K=1 TO N0
770      T(0,K)=P(1)+P(2)*X(K)^P(3)
780      T(1,K)=1
790      T(2,K)=X(K)^P(3)
800      T(3,K)=P(2)*X(K)^P(3)*LOG(X(K))
805      REM CONTINUE AS REQUIRED
810   NEXT K
820   RETURN
830   DATA 3:REM NUMBER OF UNKNOWN COEFFICIENTS IN FUNCTION
840   DATA 9:REM NUMBER OF DATA POINTS
850   DATA .1,.2,.6,1,1.5,2,3.3,3.6,4 :REM VALUES OF X
860   DATA 2.22,2.23,2.01,2.25,2.92,3.76,4.69,5.95,5.47:REM VALUES OF Y
870   DATA 1,1,1,1,1,1,1,1,1:REM WEIGHTINGS OF DATA POINTS
```

Program 8.14.1

8.15 SOME EXAMPLES OF CURVE FITTING

We will now use programs 8.12.1, 8.13.1 and 8.14.1 to find approximate functions to fit a variety of data.

Example 1 : Consider the data given in Table 8.15.1 and assume that the

(1)   The data is expected to follow a linear relationship.

(2)   For $x > 3$ it is known that the measurement accuracy is four or five times better than for values of $x \leq 3$.

(3)   Due to the physical nature of the problem it is known that when $x = 0$ then y must equal 3 although it not possible to make measurements when $x = 0$.

Program 8.13.1 has been used to fit a first degree polynomial to this data and three slightly different equations have been derived by making three different assumptions about the data. The assumptions are:

(1)   The data is unweighted: Since the program requires values for the weighting, w_i, to be specified these are set to one. In this case it is found that
$y = 3.19529569 + 1.96641281x$.

(2)   The data is weighted: Here values of $x > 3$ were given a weighting of 5, the remainder of the data was given a unit weighting. The resulting equation is
$y = 3.07366607 + 1.9753961x$.

(3)   The data is weighted and the fitting function is forced to fit $x = 0, y = 3$ exactly: In order to force the polynomial through $x = 0, y = 3$, this point is given a very high weighting, say 10^6. This results in
$y = 3.0000005 + 1.98375176x$. The constant term should be 3 and the error of 5×10^{-7} is a rounding error.

The data given in Table 8.15.1 was derived as follows. Initially values of y were obtained from the equation $y = 3 + 2x$ for given values of x. The values of y obtained were then modified by adding a random value of up to $\pm 20\%$ of y when $x < 3$ and by adding a random value of up to $\pm 5\%$ to y for values of $x > 3$.

Table 8.15.1. Data based on $f(x) = 3 + 2x$ with random
error $\pm 20\%$ $(x \le 3)$ and $\pm 5\%$ $(x > 3)$.

x	f(x)	Weight
1	5.4	1
1.5	5.4	1
2	6.2	1
2.5	8.4	1
3	10.8	1
4	10.75	5
5	13.50	5
7	16.38	5
7.5	18.00	5
10	22.09	5
11	24.58	5
12	27.61	5

Example 2 : Program 8.13.1 is used to fit a second and third degree polynomial to
the unweighted data of Table 8.15.2. The polynomials determined are

$$y = 19.3958037 - 32.9828901x + 11.4158508x^2$$

and

$$y = 9.23286486 - 0.687330246x - 5.52238178x^2 + 2.25843072x^3$$

Note that the constant term and the coefficients of x and x^2 are quite different in the
two equations; it is generally not possible to deduce the coefficients of a low degree
approximating polynomial from a higher degree one. Fig. 8.15.1 shows that the
cubic function fits the data much better than the quadratic. This should be so since the

Table 8.15.2. Data based on $f(x) = 10 + 3x - 4x^2 + 2x^3$ with random errors added.

x	f(x)
0	9.6
0.5	7.5
1	4.9
1.5	3.3
2	3.8
2.5	8.7
3	18.5
3.5	37.6
4	60.7
4.5	99.9
5	150.6

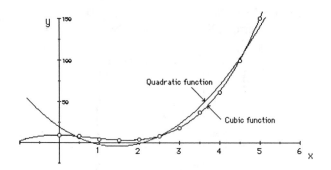

Fig. 8.15.1 Least squares approximations to the data of Table 8.15.2
using a cubic and a quadratic polynomial function.

data was calculated from the equation $y = 10 + 3x - 4x^2 + 2x^3$ with a random error
of up to $\pm 5\%$ added to each value of y. It should be noted that as the degree of the
polynomial is increased the closeness of the function to any individual point may not
improve but the sum of the squares of the errors will.

Example 3 : A series of least square approximations have been made to the data of
Table 8.15.3. The functions chosen and the programs used in the analysis were as
follows:-

(1) $y = a + b \log x$ Using program 8.12.1.

(2) $y = a + b/x + c/x^2 + d/x^3$ Using program 8.12.1.

(3) $y = a + bx + cx^2 + dx^3$ Using program 8.13.1

Table 8.15.3. Data based on $f(x) = 3 + 1.5 \log_e x$ with random error $\pm 5\%$ added.

x	f(x)
0.2	0.615
0.6	2.33
1.2	3.10
1.2	3.27
3	4.88
5	5.25
7	6.21
7	5.62
7.5	5.76
9.6	6.40

The computer outputs for these three cases are shown in Table 8.15.4 and Table 8.15.5 and the functions are plotted in Fig. 8.15.2a and Fig. 8.15.2b. Note that Fig. 8.15.2b shows the behaviour of the functions in detail in the region $0 \le x \le 2$. The logarithmic function gives a suspiciously different picture from the other functions of the behaviour of the underlying law from which the data was taken. To test if the logarithmic function is a better or worse representation of the underlying law it would be necessary to obtain more data in this region.

Table 8.15.4. Use of program 8.12.1 to fit $y = a + b \log x$ and then $y = a + b/x + c/x^2 + d/x^3$ to the data of Table 8.15.3.

```
WEIGHTED LEAST SQUARES
USING PRESCRIBED FUNCTIONS

COEFFICIENT  1  =  3.001855
COEFFICIENT  2  =  1.470954

TABLE OF DATA AND APPROXIMATING FUNCTION
   X              Y                 f(X)
  .2            .615              .6344461
  .6           2.33              2.250454
 1.2           3.1               3.270042
 1.2           3.27              3.270042
 3             4.88              4.617863
 5             5.25              5.369264
 7             6.21              5.864199
 7             5.62              5.864199
 7.5           5.76              5.965685
 9.6           6.4               6.328804
```

```
COEFFICIENT  1  =   6.741642
COEFFICIENT  2  =  -6.533745
COEFFICIENT  3  =   2.978523
COEFFICIENT  4  =  -.3833714

TABLE OF DATA AND APPROXIMATING FUNCTION
   X              Y                 f(X)
  .2            .615              .6145744
  .6           2.33              2.350876
 1.2           3.1               3.143415
 1.2           3.27              3.143415
 3             4.88              4.880475
 5             5.25              5.550967
 7             6.21              5.867918
 7             5.62              5.867918
 7.5           5.76              5.922518
 9.6           6.4               6.092929
```

Table 8.15.5. Weighted least squares fit to the data of Table 8.15.3 using a cubic polynomial.

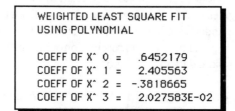

```
WEIGHTED LEAST SQUARE FIT
USING POLYNOMIAL

COEFF OF X^ 0 =    .6452179
COEFF OF X^ 1 =    2.405563
COEFF OF X^ 2 =   -.3818665
COEFF OF X^ 3 =    2.027583E-02
```

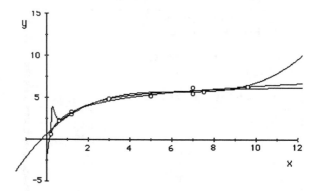

Fig. 8.15.2(a). Least square fit to the data of Table 8.15.3 using
functions (1), (2) and (3) given in the text.

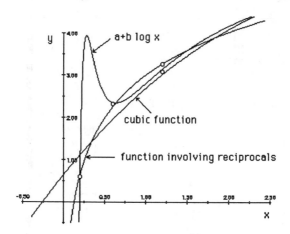

Fig. 8.15.2(b). This graph is an enlargement of Fig.8.15.2(a).
The logarithmic function behaves in an unsatisfactory manner in this region.

Table 8.15.6.

x	f(x)
0.1	2.22
0.2	2.23
0.6	2.01
1	2.25
1.5	2.92
2	3.76
3.3	4.69
3.6	5.95
4	5.47

Table 8.15.7. Non-linear least squares fit to the data of Table 8.15.6.

```
NON-LINEAR WEIGHTED LEAST SQUARE FIT

ENTER TRIAL VALUE, COEFF  1 ?  2
ENTER TRIAL VALUE, COEFF  2 ?  .5
ENTER TRIAL VALUE, COEFF  3 ?  1.5

ITERATION OF COEFFICIENTS
2              .5              1.5
 2.003285      .5580364        1.387903
 2.018018      .5374402        1.418131
 2.013239      .5442228        1.40983
 2.014486      .5424563        1.41216
 2.014131      .5429593        1.411516

X              Y               f(X)
 .1            2.22            2.035485
 .2            2.23            2.070372
 .6            2.01            2.278167
 1             2.25            2.556942
 1.5           2.92            2.976173
 2             3.76            3.458154
 3.3           4.69            4.942611
 3.6           5.95            5.325439
 4             5.47            5.856598

REPEAT WITH SAME DATA ?  N
```

Example 4 : A final illustration of least squares approximation is fitting the non-linear function $y = a + bx^c$ to the data of Table 8.15.6 using program 8.14.1. For this non-linear curve fitting procedure the partial derivatives of y with respect to a, b and c are required. These are

$$\partial y/\partial a = 1; \quad \partial y/\partial b = x; \quad \partial y/\partial c = bx^c \log x.$$

Table 8.15.7 shows that with two differing sets of starting values for a, b and c convergence is quite rapid and tends to $a = 2.014$; $b = 0.543$ and $c = 1.411$ approximately. However not all starting values lead to convergence: for example using the initial approximations 5, 1, 5 for a, b and c respectively the method fails. A graph of the approximating function, using the values of a, b and c given above, is shown together with the original data, in Fig. 8.15.3.

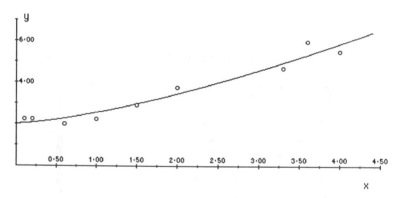

Fig. 8.15.3. Plot of the data of Table 8.15.6 and the function $y = 2.014 + 0.543x^{1.411}$.

8.16 LEAST SQUARES USING ORTHOGONAL POLYNOMIALS

In section 8.13 it was stated that problems can sometimes arise when determining the coefficients of the function (8.12.1) because (8.12.3) can represent a set of ill-conditioned equations. Here we will examine the use of orthogonal polynomials to avoid the need to solve this set of simultaneous equations. However, to use these polynomials it is necessary to stipulate that the data is equi-spaced. In section 8.19 a further example of the power and advantages of orthogonal functions will be given. Consider the following set of polynomials

$$p_0^{(n)}(v) = 1$$

$$p_1^{(n)}(v) = 1 - 2v/n$$

$$p_2^{(n)}(v) = 1 - 6v/n + 6v(v-1)/[n(n-1)]$$

$$p_3^{(n)}(v) = 1 - 12v/n + 30v(v-1)/[n(n-1)]$$
$$- 20v(v-1)(v-2)/[n(n-1)(n-2)]$$

and in general

$$p_m^{(n)}(v) = \sum_{i=0}^{m} (-1)^i \, {}^mC_i \, {}^{m+i}C_i \, v^{(i)}/n^{(i)} \tag{8.16.1}$$

where v is an integer in the range 0 to n and

$$v^{(i)} = v(v-1) \dots (v-i+1)$$
$$n^{(i)} = n(n-1) \dots (n-i+1)$$
$${}^mC_i = m!/[(m-i)! \, i!]$$

It can be proved, see Scheid (1968), that these polynomials are orthogonal over a set of equi-spaced arguments v. Thus

$$\sum_{v=0}^{n} p_r^{(n)}(v) \, p_s^{(n)}(v) = 0, \quad r \neq s \tag{8.16.2}$$

Let the polynomial that we wish to fit to the equi-spaced data (x_v, y_v), $v = 0, 1, 2, \dots, n$ be $f(x)$. This polynomial will be constructed from the orthogonal polynomials of (8.16.1). Thus

$$f(x) = b_0 p_0^{(n)}(v) + b_1 p_1^{(n)}(v) + \dots + b_m p_m^{(n)}(v) \tag{8.16.3}$$

where $v = (x - x_0)/h$ and h is the interval between data points. The sum of the differences squared, S, is

$$S = \sum_{v=0}^{n} [y_v - b_0 p_0^{(n)}(v) - b_1 p_1^{(n)}(v) - \dots - b_m p_m^{(n)}(v)]^2$$

To minimise this expression $\partial S/\partial b_j = 0$. Hence

$$\partial S/\partial b_j = -2 \sum_{v=0}^{n} [y_v - b_0 p_0^{(n)}(v) - b_1 p_1^{(n)}(v) - \dots - b_m p_m^{(n)}(v)] p_j^{(n)}(v)$$

$$= 0 \quad \text{for} \quad j = 0, 1, \dots, m \tag{8.16.4}$$

As a consequence of the orthogonality condition stated in (8.16.2), most terms in (8.16.4) are zero and (8.16.4) becomes

$$\sum_{v=0}^{n} y_v p_j^{(n)}(v) - b_j \sum_{v=0}^{n} \{p_j^{(n)}(v)\}^2 = 0 \quad j = 0, 1, \ldots, m \quad (8.16.5)$$

Unlike the general least squares formulation previously derived, the unknown coefficients are determined without the need to solve a set of simultaneous linear equations; each equation of (8.16.5) can solved individually to give

$$b_j = \sum_{v=0}^{n} y_v p_j^{(n)}(v) / \sum_{v=0}^{n} \{p_j^{(n)}(v)\}^2 \quad j = 0, 1, \ldots, m \quad (8.16.6)$$

One further advantage of orthogonal functions is now apparent. The value of the coefficients b_j do not change as we change the number of terms used in the approximating function. For example, if we compute the first five coefficients we can also examine the effect of only using one, two, three or four coefficients in the series.

There is a price to be paid for the computational advantages described. One disadvantage, already noted, is that we are restricted to using equi-spaced data. A second disadvantage is that the coefficients b_j determine the contribution of each of the orthogonal polynomials in (8.16.3). This does not allow us to obtain a direct representation of the approximating polynomial of the form

$$f(x) = a_0 + a_1 x + a_2 x^2 + \ldots + a_m x^m \quad (8.16.7)$$

Calculating the coefficients of (8.16.7) from the coefficients of (8.16.1) and (8.16.3) is tedious and not easily programmed. Program 8.16.1 uses the least squares criteria to fit five approximating orthogonal polynomials for a particular value of x. The program outputs an estimate for y for the given x using 1, 2, 3, 4 and 5 orthogonal polynomials. The number of orthogonal polynomials used in the program can be increased by including the coefficients of the higher degree polynomials.

Table 8.16.1 shows an example of the output of program 8.16.1. In this example the program is being used to fit a set of the orthogonal polynomials (8.16.1) to the data of Table 8.15.2.

```
100    REM ----------------------------------
110    REM L S FIT BY ORHOGONAL POLY JP/GL (C)
120    REM ----------------------------------
130    PRINT "LEAST SQUARE FIT USING"
140    PRINT "ORTHOGONAL POLYNOMIALS"
150    M=5
160    DIM A(M,M),B(M,M),D(M),N(M)
170    FOR K=1 TO M
180        B(K,1)=1
190    NEXT    K
200    FOR K=1 TO M
210        FOR J=1 TO K
220            READ A(K,J)
230        NEXT J
240    NEXT K
245    REM COEFFICIENTS OF FIRST FIVE ORTHOGONAL POLYNOMIALS
250    DATA 1
260    DATA 1,-2
270    DATA 1,-6,6
280    DATA 1,-12,30,-20
290    DATA 1,-20,90,-140,70
300    READ N,H,X0
310    N=N-1
320    DIM Y(N)
330    FOR T=0 TO N
340        READ Y(T)
350    NEXT T
360    FOR J=2 TO M
370        T=N:GOSUB 880
380        FOR K=J TO M
390            A(K,J)=A(K,J)/P
400        NEXT K
410    NEXT J
420    FOR K=1 TO M
430        D(K)=0: N(K)=0
440    NEXT K
450    FOR T=0 TO N
460        FOR J=2 TO M
470            GOSUB 880
480            FOR K=J TO M
490                B(K,J)=A(K,J)*P
500            NEXT K
510        NEXT J
520        FOR K=1 TO M
530            P=0
540            FOR J=1 TO K
550                P=P+B(K,J)
560            NEXT J
570            D(K)=P*P+D(K)
580            N(K)=P*Y(T)+N(K)
590        NEXT K
600    NEXT T
610    PRINT
615    REM COMPUTE COEFFICIENTS OF FITTED POLYNOMIAL
620    FOR K=1 TO M
```

Program 8.16.1(continues).

```
630      D(K)=N(K)/D(K)
640        PRINT"COEFF OF P (";K-1;")  =   ";D(K)
650   NEXT K
660   PRINT:PRINT"EVALUATION OF F(X) - ENTER  X";
670   INPUT X
680   T=(X-X0)/H
685   REM EVALUATE POLYNOMIALS
690   FOR J=2 TO M
700      GOSUB 880
710      FOR K=J  TO M
720         B(K,J)=A(K,J)*P
730      NEXT K
740   NEXT J
750   Q=0
760   FOR K=1 TO M
770      P=0
780      FOR J=1 TO K
790         P=P+B(K,J)
800      NEXT J
810      Q=P*D(K)+Q
820      PRINT"USING ";K;" TERMS, F(X)=";Q
830   NEXT K
840   PRINT:PRINT"ANOTHER EVALUATION (Y/N)";
850   INPUT A$
860   IF A$="Y" THEN 660
870   END
880   P=1
890   FOR S=1 TO J-1
900      P=P*(T-S+1)
910   NEXT S
920   RETURN
930   DATA 11,.5,0:REM NUMBER OF DATA POINTS,INTERVAL AND FIRST VALUE OF X
940   DATA9.6,7.5,4.9,3.3,3.8,8.7,18.5,37.6,60.7,99.9,150.6:REM VALUESOF Y
```

Program 8.16.1.

Table 8.16.1. Results from program 8.16.1.

```
LEAST SQUARE FIT USING
ORTHOGONAL POLYNOMIALS

COEFF OF P( 0 ) =   36.82727
COEFF OF P( 1 ) = -60.2409
COEFF OF P( 2 ) =   42.80945
COEFF OF P( 3 ) = -10.16293
COEFF OF P( 4 ) =   .4971772


EVALUATION OF F(X) - ENTER X? 5


USING  1  TERMS, F(X)=  36.82727
USING  2  TERMS, F(X)=  97.06818
USING  3  TERMS, F(X)=  139.8777
USING  4  TERMS, F(X)=  150.0406
USING  5  TERMS, F(X)=  150.5378
```

8.17 APPROXIMATIONS IN TWO DIMENSIONS

In section 8.11 we used the method of least squares to fit the best curve to a set of data points (x_i, y_i) and we determined a variety of approximating functions $y = f(x)$ for the unknown, underlying relationship between x and y. We now extend this work to deal with the problem of fitting a function $z = f(x,y)$ to data points (x_i, y_i, z_i). Because of the algebraic complexity of this task we will restrict the form of the function to a low degree unweighted polynomial in x and y.

The procedure for determining the specific functional approximation is similar to the method used in section 8.11, 8.12 and 8.13 to determine the one dimensional polynomial approximation. Because of the complexity of two variable polynomial functions it is difficult to write down a general form and so we illustrate the method by fitting a two dimensional second degree polynomial to the data (x_i, y_i, z_i). From (8.11.2), the least squares expression takes the form

$$S = \sum_{i=0}^{n} [z_i - f(x_i, y_i)]^2$$

Specifically,

$$S = \sum_{i=0}^{n} [z_i - a - bx_i - cy_i - dx_i^2 - exy_i - fy_i^2]^2$$

We require to minimise S, the sum of the squares of the errors, with respect to the coefficients a, b, c, d, e and f. Therefore, from the standard rules of calculus, we must find the values of a, b, c, d, e and f such that

$$\partial S/\partial a = 0, \ \partial S/\partial b = 0, \ \partial S/\partial c = 0, \ \partial S/\partial d = 0, \ \partial S/\partial e = 0, \ \partial S/\partial f = 0$$

This leads to the system of equations

$$na + (\Sigma x)b + (\Sigma y)c + (\Sigma x^2)d + (\Sigma xy)e + (\Sigma y^2)f = \Sigma z$$
$$(\Sigma x)a + (\Sigma x^2)b + (\Sigma xy)c + (\Sigma x^3)d + (\Sigma x^2 y)e + (\Sigma xy^2)f = \Sigma zx$$
$$(\Sigma y)a + (\Sigma x y)b + (\Sigma y^2)c + (\Sigma x^2 y)d + (\Sigma x\, y^2)e + (\Sigma y^3)f = \Sigma zy$$
$$(\Sigma x^2)a + (\Sigma x^3)b + (\Sigma x^2 y)c + (\Sigma x^4)d + (\Sigma x^3 y)e + (\Sigma x^2 y^2)f = \Sigma zx^2$$
$$(\Sigma xy)a + (\Sigma x^2 y)b + (\Sigma xy^2)c + (\Sigma x^3 y)d + (\Sigma x^2 y^2)e + (\Sigma xy^3)f = \Sigma zxy$$
$$(\Sigma y^2)a + (\Sigma xy^2)b + (\Sigma y^3)c + (\Sigma x^2 y^2)d + (\Sigma xy^3)e + (\Sigma y^4)f = \Sigma zy^2$$

where, in these equations,

$$\Sigma xy = \sum_{i=0}^{n} x_i y_i \quad \text{for example.}$$

This is a system of six linear equations in six unknowns and can be solved using the methods of Chapter 3. If desired, the reader can extend this analysis to fit any degree polynomial; a task which is straight forward but tedious. For example, fitting a third degree polynomial results in 10 equations and a fourth degree polynomial results in 15 equations.

The program 8.17.1 allows the user to fit a first, second, third or fourth degree two dimensional polynomial to data. There is no simple pattern in the coefficients of the equations and this results in a lengthy program. Thus, to extend it to fit higher degree polynomials, the user would have to determine the necessary expressions for the coefficients and modify the program to include them.

```
100   REM ------------------------------
110   REM 2D LEAST SQUARES FIT JP/GL (C)
120   REM ------------------------------
130   PRINT "LEAST SQUARES FIT TO DATA IN TWO DIMENSIONS": PRINT
140   READ M
150   PRINT"SELECT DEGREE OF POLYNOMIAL TO BE USED (1-4)";
160   INPUT CA
170   IF CA=1 THEN N=3
180   IF CA=2 THEN N=6
190   IF CA=3 THEN N=10
200   IF CA=4 THEN N=15
210   DIM C(44),A(N,N+1),D(N),Q(N),R(N)
215   REM READ DATA AND COMPUTE SUMMATIONS FOR APPROX DEGREE POLYNOMIAL
216   REM ASSIGN RESULTS TO VECTORS C AND D
220   FOR I=1 TO M
230       READ X,Y,Z
240       X1=X:Y1=Y
250       X2=X1*X1:X1Y1=X1*Y1:Y2=Y1*Y1
260       C(1)=C(1)+X1: C(2)=C(2)+Y1
270       C(3)=C(3)+X2: C(4)=C(4)+X1Y1
280       C(5)=C(5)+Y2: D(1)=D(1)+Z
290       D(2)=D(2)+Z*X1: D(3)=D(3)+Z*Y1
300       IF CA=1 THEN 660
310       X3=X2*X1:Y3=Y2*Y1
320       X2Y1=X2*Y1:X1Y2=X1*Y2
330       X4=X2*X2:Y4=Y2*Y2
340       X3Y1=X3*Y1:X2Y2=X2*Y2:X1Y3=X1*Y3
350       C(6)=C(6)+X3: C(7)=C(7)+X2Y1
360       C(8)=C(8)+X1Y2: C(9)=C(9)+Y3
370       C(10)=C(10)+X4: C(11)=C(11)+X3Y1
380       C(12)=C(12)+X2Y2: C(13)=C(13)+X1Y3
390       C(14)=C(14)+Y4: D(4)=D(4)+Z*X2
400       D(5)=D(5)+Z*X1Y1: D(6)=D(6)+Z*Y2
410       IF CA=2 THEN 660
420       X5=X3*X2:Y5=Y3*Y2
430       X6=X3*X3:Y6=Y3*Y3
440       C(15)=C(15)+X5: C(16)=C(16)+X4*Y1
450       C(17)=C(17)+X3*Y2: C(18)=C(18)+X2*Y3
460       C(19)=C(19)+X1*Y4: C(20)=C(20)+Y5
470       C(21)=C(21)+X6: C(22)=C(22)+X5*Y1
480       C(23)=C(23)+X4*Y2: C(24)=C(24)+X3*Y3
490       C(25)=C(25)+X2*Y4: C(26)=C(26)+X1*Y5
```

Program 8.17.1(continues).

```
500     C(27)=C(27)+Y6
510     D(7)=D(7)+Z*X3: D(8)=D(8)+Z*X2Y1
520     D(9)=D(9)+Z*X1Y2:D(10)=D(10)+Z*Y3
530     IF CA=3 THEN 660
540     X7=X1*X6:Y7=Y1*Y6
550     C(28)=C(28)+X7: C(29)=C(29)+X6*Y1
560     C(30)=C(30)+X5*Y2: C(31)=C(31)+X4*Y3
570     C(32)=C(32)+X3*Y4: C(33)=C(33)+X2*Y5
580     C(34)=C(34)+X1*Y6: C(35)=C(35)+Y7
590     C(36)=C(36)+X7*X1: C(37)=C(37)+X7*Y1
600     C(38)=C(38)+X6*Y2: C(39)=C(39)+X5*Y3
610     C(40)=C(40)+X4*Y4: C(41)=C(41)+X3*Y5
620     C(42)=C(42)+X2*Y6: C(43)=C(43)+X1*Y7
630     C(44)=C(44)+Y1*Y7: D(11)=D(11)+Z*X4
640     D(12)=D(12)+Z*X3Y1: D(13)=D(13)+Z*X2Y2
650     D(14)=D(14)+Z*X1Y3: D(15)=D(15)+Z*Y4
660   NEXT I
665   REM ALLOCATE SUMMATION VALUES TO SYSTEM MATRIX
670   A(1,1)=M
680   FOR I=2 TO N
690       A(1,I)=C(I-1)
700   NEXT I
710   A(2,2)=C(3):A(2,3)=C(4)
720   A(3,3)=C(5)
730   IF CA=1 THEN 1210
740   FOR I=2 TO 3
750       FOR J=4 TO 6
760           A(I,J)=C(I+J)
770       NEXT J
780   NEXT I
790   FOR I=4 TO 6
800       FOR J=I TO 6
810           A(I,J)=C(I+J+2)
820       NEXT J
830   NEXT I
840   IF CA=2 THEN 1210
850   FOR I=2 TO 3
860       FOR J=7 TO 10
870           A(I,J)=C(I+J+1)
880       NEXT J
890   NEXT I
900   FOR I=4 TO 6
910       FOR J=7 TO 10
920           A(I,J)=C(I+J+4)
930       NEXT J
940   NEXT I
950   FOR I=7 TO 10
960       FOR J=I TO 10
970           A(I,J)=C(I+J+7)
980       NEXT J
990   NEXT I
1000  IF CA=3 THEN 1210
```

Program 8.17.1(continues).

```
1010 FOR I=2 TO 3
1020     FOR J=11 TO 15
1030         A(I,J)=C(I+J+2)
1040     NEXT J
1050 NEXT I
1060 FOR I=4 TO 6
1070     FOR J=11 TO 15
1080         A(I,J)=C(I+J+6)
1090     NEXT J
1100 NEXT I
1110 FOR I=7 TO 10
1120     FOR J=11 TO 15
1130         A(I,J)=C(I+J+10)
1140     NEXT J
1150 NEXT I
1160 FOR I=11 TO 15
1170     FOR J=I TO 15
1180         A(I,J)=C(I+J+14)
1190     NEXT J
1200 NEXT I
1210 FOR I=1 TO N
1220     FOR J=I TO N
1230         A(J,I)=A(I,J)
1240     NEXT J
1250     A(I,N+1)=D(I)
1260 NEXT I
1265 REM SUB 1500 SOLVES SIMULTANEOUS EQUNS (NOT LISTED).
1266 REM BASED ON PROG.3.6.1. RETURNS SOLUTION IN VECTOR Q
1270 GOSUB 1500
1280 FOR I=1 TO N
1290 IF I=1 THEN PRINT"Constant term   ";Q(1)
1300 IF I=2   THEN PRINT"Coeff of x      ";Q(2)
1310 IF I=3   THEN PRINT"Coeff of y      ";Q(3)
1320 IF I=4   THEN PRINT"Coeff of x^2    ";Q(4)
1330 IF I=5   THEN PRINT"Coeff of xy     ";Q(5)
1340 IF I=6   THEN PRINT"Coeff of y^2    ";Q(6)
1350 IF I=7   THEN PRINT"Coeff of x^3    ";Q(7)
1360 IF I=8   THEN PRINT"Coeff of x^2y   ";Q(8)
1370 IF I=9   THEN PRINT"Coeff of xy^2   ";Q(9)
1380 IF I=10  THEN PRINT"Coeff of y^3    ";Q(10)
1390 IF I=11  THEN PRINT"Coeff of x^4    ";Q(11)
1400 IF I=12  THEN PRINT"Coeff of x^3y   ";Q(12)
1410 IF I=13  THEN PRINT"Coeff of x^2y^2 ";Q(13)
1420 IF I=14  THEN PRINT"Coeff of xy^3   ";Q(14)
1430 IF I=15  THEN PRINT"Coeff of y^4    ";Q(15)
1440 NEXT I
1450 END
1460 DATA 25:REM NUMBER OF DATA POINTS FOLLOWED BY X,Y,Z VALUES
1470 DATA 1,1,1,2,1,4,3,1,9,4,1,16,5,1,25,1,2,2,2,2,8,3,2,18,4,2,32
1480 DATA 5,2,50,1,3,3,2,3,12,3,3,27,4,3,48,5,3,75,1,4,4,2,4,16,3,4,36
1490 DATA 4,4,64,5,4,100,1,5,5,2,5,20,3,5,45,4,5,80,5,5,125
```

Program 8.17.1.

The set of equations resulting from a least squares problem may often be ill-conditioned and consequently it is necessary to work to a high level of precision to obtain accurate results. To illustrate this, program 8.17.1 was modified so that the data was generated from a polynomial function in x and y defined within the program. A comparison could then be made between the actual coefficients of the function and the coefficients computed by the program. Any errors in this process arise because the set of equations defined by (8.17.1) are ill-conditioned. These errors in table 8.17.1 and it is seen for the larger systems of equations double precision working becomes imperative.

Table 8.17.1 Errors in determining the coefficients of the polynomials
$$f(x,y) = -6 + 2x + y + x^2 - 5xy + 2y^2$$
$$g(x,y) = f(x,y) + x^3 - 3x^2y + xy^2 + 2.5y^3$$
$$h(x,y) = g(x,y) - 4x^3y + 4x^2y^2 + 2xy^3 + y^4$$
50 values were sampled from the polynomials in the range $0 < x < 1$ and
$0 < y < 1$ and thus the least squares fit should be exact.
The errors are due to ill-conditioning.

	Maximum error in the determined coefficients	
Degree of polynomial	Single precision	Double precision
2	1×10^{-4}	9×10^{-14}
3	4×10^{-3}	9×10^{-12}
4	0.285	7×10^{-11}

8.18 THE MINIMAX CRITERION

We will conclude this discussion on approximating functions by introducing the minimax criterion for curve fitting. A function generated according to this criterion is such that the maximum difference between the function and the data points is minimised: hence the name "minimax". In this section a procedure to generate minimax polynomials will be described and justified intuitively and by example.

In order to introduce the underlying principles of the procedure we will initially examine the simplest situation; that of fitting a straight line to three distinct data points. The diagram given in Fig. 8.18.1 illustrates that the difference between the approximating straight line and the data points must alternate in sign if the three differences are to be equal. A line drawn in such a way as to give differences of the same sign will always cause one difference to be larger or smaller than the other two. In addition, Fig. 8.18.2 shows that the so called equal error line is unique. If the slope or position of the line is perturbed from the equal error position then one

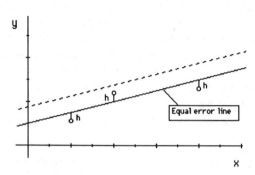

Fig. 8.18.1. Equal error line for three points. Other lines such as the dotted line shown
misses some points by a distance greater than the minimum h.

or more of the differences increase and one or more decrease; another equal error line
will not be found.

Thus we may summarise by stating that for any three points with distinct values
of x, there is only one straight line which misses all three points by equal amounts
and with alternating sign. Such a line is called an equal error or Chebyshev line.
Furthermore, any other line will miss at least one point with a larger difference and
thus for three points the equal error line is also the minimax line.

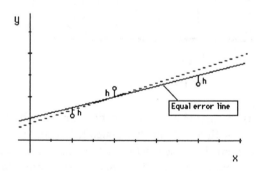

Fig. 8.18.2. As Fig.8.18.1, but the dotted line shown passes closer to one point
at the expense of missing another point by a distance greater than h.

A problem arises if two of the three data points have the same value of x; the
minimax line is no longer unique and a family of possible minimax lines exist as
shown in Fig. 8.18.3. Consider three points $x_i, y_i, i = 0, 1, 2$. Let the minimax line
be given by

$$f(x) = a_0 + a_1 x$$

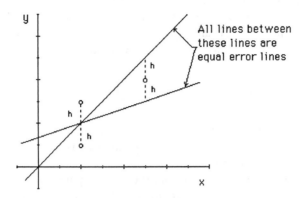

Fig. 8.18.3. In this illustration the equal-error line is not unique.

Accepting that the minimum difference, h, between the points and the minimax line alternate in sign we have

$$a_0 + a_1 x_1 - y_1 = h$$
$$a_0 + a_1 x_2 - y_2 = -h$$
$$a_0 + a_1 x_3 - y_3 = h$$

Thus

$$a_0 + a_1 x_1 - h = y_1$$
$$a_0 + a_1 x_2 + h = y_2$$
$$a_0 + a_1 x_3 - h = y_3$$

These three equations may be solved to determine a_0, a_1 and h.

Generally we will wish to fit a minimax line to more than three data points and to accomplish this task the exchange algorithm is used. Three arbitrary points, usually three adjacent points at the beginning of the sequence of data, are chosen and the equal error line for these three points is established. A search is then carried out to determine if this equal error line misses any of the remaining points by a greater amount than it misses the three chosen points. Assuming that the search does reveal such points then a new equal error line is constructed, using the point where the greatest difference occurs and two of the original three points. The process of searching, exchanging one point and constructing a new equal error line is continued until no point is missed by more than the amount that the three currently "best" points are missed. The equal error line for these three points is then the min-max line for the complete set of data.

The details of the exchange algorithm will now be described with the aid of an example. Consider thirty data points, $(x_0, y_0)...(x_{29}, y_{29})$. Suppose a stage is reached where, for example, the current working points (x_4, y_4), (x_9, y_9) and (x_{14}, y_{14}) which we shall now denote by P_4, P_9 and P_{14}. An equal error line is computed for these three points and it is found that the sign of the differences are as follows:

Data points:	P_4	P_9	P_{14}
Sign of difference:	+	−	+

The point of maximum difference is now calculated and this will either lie inside or outside of the range of the three original points. Let us consider the case where the point of maximum difference lies inside the range. For example, let the maximum difference be a negative quantity at P_7. This point will now be introduced into the three point sequence by deleting one of the adjacent points in the current sequence so that the pattern of the signs of the differences remains the same. Thus in the example shown both points P_4 and P_9 are adjacent to point P_7 but P_9 is replaced by virtue of the sign of its difference to give

Data points:	P_4	P_7	P_{14}
Sign of difference:	+	−	+

Reverting to the original set of points, P_4, P_9 and P_{14}, suppose the point of maximum difference is outside this current range, say at P_{23}. If the difference at P_{23} is the same sign as the difference associated with the highest numbered member of the current group, i.e. in this example positive, then P_{23} replaces this point in the sequence to give

Data points:	P_4	P_9	P_{23}
sign of difference:	+	−	+

If, in contrast, the difference at P_{23} is negative then the member of the original group at the other end of the sequence is discarded to give

Data points:	P_9	P_{14}	P_{23}
Sign of difference:	−	+	−

Fig. 8.18.4 shows the exchange algorithm operating on a sequence of 6 points to find the minimax line after two exchanges.

So far the discussion has been limited to the process of fitting a straight line through three points to obtain an equal error line and then to use the exchange algorithm to converge to the minimax line for all the data points. This process can be generalised to fit an equal error polynomial of degree m through $m + 2$ distinct points by solving $m + 2$ equations. The signs of the differences will alternate and

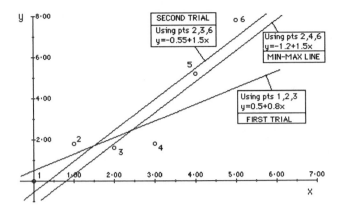

Fig. 8.18.4. Steps required to find the Minimax line for the data shown.

the polynomial will be unique. Thus if

$$p(x) = a_0 + a_1x + a_2x^2 + ... + a_mx^m$$

then

$$p(x_i) - y_i = \pm h \text{ (in an alternating sequence)} \quad \text{for } i = 1, 2, ... , m + 2$$

Solving this set of $m + 2$ linear equations yields $a_0, a_1, a_2, ... , a_m$ and h. The exchange algorithm may then be used to find the minimax polynomial of degree m that fits n data points. As will be seen in the examples that follow, the exchange algorithm is remarkably efficient, usually reaching the required solution in a few steps.

Program 8.18.1 determines the minimax polynomial using the exchange algorithm. The program has been used to fit a first degree minimax polynomial through the data of Table 8.15.1 and the output of the program is shown in Table 8.18.1. It should be noted that only 5 exchanges were required to determine the minimax solution. The result, $y = 3.534375 + 1.98625x$ may be compared with the unweighted least squares fit to the same data, i.e. $y = 3.19529569 + 1.966412811x$, as shown in Fig. 8.18.5.

As a further illustration, a third degree polynomial has been fitted to the data of Table 8.15.2. The result is shown below. Again a comparison is made with the least squares solution thus

Minimax: $y = 8.0780 + 2.7733x - 6.8880x^2 + 2.3947x^3$
Least squares: $y = 9.2329 - 0.6873x - 5.5224x^2 + 2.2258x^3$

```
100   REM ----------------------
110   REM MINIMAX POLY   JP/GL (C)
120   REM ----------------------
130   PRINT"POLYNOMIAL MINIMAX FIT"
140   PRINT
150   READ M,N0
160   N=N0+2
170   DIM X(M),Y(M),H(M),Z(N),A(N,N+1),C(N),R(N)
180   FOR I=1 TO N
190       Z(I)=I
200   NEXT I
210   PRINT"INITIAL GROUP  -";
220   FOR I=1 TO N
230       PRINT"X(";Z(I);")  ";
240   NEXT I
250   PRINT
260   FOR I=1 TO M
270       READ X(I)
280   NEXT I
290   FOR I=1 TO M
300       READ Y(I)
310   NEXT I
315   REM START ITERATION
320   A(1,1)=1:A(1,N)=-1
330   FOR I=2 TO N
340       A(I,1)=1: A(I,N)=-A(I-1,N)
350   NEXT I
360   FOR I=1 TO N
370       A(I,2)=X(Z(I)): A(I,N+1)=Y(Z(I))
380       IF N=3 THEN 420
390       FOR  J=3 TO N-1
400           A(I,J)=A(I,J-1)*A(I,2)
410       NEXT J
420   NEXT I
425   REM SUB 910 SOLVES SIMULTANEOUS EQUATIONS (NOT LISTED)
426   REM BASED ON PROG 3.6.1.  RETURNS SOLUTION IN VECTOR C
430   GOSUB 910
440   H=C(N)
450   Hmax=0
460   FOR I=1 TO M
470       S=C(N-1)
480       FOR J=N-2 TO 1 STEP -1
490           S=C(J)+X(I)*S
500       NEXT J
510       H(I)=Y(I)-S
520       IF ABS(H(I))>Hmax THEN Hmax=ABS(H(I)):I0=I
530   NEXT I
540   PRINT"Max difference = ";ABS(H(I0));" at X(";I0;")"
550   IF ABS(Hmax)-ABS(H)<.001 THEN 820
560   X0=X(I0): H0=H(I0)
570   IF Z(1)>I0 THEN 670
580   FOR I=2 TO N
590       IF Z(I)>I0 THEN 730
600   NEXT I
610   IF SGN(H0)=SGN(H(Z(N))) THEN 650
```

Program 8.18.1 (continues).

```
620   FOR I=1 TO N-1
630       Z(I)=Z(I+1)
640   NEXT I
650   Z(N)=I0
660   GOTO 750
670   IF SGN(H0)=SGN(H(Z(1))) THEN 710
680   FOR I=N TO 2 STEP-1
690       Z(I)=Z(I-1)
700   NEXT I
710   Z(1)=I0
720   GOTO 750
730   IF SGN(H0)=SGN(H(Z(I))) THEN Z(I)=I0:GOTO 750
740   Z(I-1)=I0
750   PRINT
760   PRINT"NEXT GROUP - ";
770   FOR I=1 TO N
780       PRINT"X(";Z(I);")   ";
790   NEXT I
800   PRINT
810   GOTO 320
820   PRINT
830   FOR I=1 TO N-1
840       PRINT"COEFF OF X ^";I-1;"  =";C(I)
850   NEXT I
860   END
870   DATA 11,3:REM NUMBER OF DATA POINTS AND DEGREE OF POLYNOMIAL
880   DATA 0,.5,1,1.5,2,2.5,3,3.5,4,4.5,5: REM VALUES OF X
890   DATA 9.6,7.5,4.9,3.3,3.8,8.7,18.5,37.6,60.7,99.9,150.6:REM Y VALUES
```

Program 8.18.1

Table 8.18.1. Generation of the minimax line to fit the data of Table 8.15.1.

```
POLYNOMIAL MINIMAX FIT

INITIAL GROUP  -X( 1 ) X( 2 ) X( 3 )
Max difference =   13.61   at X( 12 )

NEXT GROUP - X( 1 ) X( 2 ) X( 12 )
Max difference =   1.86659   at X( 5 )

NEXT GROUP - X( 1 ) X( 2 ) X( 5 )
Max difference =   7.145   at X( 11 )

NEXT GROUP - X( 2 ) X( 5 ) X( 11 )
Max difference =   1.656843   at X( 10 )

NEXT GROUP - X( 2 ) X( 5 ) X( 10 )
Max difference =   1.409118   at X( 3 )

NEXT GROUP - X( 3 ) X( 5 ) X( 10 )
Max difference =   1.306875   at X( 3 )

COEFF OF X ^ 0   = 3.534375
COEFF OF X ^ 1   = 1.98625
```

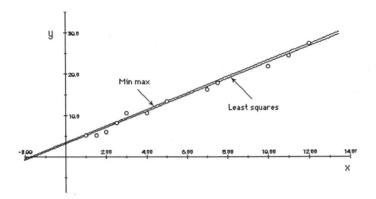

Fig. 8.18.5. Least squares and minimax fits to the data of Table 8.15.1.

Fig. 8.18.6 shows that these functions when plotted differ little from each other for the range data considered.

Provided that all values of the independent variable are distinct a unique minimax solution will exist. However, it has already been stated that if the data contains repeated values of the independent variable the minimax solution may not be unique. For example, using the data of Table 8.18.2 it can be shown that

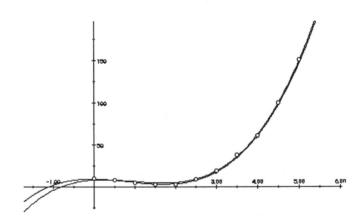

Fig.8.18.6. Cubic fits to the data of Table 8.15.2. Approximations given by the least squares and minimax methods produce results which are very similar.

Table 8.18.2.

x	f(x)
1	0.5
1	0.9
2	2.1
3	2.9
3	3.4
4	4.3
5	5.5
5	5.6
6	6.7

$$y = -0.60 + 1.25x$$

and

$$y = -0.45 + 1.20x$$

are both minimax solutions with a maximum difference of 0.25.

8.19 FOURIER ANALYSIS OF DISCRETE DATA

In this section Fourier analysis of discrete data is briefly introduced. This type of analysis is an important tool for the engineer and scientist engaged in the interpretation of data that has an underlying periodic nature. Because of the advances in computer hardware and the development of new and powerful algorithms the application of these techniques and the associated literature has grown enormously over the passed twenty years. A detailed description of these is given by Brigham (1974) and a straightforward introduction which emphasises the practical problems in using this type of analysis is given by Ramirez (1986).

   Suppose we wish to fit a finite set of trigonometric functions to n equi-spaced data points (x_r, y_r) where $r = 0, 1, 2, \ldots, n-1$. The data may be complex although in most practical situations it will be real. The finite Fourier series is defined by

$$y = A_0/2 + \sum_{k=1}^{m-1} [A_k \cos(2\pi kx/L) + B_k \sin(2\pi kx/L)]$$

$$+ (A_m/2)\cos(2\pi mx/L) \qquad (8.19.1)$$

In (8.19.1) n is assumed to be even, $m = n/2$, L is the range of the data and A_0, A_m, A_k and B_k (where $k = 1, 2, \ldots, m-1$) are n unknown coefficients. The fact that the equations are defined in terms of $A_0/2$ and $A_m/2$ simplifies the mathematics. The coefficients B_0 and B_m are absent from (8.19.1) for reasons which will become apparent. Since the n data values are equally spaced in the range L we may define x_r as

$$x_r = rL/n \quad r = 0, 1, 2, \dots, n-1 \tag{8.19.2}$$

Substituting for (x_r, y_r) into (8.19.1) and using (8.19.2) we have

$$y_r = A_0/2 + \sum_{k=1}^{m-1} [A_k \cos(2\pi kr/n) + B_k \sin(2\pi kr/n)]$$

$$+ (A_m/2)\cos(\pi r) \tag{8.19.3}$$

where $r = 0, 1, 2, \dots, n-1$

It is now clear that the coefficients B_0 and B_m are not required in (8.19.1) since $\sin(0)$ and $\sin(\pi r)$ are zero.

The method of undetermined coefficients could be used to find the values of the unknown coefficients of (8.19.3). However, this procedure is not necessary because the set of trigonometric functions used in (8.19.3) have the property that for n equi-spaced samples they form an orthogonal set. Thus

$$\sum_{r=0}^{n-1} \cos(2\pi rj/n)\cos(2\pi rk/n) = \begin{cases} 0 & \text{if } j \neq k \\ n/2 & \text{if } j = k \neq (0 \text{ or } n/2) \\ n & \text{if } j = k = (0 \text{ or } n/2) \end{cases} \tag{8.19.4}$$

$$\sum_{r=0}^{n-1} \sin(2\pi rj/n)\sin(2\pi rk/n) = \begin{cases} 0 & \text{if } j \neq k \\ n/2 & \text{if } j = k \neq (0 \text{ or } n/2) \\ 0 & \text{if } j = k = (0 \text{ or } n/2) \end{cases} \tag{8.19.5}$$

$$\sum_{r=0}^{n-1} \sin(2\pi rj/n)\cos(2\pi rk/n) = 0 \tag{8.19.6}$$

These properties are analogous to the orthogonality properties of continuous sine and cosine functions when integrated over one cycle. Multiplying (8.19.3) by the terms $\cos(2\pi rj/n)$ and $\sin(2\pi rj/n)$ in turn, summing over the n values of r and then using (8.19.4), (8.19.5) and (8.19.6), expressions for the unknown coefficients can be found thus:

$$A_k = (2/n)\sum_{r=0}^{n-1} y_r \cos(2\pi rk/n); \quad k = 0, 1, 2, \dots, n/2 \tag{8.19.7}$$

and

$$B_k = (2/n)\sum_{r=0}^{n-1} y_r \sin(2\pi rk/n); \quad k = 1, 2, \dots, (n/2) - 1 \tag{8.19.8}$$

8.20 FINITE FOURIER SERIES IN COMPLEX EXPONENTIAL FORM

It is sometimes convenient to express the finite Fourier series in terms of complex exponentials. Using the fact that

$$\exp(i2\pi kr/n) = \cos(2\pi kr/n) + i \sin(2\pi kr/n)$$

and that

$$\cos(2\pi(n - k)r/n) = \cos(2\pi kr/n)$$

$$\sin(2\pi(n - k)r/n) = -\sin(2\pi kr/n)$$

where $k = 1, 2, \dots, (n/2) - 1$, then it can be shown from (8.19.7) and (8.19.8) that

$$Y_k = (1/n)\sum_{r=0}^{n-1} y_r \exp(-i2\pi kr/n) \quad k = 0, 1, 2, \dots, n - 1 \tag{8.20.1}$$

where

$$Y_0 = A_0/2; \; Y_m = A_m/2 \text{ where } m = n/2$$

$$Y_k = (A_k - iB_k)/2 \quad k = 1, 2, \dots, (n/2) - 1$$

$$Y_p = (A_p + iB_p)/2 \quad \text{where } p = n - k, \; k = 1, 2, \dots, (n/2) - 1$$

Substituting for A_k and B_k in terms of Y_k in (8.19.3) we have

$$y_r = \sum_{k=0}^{n-1} Y_k \exp(i2\pi kr/n) \quad r = 0, 1, 2, \dots, n - 1 \tag{8.20.2}$$

We can obtain the coefficients Y_k from the equi-spaced data (x_r, y_r) by using (8.20.1). Equation (8.20.1) is called the discrete Fourier transform (or the DFT). The reverse of this process is called the inverse discrete Fourier transform (IDFT) and is implemented by (8.20.2). This equation is identical to the finite Fourier series, (8.19.1) but it is expressed in complex form. Both y_r and Y_k may be complex although, as has been previously stated, the samples y_r are usually real. Note also that these transforms constitute a pair: if the data is transformed by the discrete Fourier transform to determine the coefficients Y_k then it can be recovered exactly by means of the inverse transform.

8.21 THE FAST FOURIER TRANSFORM

To evaluate the coefficients of the DFT it would appear convenient to use (8.19.7) and (8.19.8). Although using these equations is satisfactory for small quantities of data, calculating the DFT for n real data points requires in excess of $2n^2$ multiplications. Thus, to transform a sequence of 4096 data points would require approximately 33 million multiplications; a task that would be a very time consuming and expensive operation using even a powerful main frame computer and impractical using a microcomputer. For this reason, when engineers required a continuous signal to be transformed into the Fourier domain it was the practice to carry out the operation using electronic equipment involving amplifiers and filters rather than by attempting to evaluate the coefficients A_k and B_k mathematically from discrete samples of the continuous data.

In 1965 this situation was dramatically changed with the development of the fast Fourier transform (FFT) algorithm (Cooley & Tukey 1965) The FFT algorithm is extremely efficient and only about $2n \log_2 n$ multiplications are required to compute the DFT for real data. With this development it became feasible to compute the DFT of a sampled waveform using a digital computer. Allied with the developments in computing hardware that have occurred in the past 20 years, it is now possible to compute the DFT for a relatively large number of data points using the FFT algorithm on a microcomputer.

Since it was first formulated, many refinements have been made to the basic FFT algorithm and several variants have been introduced. Here one of the simpler forms of the algorithm will be described. A rigorous algebraic development will not be presented; the details of the algorithm will be described and partially justified with the aid of diagrams.

Most versions of the FFT algorithm place one further restriction on our data. In addition to using equi-spaced values of x, the number of data points is restricted to a value that can be expressed as an integer power of 2. Here, for simplicity, a description of the algorithm is presented for $n = 2^3 = 8$.

It is easier to explain the algorithm in terms of the complex exponential form of the DFT, (8.20.1). This equation can be written in matrix notation thus

$$\mathbf{Y} = (1/n)\mathbf{W}\mathbf{y} \qquad (8.21.1)$$

where

$$w_{kr} = \exp(-i2\pi kr/n) \qquad k, r = 0, 1, 2, \dots, n-1$$

Note that $\mathbf{W}$ is an n by n array but $\mathbf{Y}$ is a *vector* of the Fourier coefficients and in this instance we are departing from our usual convention that emboldened upper case letters represent arrays.

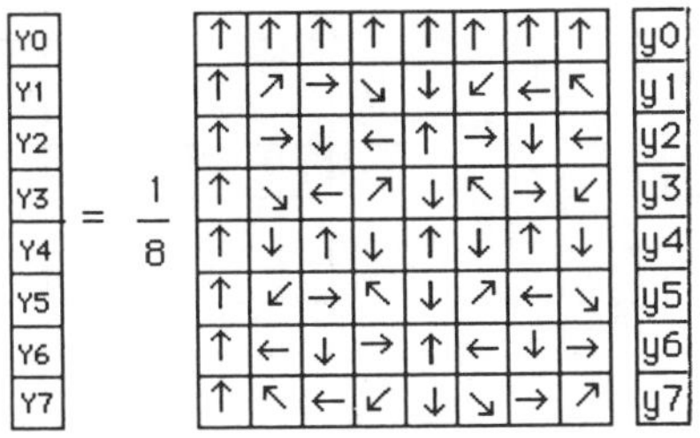

Fig. 8.21.1. Representation of equation (8.21.1) for 8 points.

Each element of **W** is a unit vector in the complex plane. Thus Fig. 8.21.1 illustrates (8.21.1) for $n = 8$ by showing an arrow to represent the unit vector; it is seen that the unit vectors may point in any one of 8 directions. For example, an upward pointing arrow represents exp(0) which equals 1, a downward pointing arrow represents exp(π) which equals -1. The arrows at other angles represent complex quantities.

The first step in the procedure is to reorder the rows of (8.21.1) as shown in Fig. 8.21.2. This reordering is based on a "bit reversed" order. In this process the subscripts of **Y** are converted to a binary number, the order of the binary digits is reversed and converted back to decimal. For example, it can be seen in Fig. 8.21.2 that the subscript 6 converts to binary 110. Reversing this binary number gives 011, equivalent to decimal 3. Thus equation 6 is moved from row 6 and placed in row 3. Similarly the original equation 3 is moved to row 6. At first sight little has been achieved by this reordering but it can be shown that the new array, illustrated in

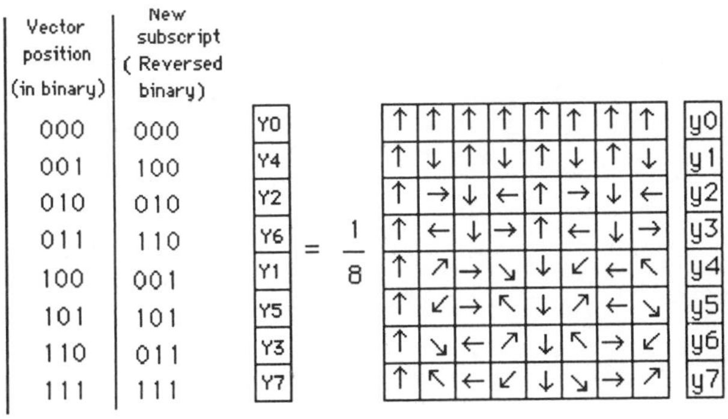

Fig. 8.21.2. 8-point DFT showing the effect of re-ordering **Y** in bit-reversed order.

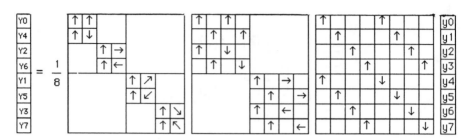

Fig. 8.21.3. 8-point DFT showing factorising into the product of $\mathbf{W}_1$, $\mathbf{W}_2$ and $\mathbf{W}_3$.

Fig. 8.21.2, can now be factorised into three arrays as shown in Fig. 8.21.3. Thus

$$\mathbf{Y}^* = \mathbf{W}_1 \mathbf{W}_2 \mathbf{W}_3 \, \mathbf{y} \qquad\qquad (8.21.2)$$

where $\mathbf{Y}^*$ is the vector $\mathbf{Y}$ in bit-reversed order. The rearranged form of $\mathbf{W}$ has been factored into three matrices and in general the number of matrix factors is equal to $\log_2 n$. In this case $n = 8$ and $\log_2 8 = 3$.

The details of the procedure used to factorise $\mathbf{W}$ are not given, indeed they are not required. Instead we will concentrate on the patterns that have emerged from the process. These are

(1)   In each of $\mathbf{W}_1, \mathbf{W}_2$ and $\mathbf{W}_3$ there are only two non-zero entries per row and one of these is unity.

(2)   In $\mathbf{W}_3$ the vectors turn by steps of π only; in $\mathbf{W}_2$ the vectors turn by steps of $\pi/2$; in $\mathbf{W}_1$ the vectors turn in steps of $\pi/4$.

(3)   Some of the patterns of non-zero coefficients in one array can be seen to repeat in another array. These similarities are highlighted in Fig. 8.21.4. Referring to this diagram and also Fig 8.21.3 it can be seen that the pattern called "A" is an 8 x 8 array in $\mathbf{W}_3$, it retains its essential characteristics but is compressed to a 4 x 4 array in $\mathbf{W}_2$ and still further compressed to a 2 x 2 array in $\mathbf{W}_1$. Pattern "B" is a 4 x 4 array in $\mathbf{W}_2$ and a 2 x 2 array in $\mathbf{W}_1$ and patterns "C" and "D" are unique to $\mathbf{W}_1$. Thus it is reasonable to infer that the task of determining the elements of $\mathbf{W}_1, \mathbf{W}_2$ and $\mathbf{W}_3$ is relatively straight forward and is based on a knowledge of these simple patterns.

Equation (8.21.2) can be broken down into stages so that

$$\mathbf{u} = \mathbf{W}_3 \mathbf{y} \qquad\qquad (8.21.3)$$
$$\mathbf{v} = \mathbf{W}_2 \mathbf{u} \qquad\qquad (8.21.4)$$
$$\mathbf{Y}^* = \mathbf{W}_1 \mathbf{v} \qquad\qquad (8.21.5)$$

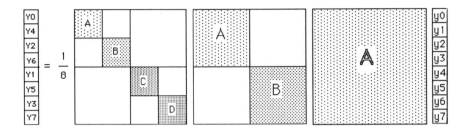

Fig. 8.21.4. The diagram shows the underlying pattern of coefficients in the
factorised matrices of the 8-point DFT.

Let us examine (8.21.3). With reference to Fig. 8.21.3 we have

$$u_0 = (\uparrow) y_0 + (\uparrow) y_4 = y_0 + y_4$$
$$u_4 = (\uparrow) y_0 + (\downarrow) y_4 = y_0 - y_4$$
$$u_1 = (\uparrow) y_1 + (\uparrow) y_5 = y_1 + y_5$$
$$u_5 = (\uparrow) y_1 + (\downarrow) y_5 = y_1 - y_5$$
$$u_2 = (\uparrow) y_2 + (\uparrow) y_6 = y_2 + y_6$$
$$u_6 = (\uparrow) y_2 + (\downarrow) y_6 = y_2 - y_6$$
$$u_3 = (\uparrow) y_3 + (\uparrow) y_7 = y_3 + y_7$$
$$u_7 = (\uparrow) y_3 + (\downarrow) y_7 = y_3 - y_7$$

This shows the simplicity of the actual operations. In each equation it can be seen that the lowest numbered y_r is always multiplied by unity. Furthermore this is true at every stage of the matrix multiplication since the left-most non-zero element in each of $\mathbf{W}_1$, $\mathbf{W}_2$ and $\mathbf{W}_3$ is unity. Also each equation only has two terms because each of $\mathbf{W}_1$, $\mathbf{W}_2$ and $\mathbf{W}_3$ has only two non-zero entries per row. Finally, once a pair of y_r values has been used twice they are not required again and so a pair of u_r values can be stored in the same location in the computer memory. For example, y_0 and y_4 are used to compute u_0 and u_4 but since they are not used again then u_0 and u_4 can replace y_0 and y_4 in the memory of the computer. This is called "computation in place". The stages in the matrix multiplication can be illustrated using a "signal flow graph" as shown in Fig. 8.21.5.

The extreme left-hand column gives the values of y. At each stage of the calculations we move across the diagram towards the right. The diagram shows that a very clear pattern of operations emerges and it is not difficult to imagine how this diagram could be extended to cope with $n = 16, 32, 64$ and so on. Finally to complete the process, $\mathbf{Y}^\ast$ must be rearranged into sequential order.

Many variations of this procedure have been developed. One variation is to rearrange the data y into 'bit reversed order' rather than producing the solution in 'bit reversed order' i.e. $\mathbf{Y}^\ast$. This does not change the essential features of the signal flow

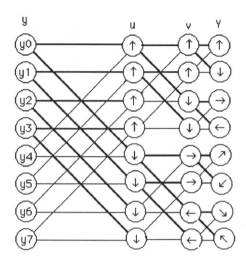

Fig.8.21.5. Signal flow or "butterfly" diagram for the FFT algorithm. The procedure begins with the values of y at the left-hand side of the diagram and progresses to the right. As an example, follow the emboldened line down from y3 to the point where it intersects the line from y7. The result of this flow is u7 = y3 + (↓)y7.

graph geometry nor does it change the total number of operations. However, the order and detail of the arithmetic are changed so that the Fourier coefficients are in sequential order rather than in bit reversed order in the resulting vector. Program 8.21.1 implements this variation of the FFT algorithm.

The procedure is further modified because when coding the algorithm in BASIC it is necessary to work with trigonometric functions rather than with complex exponentials: unlike the programming language FORTRAN, complex arithmetic is not defined in BASIC. The program could, with little modification, be used to analyse complex data and also determine the IDFT using the FFT algorithm. It is useful to mention an important practical point here. The process of calculation can be further speeded up by using look-up tables which provide precomputed values for the required sine and cosine values within the program structure.

```
100    REM ----------------------
110    REM FFT ALGORITHM JP/GL (c)
120    REM ----------------------
130    PRINT "COEFFICIENTS OF THE FINITE FOURIER SERIES"
140    PRINT "(FFT ALGORITHM)": PRINT
150    READ Q
160    P=Q-1: Ep=.001: N=INT(2^Q+Ep)
170    R=N-1: PI=3.141593: K=2*PI/N
180    DIM Re(R),Im(R),Si(N),Co(N)
190    FOR X=0 TO R
200       Im(X)=0
210       READ Re(X)
220       Re(X)=Re(X)/N
230    NEXT X
235    REM SUBROUTINE TO GENERATE SINE/COS TABLE
240    GOSUB 360
250    Flag=1
255    REM SUBROUTINE 500 IMPLEMENTS FFT ALGORITHM
260    GOSUB 500
270    PRINT "COEFFICIENTS OF COS(2*PI*K*X/L) AND SIN(2*PI*K*X/L)
280    PRINT: PRINT"K";TAB(10);"A(K)";TAB(25);"B(K)"
290    PRINT 0;TAB(10);Re(0)
300    FOR X=1 TO R/2
310      PRINT X;TAB(10);Re(X)*2;TAB(25);Im(X)*2
320    NEXT X
330    PRINT R2;TAB(10);Re(R2)
340    PRINT:PRINT"NOTE: A(0) AND A(";R2;") HAVE BEEN DIVIDED BY 2."
350    END
355    REM SUBROUTINE GENERATES SINE AND COS TABLE
360    R1=N/4
370    R2=N/2
380    R3=3*R1
390    FOR X=0 TO R1
400       Si(X)=SIN(K*X)
410       Si(R2-X)=Si(X)
420       Si(R2+X)=-Si(X)
430       Si(N-X)=-Si(X)
440       Co(R1-X)=Si(X)
450       Co(R1+X)=-Si(X)
460       Co(R3-X)=-Si(X)
470       Co(R3+X)=Si(X)
480    NEXT X
490    RETURN
495    REM FFT SUB, BEGIN BIT REVERSAL
500    J=0
510    FOR L=0 TO N-2
520       IF L<=J THEN 550
530       Re=Re(J): Re(J)=Re(L): Re(L)=Re
540       Im=Im(J): Im(J)=Im(L): Im(L)=Im
550       K=R2
560       IF K<J+1 THEN J=J-K:K=K/2:GOTO 560
570       J=J+K
580    NEXT L
```

Program 8.21.1. (continues).

```
585   REM FFT ALGORITHM, Flag=1, INVERSE FFT ALGORITHM Flag<>1
590   IF Flag=1 THEN 630
600   FOR X=0 TO N
610       Si(X)=-Si(X)
620   NEXT X
630   FOR S=0 TO P
640       T=INT(2^S+Ep)
650       D=INT(2^(P-S)+Ep)
660       FOR Z=0 TO T-1
670           L=INT(D*Z+Ep)
680           FOR I=0 TO D-1
690               A=2*I*T+Z: B=A+T
700               F1=Co(L)*Re(B)-Si(L)*Im(B)
710               F2=Si(L)*Re(B)+Co(L)*Im(B)
720               Re(B)=Re(A)-F1
730               Im(B)=Im(A)-F2
740               Re(A)=Re(A)+F1
750               Im(A)=Im(A)+F2
760           NEXT I
770       NEXT Z
780   NEXT S
790   RETURN
800   DATA 3: REM NUMBER OF DATA POINTS EXPRESSED AS A POWER OF 2
810   DATA 0,1.75,3,3.75,4,3.75,3,1.75: REM VALUES OF Y(K)
```

Program 8.21.1.

8.22 TRUNCATING THE FINITE FOURIER SERIES

In the previous analysis n coefficients have been adjusted to make a combination of n trigonometric or complex exponential functions pass through n equi-spaced data points. If the finite Fourier series is truncated then it can be shown that it fits the data approximately according to the unweighted least squares criterion. This property is in marked contrast to the behaviour of a general, non-orthogonal, set of functions: if a non-orthogonal collocating polynomial is truncated it does not satisfy the least squares criterion.

8.23 USING THE FINITE FOURIER SERIES

We will conclude this chapter with two examples of curve fitting using the discrete Fourier series. Program 8.21.1 has been used to determine the coefficients of the finite Fourier series appropriate to the data of Table 8.23.1. The results confirm the fact that the data was generated from the expression

$$y = 1 + 3\cos(2\pi x) + \cos(4\pi x) - 2\cos(6\pi x) + \cos(8\pi x)$$

$$+ 2\sin(2\pi x) - 3\sin(4\pi x) + \sin(6\pi x) \quad (8.23.1)$$

Table 8.23.1.

x	y
0	4
0.125	2.65685425
0.25	2
0.375	1.58578643
0.5	2
0.625	-8.65685425
0.75	0
0.875	4.41421357

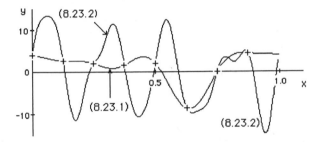

Fig. 8.23.1. The eight data points of Table 8.23.1 and the functions defined
by equations (8.23.1) and (8.23.2).

Consider the following equation:

$$y = \cos(2\pi x) + \cos(4\pi x) + \cos(8\pi x) - 2\cos(10\pi x)$$
$$+ \cos(16\pi x) + 2\cos(18\pi x) + 2\sin(2\pi x)$$
$$+ \sin(6\pi x) + 5\sin(8\pi x) + 3\sin(12\pi x) \qquad (8.23.2)$$

Fig. 8.23.1 shows that the two curves defined by (8.23.1) and (8.23.2) are coincident
at the values of x chosen in Table 8.23.1. However, using this data, program 8.21.1
will *always* determine the coefficients of the series (8.23.1).

From the point of view of curve fitting the function given by (8.23.1) fits the given
data satisfactorily, irrespective of how the data was originally generated. However,
if we wish to obtain an insight into the mechanism of the data generation in a physical
system, then determining the coefficients of (8.23.1) from data which, in reality, was
sampled from the function given by (8.23.2) would lead to a serious misunderstand-
ing of the system. This is an example of a phenomena called "aliasing". Aliasing
can only be avoided by taking additional samples from the continuous function from

Table 8.23.2.

x		y
0	4	0
0.5	3.5	1.75
1	3	3
1.5	2.5	3.75
2		4

Table 8.23.3. Discrete Fourier coefficients for the data of Table 8.23.2 determined using program 8.21.1.

```
COEFFICIENTS OF THE FINITE FOURIER SERIES
(FFT ALGORITHM)

COEFFICIENTS OF COS(2*PI*K*X/L) AND SIN(2*PI*K*X/L)

K         A(K)              B(K)
  0        2.625
  1       -1.707107           0
  2       -.5                 0
  3       -.2928932           0
  4       -.125

NOTE: A(0) AND A( 4 ) HAVE BEEN DIVIDED BY 2.
```

which the data is taken. The optimum criteria which determines the number of samples is called the Nyquist Criteria, see Remirez (1986). A classic example of aliasing is seen when old films attempt to depict the continuous cyclic motion of the spokes of a wheel; the wheel can appear to be rotating backward whilst the vehicle is going forward. This strange result arises because the film attempts to capture continuous motion by a sequence of discrete pictures or samples and aliasing may give a false impression of the cyclic motion.

Considering a second example of the use of the Discrete Fourier Transform, program 8.21.1 has been used to calculate the coefficients of the finite Fourier series using 8 equi-spaced data points taken from Table 8.23.2. This data of has been generated from $y = x(4 - x)$ in the range $x = 0$ to $x = 4$ and the output of the program is shown in Table 8.23.3.

Note that the coefficients of the sine terms are zero. Using these coefficients we can generate the finite Fourier series and Fig. 8.23.2 shows that this series fits the original data of Table 8.23.2 exactly and is a close approximation to the continuous function from which the data was sampled.

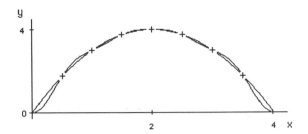

Fig. 8.23.2. The 8 data points of Table 8.23.2, together with the function $y = x(4 - x)$ and the Discrete Fourier Approximation.

8.24 CONCLUSIONS AND SUMMARY

Methods have been described for fitting functions to exact and approximate data. These have included Aitken's method and the use of spline functions for exact data. For fitting approximate data we have given the least squares method, for one and two variable problems, and the minimax method. For fitting harmonic functions to data with an underlying periodic nature we have described the fast Fourier transform algorithm. Table 8.24.1 summarises all the methods and indicates the problems to which they are most efficiently applied.

Table 8.24.1. Methods of fitting functions to discrete data.

Method	Program	Type of fit.	Data interval	Function given explicitly	Function Type
Undetermined coefficients	None	Exact	Any	Yes	Polynomial
Lagrange/Aitken	8.6.1	Exact	Any	No	Polynomial
Cubic spline	8.8.1	Exact	Any	Yes	Piecewise poly.
General l. s.	8.12.1	Approx.	Any	Yes	Any
Polynomial l. s.	8.13.1	Approx.	Any	Yes	Polynomial
Non-linear l. s.	8.14.1	Approx.	Any	Yes	Any
Orthog poly.	8.16.1	Approx.	Even	No	Orthogonal poly.
2-dimension l. s.	8.17.1	Approx.	Any	Yes	2 D polynomial
Minimax	8.18.1	Approx.	Any	Yes	Polynomial
Fast Fourier	8.21.1	Exact	Even	Yes	Trigonometric

PROBLEMS

8.1   It is necessary to generate a curve described by the polynomial $y = f(x)$ that touches the line $y = 3$ tangentially at $x = 2$, touches the line $2x = 3y$ tangentially at $x = 7.5$ and passes through the point $(5, 3.5)$, as shown in the diagram below. Set up the equations necessary to satisfy these requirements and use program 3.6.1 to solve them to determine the coefficients of $f(x)$.

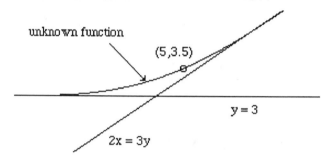

8.2   A definition of the error function erf(z) is given by

$$\text{erf}(z) = (2/\sqrt{\pi}) \int_0^z \exp(-t^2)\, dt$$

The following table gives the values of erf(z) for specific values of z.

z	erf(z)	z	erf(z)
0.5	0.5204999	1.1	0.8802050
0.6	0.6038561	1.2	0.9103141
0.7	0.6778011	1.3	0.9340079
0.8	0.7421010	1.4	0.9522850
0.9	0.7969082	1.5	0.9661053
1.0	0.8427007		

Using Aitken's interpolation method, program 8.6.1, to evaluate erf(z) for $z = 0.76, 1.35$.

8.3   Generate a table of v and $f(v) = v^3 + 14v^2 + 34v - 50$ for $v = 0(0.2)2$
Use Aitken's interpolation method, program 8.6.1, to find the value of v for which $f(v) = 0$, thereby obtaining an approximation to the root of the equation $f(v) = 0$. *Hint.* In Aitken procedure input $f(v)$ as x and v as y. This is an example of inverse interpolation.

8.4 The table below gives values for x and $\cos(\pi x/2)$. Use the methods of (i) least squares, program 8.13.1 and (ii) minimax program 8.18.1 to generate polynomial of order 2 and 4 to fit this data.

x	$\cos(\pi x/2)$	x	$\cos(\pi x/2)$
−1.0	0		
−0.8	0.3090170	0.2	0.9510565
−0.6	0.5877852	0.4	0.8090169
−0.4	0.8090169	0.6	0.5877852
−0.2	0.9510565	0.8	0.3090170
0	1	1.0	0

8.5 Fit the data given in the table below to the following functions
  (i) $f(x) = a + bx + cx^2$ using program 8.12.1
  (ii) $f(x) = a + b/(x + 1) + c/(x + 1)^2$ using program 8.12.1
  (iii) $f(x) = a + be^x + ce^{2x}$ using program 8.12.1

x	y	x	y	x	y
0	1.0	1.25	1.32	2.25	6.2
0.25	0.707	1.5	1.84	2.50	9.88
0.5	0.707	1.75	2.66	2.75	16.1
0.75	0.806	2.0	4.0	3.0	27.0
1.0	1.0				

8.6 A theoretical study has shown that the following experimental data should fit the function $y = a \exp(b(x − c)^2)$, where, the constants a, b and c are expected to have values in the region of 3, 1.5 and 1 respectively.

x	y	x	y	x	y
0	13	0.8	3	1.6	9
0.2	7	1.0	3	1.8	18
0.4	5	1.2	4	2.0	40
0.6	3	1.4	5		

Fit the function to the data to determine the actual values of a, b and c by
    (i) by using program 8.14.1. The convergence is slow.
    (ii) linearising y with respect to the coeficients a, b and c and then use program 8.13.1.

8.7 Fit a third order polynomial to the following table of data (i) by using the polynomial least square program, 8.13.1, (ii) by defining $z = x − 12$ and determining a polynomial in z by using the same program.

x	y	x	y	x	y
10	4.2	11.5	0.1	13	2.8
10.5	−0.1	12	2.2	13.5	−0.4
11	−0.9	12.5	3.4	14	−8.3

Compare the values of the polynomials obtained from (i) and (ii) for $x = 10.6$ with the result obtained using orthogonal polynomials, program 8.16.1, with the same data.

8.8   Fit a cubic spline function using program 8.8.1 and a 5th degree polynomial using program 8.13.1 to the following data.

x	y	x	y	x	y
−2	4	2	−4	4	−40
0	0	3	−30	5	−50

Evaluate the least squares polynomial and the cubic spline for selected values of x within the in the range $-2 \leq x \leq 5$ and decide which gives the better representation of the data.

8.9   Use program 8.21.1 to determine the coefficients of the discrete Fourier series for the following data:

t	y	t	y
0	0	1	1
0.25	0.0625	1.25	1.5625
0.5	0.25	1.5	2.25
0.75	0.5625	1.75	3.0625

Using program 8.12.1 (general least squares), demonstrate that if a truncated version of the discrete Fourier series is used as the basis of a least squares fit then the values of the coefficients determined will be unchanged, even when fewer terms are used in the least squares approximation.

8.10 The data for problem 8.9 was obtained by sampling the function $y = t^2, 0 \leq t \leq 2$. The Fourier series for this function is

$$f(t) = \frac{4}{3} + \sum_{n=1}^{\infty} \frac{16(-1)^n}{(n\pi)^2} \cos\left(\frac{n\pi t}{2}\right)$$

(i) Show that if this series is truncated to 8 terms it does not fit the data of problem 8.9 accurately. In contrast, show that the finite series approximation computed using program 8.21.1 does give an exact answer. Use $t = 1$ and $t = 1.5$ as test values.

(ii) Show that for values of t not sampled in the data of problem 8.9 the finite series approximation does not fit the function $y = t^2$.

9

Monte Carlo integration

9.1 INTRODUCTION

Most versions of BASIC used in microcomputers have a function that returns random numbers for use within a program. This facility is provided in low-cost home computers so that computer games can include an element of chance. More expensive microcomputers used in business also require the facility because sequences of random numbers are used in certain business games and simulations. Random numbers are also used in problems of simulation in science, engineering and statistics.

Perhaps surprisingly, sequences of random numbers also have a role to play in numerical analysis. They can be used to determine solutions for such apparently diverse problems as systems of linear algebraic equations and boundary-value problems. In this text, only the application to definite integrals will be described.

It must be stressed that integration based on the use of random numbers, known as Monte Carlo integration, is only suitable for applications where low accuracy is sufficient. In contrast to the methods of numerical integration described in Chapter 4, Monte Carlo integration does not yield results of high accuracy unless tens or hundreds of thousands of random numbers are used. However, it does have the advantage of offering very simple algorithms which in turn lead to very simple programs. Furthermore, the algorithms are such that the programs used to evaluate repeated integrals are only marginally more complicated than those used to evaluate a single integral. This is in marked contrast to the programs that result from the use of conventional methods.

Before Monte Carlo integration methods are discussed in detail a series of tests are presented which allow the user to attempt to verify that the random number function of the computer is generating numbers that satisfy some of the accepted criteria of randomness. This is important if the results of the Monte Carlo integration are to be free from bias. The user should be cautious in accepting any random number generator for applications that are critical and in recent years concern has been expressed as to the quality and implementation of some random number gereators, particularly those used in microcomputers. (Modianos *et al* 1987 and Sawitzki 1985). If the random number function provided in BASIC is considered to be unsatisfactory then a suitable alternative algorithm may be programmed. For guidance on this see Ripley (1983 and 1987)

9.2 RANDOM NUMBERS

It is difficult to verify that a group of numbers constitutes a set or sequence of random numbers. Random numbers are required to be statistically independent and to have a uniform probability density in a particular range. This implies that each member of the set of numbers has an equal likelihood of taking any value in the range so that the value of each member of the set cannot be predicted. No single test has been devised to verify that a group of numbers constitutes a set or sequence of random numbers and it is necessary to perform several different tests before such a statement can be made with some degree of confidence.

The fact that random number functions in BASIC gives numbers that are reasonably random can be nicely illustrated on computers with medium or high resolution graphics. The random number function, suitably scaled, is used to produce pairs of x and y coordinates confined within the area of the computer graphics display and the screen element corresponding to these coordinates is changed from, say, white to black. The screen will be seen to change colour; rapidly at first and then more slowly as the number of unchanged elements decreases. Fig. 9.2.1 shows this

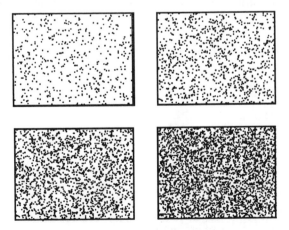

Fig 9.2.1. Randomly selected groups of 500, 1000, 2000 and 4000 points.

effect using 500, 1000, 2000 and 4000 points.

If this test is performed several times and no particular pattern of change is discernible then it is an indication that the random number function is indeed producing random numbers. However, suppose it is noted that areas near the edge of the screen are more often than not the last to change colour. This might indicate that the random number function is producing too few numbers at the extreme values of the range. Whether this is statistically significant is another matter and a visual inspection of the display is unlikely to give this information: more rigorous and quantitative testing methods are required.

9.3 TESTS   FOR   RANDOMNESS

Six tests for randomness are now presented. In each case a particular property of a random number sequence is examined in order that a statistical comparison can be made between the theoretical expectation and the observed results, the latter calculated from the numbers produced by the random number function. To measure the significance of each of these comparisons, the chi-squared test must be used but details of this test are omitted. Without loss of generality the tests will be described on the assumption that the numbers in the sequence are in the range 0 to 1.

(1) *The Frequency Test.* In this test the range 0 to 1 is divided up into non-overlapping sub-intervals, typically 10 in number, and the frequency of numbers in the sequence falling into any sub-interval is compared with the expected frequency. If there are N numbers in the sequence and p sub-intervals are used then the expected frequency in each sub-interval is N/p. A minimum of 50 numbers should be used. Note that this test by itself is insufficient to establish randomness since a sequence of numbers such as 0.01, 0.02, ... , 0.99 and 1 would give an excellent agreement with the expected distribution in the sub-intervals, although clearly this sequence is not random.

(2) *Run Up and Down Test.* Consider the following sequence of numbers:

$$0.0823$$
$$0.1787$$
$$0.2444$$
$$0.3612$$
$$0.2101$$
$$0.0787$$
$$0.2553$$

From the start of the sequence, there is a "run up" of three numbers. This is based on the fact that $0.0823 < 0.1787 < 0.2444 < 0.3612$ but 0.3612 is not less than the fifth number, 0.2101, and so the run ends at the fourth number. This "run up" is followed by a "run down" of two and finally a run up of one. For N numbers, the expected number of runs (both up and down) of length r is given by

$$E(r) = 2[(r^2 + 3r + 1)N - (r^3 + 3r^2 - r - 4)]/(r + 3) \quad r = 1, 2, 3, \ldots$$

and the expected number greater or equal to r is approximately

$$S(r) = 2[(r + 1)N - (r^2 + r - 1))]/(r + 2)$$

A minimum of 50 numbers should be used.

(3) *The Gap Test*. In this test only the first digit of each random number is examined. Consider the following sequence which consists of the first digit of each number of a sequence of numbers which are to be tested for randomness.

2 5 3 7 9 4 1 3 5 7 8 6 5 3 1 9 2 5

From the beginning of the sequence there are, between successive occurrences of the digit "5", gaps of 1, 6, 3 and 4 digits Theoretically the expected gap size should follow a geometric distribution given by

$$E(r) = N(0.9)^r (0.1) \quad \text{for } r = 0, 1, 2, \ldots$$

where N is the number of digits and r is the gap size. At least 100 digits should be used.

(4) *The d-squared Test*. In this test the parameter d^2 is calculated as follows

$$d^2 = (x_1 - x_3)^2 + (x_2 - x_4)^2$$

where x_1, x_2, x_3 and x_4 are four consecutive numbers in a sequence of supposedly random numbers. If x_1 and x_2 represent the x, y coordinates of a point and x_3 and x_4 are the x, y coordinates of another point then d^2 represents the square of the distance between these two randomly chosen points lying in a unit square. The test compares the theoretical and actual distribution of d. Now

$$F(d) = \pi d^2 - 8d^3/3 + d^4/2 \quad \text{for } 0 \leq d \leq 1$$

and

$$F(d) = 1/3 + (\pi - 2)d^2 + 4(d^2 - 1)^{1/2} +$$

$$(8/3)(d^2 - 1)^{3/2} - d^4/2 - 4d^2 \sec^{-1} d \quad \text{for } 1 < d \leq \sqrt{2}$$

Having determined $F(d)$, the number of occurrences between a and b is given by N times the integral of $F(d)$ with respect to d between a and b. At least 200 numbers are required.

(5) *Yule's Test.* Like the gap test, this test only examines the first digit of each random number. Supposing $5N$ numbers are sampled and their first digits are added in groups of 5. Thus N sums are obtained and Yule's test compares these sums with the expected values. Letting the sum of the group of five digits be r, the expected value of r is given by

$$E(r) = (N/10^5)\sum_{j=0}^{5} (-1)^j \{{}^5C_j \; {}^{(r-10j+4)}C_4\}$$

where ${}^nC_r$ represents the number of ways of choosing r things from n without regard to order. The test should be used with a minimum of 500 digits.

(6) *The Serial Test.* This is an extension of the frequency test but it requires many more numbers: at least 1000. The range is again divided into k non-overlapping sub-intervals and a count is kept of the number of times that a number in the ith interval is followed by a number in the jth interval. Letting this count be c_{ij}, then we can construct a $k \times k$ matrix of counts and it can readily be seen that if the numbers in the sequence are independent then the elements of the count matrix will all have the expected value N/k^2 where N is the number of pairs of random numbers used. The test compares the actual and expected values of c_{ij}.

　　　The value of chi-squared that marks the threshold between passing or failing any one of the above tests is, to some extent, arbitrary, although typically the threshold might be set at 5%. Using this figure it is to be expected that, over many tests, a random number generator will fail approximately one test in 20, even if it is normally producing sequences of numbers that satisfy all the tests for randomness. A higher failure rate would indicate an unsatisfactory performance.

　　　These tests for randomness can be fairly easily programmed. A set of supposedly random numbers should be generated using the random statement and assigned to an array; this allows each test to be applied to the same group of random numbers. However, such a program is not given here because it would be a relatively long.

　　　We will now consider some integration techniques using random numbers.

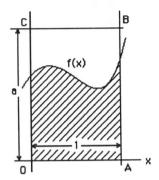

Fig. 9.4.1. The area of the rectangle OABC = a. Consequently the shaded area, which represents the integral of $f(x)$ from 0 to 1 is an unknown fraction of a.

9.4 MONTE CARLO INTEGRATION

Consider the evaluation of

$$\int_0^1 f(x)dx \qquad\qquad (9.4.1)$$

Let us assume that the value of $f(x)$ lies in the range 0 to a as shown in Fig. 9.4.1. The process of integration determines the area of the shaded part of that diagram. If points are chosen at random within the rectangle OABC then the probability that a particular point will fall within the shaded region is equal to the ratio of the area of the shaded region to the area of the rectangle OABC. Thus, by considering a large number of such randomly chosen points, the proportion of the total falling within the shaded region approximates to the ratio of the shaded to total areas. Since the area of the rectangle OABC is known, an approximation to the integral may be obtained. This method, although easy to understand, is not an efficient procedure and will not be discussed in any further detail.

An alternative method for estimating integrals using random numbers will now be presented. An integral may be interpreted as the mean value of the function within the range of integration multiplied by the difference between the upper and lower limits of integration. Thus, if the function $f(x)$ is sampled at a very large number of points the average of the function values will approximate to the mean value of the function and hence an estimation of the integral can be obtained. The only question to be answered is how should the sample values of the function be chosen? Intuitively one might consider sampling $f(x)$ at equal increments of x, but it is been shown in Chapter 4 that equal spacing of the samples can, in adverse circumstances,

lead to a very poor or even grossly erroneous result. An alternative strategy is
to sample the function at randomly chosen values of x. Using probability theory it
can be shown that if points $x_1, x_2, \ldots, x_n$ are selected at random in the range 0 to 1
then an unbiased estimator of the mean of $f(x_1), f(x_2), \ldots, f(x_n)$ is given by

$$f* = [\sum_{i=1}^{n} f_i]/n \qquad (9.4.2)$$

where $f_i = f(x_i)$. Here $f*$ is an unbiased estimator of the mean value of the function
$f(x)$ in the range 0 to 1 and an estimate of the integral of the function in the same
range. Furthermore, the maximum likely error in the approximation can be predicted.

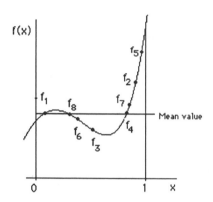

Fig. 9.4.2. Function $f(x)$ sampled at eight randomly selected points.

Referring to Fig. 9.4.2, it can be seen intuitively that the likely error in the
estimate of the mean value varies in the same fashion as the deviation of the function
from its mean value. For example, in the extreme case when the function has a
constant value over the range of integration, the sampled values of the function and
the mean value of the function are equal. Hence sample values of the function will
give an exact estimate for the integral. In general, this will not be the case and we
define a measure of the variation of $f(x)$ as the variance of $f(x)$ which is given by

$$\sigma^2 = \int_0^1 [f(x) - I]^2 \, dx \qquad (9.4.3)$$

where

$$I = \int_0^1 f(x)dx$$

On expanding (9.4.3)

$$\sigma^2 = \int_0^1 [f(x)]^2 dx - I^2 \qquad (9.4.4)$$

Of course, we would not normally know the value of I, otherwise there would be no purpose in attempting to estimate it, so it is not possible to evaluate (9.4.4) to obtain σ^2. However, if n random samples are taken of $f(x)$, then the equivalent of (9.4.4) is

$$s^2 = [\sum_{i=1}^n f_i^2 - nf*^2]/(n-1) \qquad (9.4.5)$$

where $f_i = f(x_i)$. The quantity s^2 is an estimate of σ^2. If σ is known then the standard error in the estimation of f is exactly $\sigma/\sqrt{n}$; if σ is not known then the standard error is approximately $s/\sqrt{n}$. It can be stated with a confidence of 95% that the true integral I is bounded as follows

$$f* - 2\sigma/\sqrt{n} < I < f* + 2\sigma/\sqrt{n} \qquad (9.4.6)$$

A close approximation to these bounds is given by

$$f* - 2s/\sqrt{n} < I < f* + 2s/\sqrt{n} \qquad (9.4.7)$$

The bounds given by (9.4.7) can be used in practice since they are in terms of the approximate standard error s rather than the exact standard error σ.

So far only integrals in the range 0 to 1 have been considered but we may require to evaluate integrals in any range (a,b). Let

$$I = \int_a^b g(y) dy$$

To transform the integral (9.4.1) from the range $(0, 1)$ to (a,b) we let $y = rx + a$, where $r = b - a$. Thus $x = (y - a)/r$ and $dx = dy/r$. Substituting for x and dx in (9.4.1) we have

$$I = \int_a^b f[(y-a)/r](dy/r) = \int_a^b g(y) dy \qquad (9.4.8)$$

where $f[(y - a)/r]/r = f(x)/r = g(y)$.

Thus an estimate for integral, I_e, is given by

$$I_e = f^* = [\sum_{i=1}^{n} f_i]/n = [\sum_{i=1}^{n} rg_i]/n = [r\sum_{i=1}^{n} g_i]/n = rg^* \qquad (9.4.9)$$

where $g_i = g(y_i)$ and g^* is an unbiased estimator of the mean value of $g(y)$ in the range (a,b). The variance, σ^2, is given by

$$\sigma^2 = \int_0^1 [f(x)]^2 dx - I^2 = \int_a^b [f\{(y-a)/r\}]^2 (dy/r) - I^2$$

$$\sigma^2 = r \int_a^b [g(y)]^2 dy - I^2 \qquad (9.4.10)$$

and the estimate of variance, s^2, can be shown to be

$$s^2 = r^2 [\sum_{i=1}^{n} g_i^2 - ng^{*2}]/(n-1) \qquad (9.4.11)$$

where g^* is an unbiased estimator of the mean value of the function $g(x)$ in the range a to b. Alternatively, (9.4.11) can be expressed in the form

$$s^2 = [r^2 \sum_{i=1}^{n} g_i^2 - nI_e^2]/(n-1)$$

The computer program 9.4.1 implements (9.4.9) and (9.4.11) to estimate integrands. For example, consider the evaluation of

$$\int_1^3 x^3 dx \qquad (9.4.12)$$

Using program 9.4.1 the results shown in Table 9.4.1 were obtained. It can be seen that as the sample size increases so the estimate generally becomes closer to the exact value of 20 and the bounds (with a confidence of 95%) become closer and encompass the exact value. Note that, in contrast to the examples given in the remainder of this text, the computer outputs given in this chapter are not exactly repeatable. This is, of course, due to the fact that the results are dependent on the generation of sequences of random numbers. Some models of computer always produce the same sequence of random numbers each time a program is run whilst some allow the sequence to be changed by using an appropriate statement in the program. However, different models of computer are likely to produce different sequences of numbers.

```
100    REM ------------------------------------
110    REM MONTE CARLO INTEGRATION JP/GL (C)
120    REM ------------------------------------
130    PRINT "MONTE CARLO INTEGRATION"
135    REM DEFINE FUNCTION IN STATEMENT 140
140    DEF FNA(X)=X^3
150    PRINT
160    PRINT "HOW MANY DIVISIONS OF EVALUATION REQUIRED";
170    INPUT D
180    PRINT "ENTER UPPER LIMIT FOR INTEGRAL";
190    INPUT U
200    PRINT "ENTER LOWER LIMIT FOR INTEGRAL";
210    INPUT L
220    R=U-L:S=0:T=0
230    PRINT
240    PRINT "SAMPLE SIZE";TAB(13);"ESTIMATE";TAB(25);"ST ERROR";
250    PRINT TAB(36);"(EST-2*ST ERR)";TAB(49);"(EST+2*ST ERR)"
255    REM START OF EVALUATION OF INTEGRAL
260    B=1
270    FOR J=1 TO D
275        REM N IS THE NUMBER OF SAMPLES
280        N=4^J
295        REM NOTE WE ADD TO THE CURRENT SET OF B-1 RANDOM NUMBERS
290        FOR I=B TO N
300            X=RND(1)*R+L: F=FNA(X)
310            S=S+F: T=T+F*F
320        NEXT I
325        REM STANDARD ERROR AND ESTIMATE CALCULATIONS
330        EST=R*S/N: T1=T*R*R
340        S2=(T1-EST*EST*N)/(N-1):  SE=SQR(S2/N)
350        PRINT N;TAB(12);EST;TAB(24);SE;TAB(38);EST-2*SE;TAB(50);EST+2*SE
360        B=N+1
370    NEXT J
380    END
```
Program 9.4.1.

Table 9.4.1. Estimate of the integral (9.4.12). The two right-most columns give lower and upper bounds for the estimate with a confidence of 95%.

MONTE CARLO INTEGRATION

HOW MANY DIVISIONS OF EVALUATION REQUIRED? 7
ENTER UPPER LIMIT FOR INTEGRAL? 3
ENTER LOWER LIMIT FOR INTEGRAL? 1

SAMPLE SIZE	ESTIMATE	ST ERROR	(EST-2*ST ERR)	(EST+2*ST ERR)
4	24.77157	7.787768	9.196037	40.34711
16	26.30975	4.383103	17.54354	35.07595
64	21.79399	1.810638	18.17272	25.41527
256	20.58074	.9297948	18.72115	22.44033
1024	20.22226	.4638402	19.29458	21.14994
4096	20.08105	.2325633	19.61592	20.54618
16384	19.94053	.1165297	19.70748	20.17359

Unless stated to the contrary, the standard error referred to in Table 9.4.1 and all subsequent tables is an estimate of the standard error, since the true standard error can only be computed if the exact value of the integral is known. In this example, since the exact value of the integral of (9.4.12) can readily be shown to be 20, σ^2 can be computed from (9.4.10) and is equal to 224.57. Thus the true standard error can be calculated and a comparison between the true and the estimated standard error is given in Table 9.4.2. The table shows a close agreement between the standard error and its estimate: the standard error halves as the sample size is increased by a factor of four.

A series of 1000 estimates for the previous integral have been made using a sample size of 64 for each estimate. The errors between the estimates and the true value of the integral were then determined and these errors were normalised by dividing them by the standard error $\sigma/\sqrt{n}$. A summary of these results is shown in Table 9.4.3. The table shows that the distribution of the normalised errors is close to

Table 9.4.2. Comparison of exact and estimated standard error in the evaluation of the integral (9.4.12).

Sample size	Standard error	
	Estimated	Exact
4	7.787768	7.4928
16	4.383103	3.7464
64	1.810638	1.8732
256	0.929794	0.9366
1024	0.463840	0.4683
4096	0.232563	0.2342
16384	0.116529	0.1171

Table 9.4.3. Comparison of Normal and actual distribution of standard error. By comparing 1000 estimations of (9.4.12) with the exact value the actual distribution was found. Above the dashed line values are within two standard deviation.

Standard Error	Distribution	
	Actual	Normal
±0.5	377	383
±1.0	346	300
±1.5	143	184
±2.0	89	88
±2.5	28	33
±3.0	14	10
±3.5	3	2

a Normal distribution. In particular only 45 out of 1000 estimations gave an error more than twice the standard error. Thus 95.5% of the evaluations gave a result within two standard errors of the exact result, very close to the 95% value bounded theoretically by (9.4.6).

It has been shown that the standard error is inversely proportional to the square root of the sample size so that it is often necessary to take a very large sample size in order to reduce the standard error to an acceptable level. There are, however, several techniques which can be used to reduce the variance and hence the standard error. Two of the simplest of these are now described.

9.5 REDUCING VARIANCE BY MIRROR IMAGE FUNCTIONS

Consider again the integration defined by (9.4.12). Plots of $y = x^3$ and $y = (4-x)^3$ are shown in Fig. 9.5.1. This latter function is the mirror image of $y = x^3$ over the range of integration and will, of course, evaluate to the same value as the original function when integrated from 1 to 3. The mean of these two functions has a smaller variation than the original function and will also integrate to the same value in the range of interest. Thus, by performing a Monte Carlo integration on the mean of the two functions, a smaller standard error will result by virtue of the fact that this function has a smaller variance than the original one.

Using program 9.4.1 to integrate the modified function gave the results shown in Table 9.5.1. Comparing these results with those of Table 9.4.1 it is seen that estimated standard error is reduced by about a factor of four.

The notion of creating an equivalent function as described above is not helpful if the function to be integrated is nearly symmetrical about the mid-ordinate in the

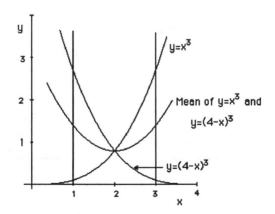

Fig. 9.5.1. Graph of $y = x^3$, $y = (4-x)^3$. Each is a mirror image of the other in the range 1 to 3. The mean of these functions is also shown.

Table 9.5.1. Estimate of integral (9.4.12). The standard error is reduced compared with standard error shown in Table 9.4.1 by using a mirror image function.

```
MONTE CARLO INTEGRATION

HOW MANY DIVISIONS OF EVALUATION REQUIRED?  6
ENTER UPPER LIMIT FOR INTEGRAL?  3
ENTER LOWER LIMIT FOR INTEGRAL?  1

SAMPLE SIZE    ESTIMATE     ST ERROR     (EST-2*ST ERR) (EST+2*ST ERR)
   4          20.26344     1.441997        17.37945       23.14744
  16          21.0376       .9949621       19.04768       23.02752
  64          19.43652      .4386616       18.5592        20.31384
 256          19.80784      .2281779       19.35149       20.2642
1024          19.88526      .1100012       19.66526       20.10526
4096          19.90478     5.533848E-02    19.7941        20.01546
```

range of integration. For example the integral

$$\int_0^1 x(1-x)^{1.1}\, dx$$

is almost symmetrical in the range of integration and so also is its mirror image, $x^{1.1}(1-x)$. This is shown in Fig. 9.5.2. Thus in the range 0 to 1 these two functions and their mean are very similar so that the variance of the mean function differs little from the variance of the original function.

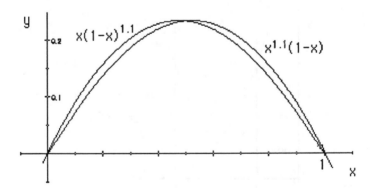

Fig.9.5.2. Two functions that are mirror images of each other. However, the mean of these two functions differs little from the functions themselves.

9.6 REMOVAL OF PART OF THE FUNCTION

Consider the integral

$$I = \int_a^b f(x)dx$$

We may write

$$I = \int_a^b [f(x) - p(x)]dx + \int_a^b p(x)dx \qquad (9.6.1)$$

where $p(x)$ is any function. Let $p(x)$ now be chosen so that it can be integrated analytically and is sufficiently close to $f(x)$ to ensure that $f(x) - p(x)$ has a much smaller variance than $f(x)$ alone. Monte Carlo integration of $f(x) - p(x)$ will give a result with a smaller standard error than integrating $f(x)$ as it stands. The integral of $f(x)$ is then determined by adding the integral $f(x) - p(x)$ (found using Monte Carlo methods) to the integral of $p(x)$ (determined analytically).

As an example of this technique, consider the following integral

$$\int_0^{\pi/2} e^{\sin x}dx \qquad (9.6.2)$$

Table 9.6.1 shows the results obtained when this integral is estimated using program 9.4.1. A function that is readily integrated analytically and is close to $e^{\sin x}$ in the range 0 to $\pi/2$ is

$$p(x) = 1 + x + x^2/2 - x^4/8$$

The integral of $p(x)$ between 0 and $\pi/2$, obtained analytically, is 3.2113831 and Table 9.6.2 shows the result of integrating $f(x) - p(x) + 2.04429978$ by random evaluations of this expression. The constant 2.04429978 is the mean value of $p(x)$ over the range of integration, i.e. $3.2113831/(\pi/2)$. It has been added to the function to be integrated so that the output from program 9.4.1 gives an estimate of the integral of $f(x)$ directly: The variance is unaffected by the addition of this constant term. This has been done for ease of presentation but it is not essential, if this constant term is not added to $f(x) - p(x)$ then the value of the integral of $p(x)$ must be added to the output of program 9.4.1. Comparison of Tables 9.6.1 and 9.6.2 shows that a skilful choice of $p(x)$ significantly reduces the estimated variance.

Table 9.6.1. Estimate of integral (9.6.2).

```
MONTE CARLO INTEGRATION

HOW MANY DIVISIONS OF EVALUATION REQUIRED?  6
ENTER UPPER LIMIT FOR INTEGRAL?  1.570796
ENTER LOWER LIMIT FOR INTEGRAL?  0

SAMPLE SIZE   ESTIMATE     ST ERROR      (EST-2*ST ERR) (EST+2*ST ERR)
    4         3.419173     .5167234        2.385727       4.452621
   16         3.390547     .2307575        2.929031       3.852062
   64         3.257344     9.968974E-02    3.057965       3.456724
  256         3.15409      5.311557E-02    3.047859       3.260322
 1024         3.126978     2.701744E-02    3.072943       3.181013
 4096         3.115789     1.352613E-02    3.088737       3.142841
```

Table 9.6.2. Estimate of integral (9.6.2). The standard error has been reduced compared with that given in Table 9.6.1 by removing part of the function.

```
MONTE CARLO INTEGRATION

HOW MANY DIVISIONS OF EVALUATION REQUIRED?  6
ENTER UPPER LIMIT FOR INTEGRAL?  1.570796
ENTER LOWER LIMIT FOR INTEGRAL?  0

SAMPLE SIZE   ESTIMATE     ST ERROR      (EST-2*ST ERR) (EST+2*ST ERR)
    4         3.069046     6.625049E-02    2.936545       3.201547
   16         3.038221     4.547399E-02    2.947273       3.129169
   64         3.095382     1.830499E-02    3.058772       3.131992
  256         3.101591     9.149326E-03    3.083292       3.119889
 1024         3.103694     4.46704E-03     3.09476        3.112628
 4096         3.104454     2.237495E-03    3.099979       3.108928
```

9.7 REPEATED INTEGRALS

The particular advantage of estimating definite integrals using random samples of the function is seen when evaluating repeated integrals. Integrals of this type are often difficult to evaluate approximately using conventional methods of numerical analysis, but Monte Carlo integration presents few conceptual or programming difficulties. Consider the integral

$$\int_{x_1=a_1}^{x_1=b_1} \int_{x_2=a_2}^{x_2=b_2} \cdots \int_{x_m=a_m}^{x_m=b_m} f(x_1, x_2, \ldots, x_m)\, dx_m \cdots dx_2\, dx_1 \qquad (9.7.1)$$

where the limits of integration may be constants or functions of the other variables. This integral may be written more concisely thus

$$\int_{a_1}^{b_1} dx_1 \int_{a_2}^{b_2} dx_2 \cdots \int_{a_m}^{b_m} f(x_1, x_2, \ldots, x_m)\, dx_m$$

Initially the estimation of a repeated integral with constant limits of integration will be examined.

Consider the integral defined by (9.7.1) where $a_1, a_2, \ldots, a_m$ and $b_1, b_2, \ldots, b_m$ are constant. Let $r_i = b_i - a_i$ $i = 1, 2, \ldots, m$. By similar reasoning to that used to develop (9.4.9), the estimate for the integral, I_e, is given by

$$I_e = (r_1 r_2 \cdots r_m)[\sum_{i=1}^{n} f_i]/n = (r_1 r_2 \cdots r_m)f^* \qquad (9.7.2)$$

where $f_i = f(x_{1i}, x_{2i}, \ldots, x_{mi})$, f^* is the mean of the function over the required domain and n is the number of random evaluations of the function, *not* the number of random numbers that must be used in these evaluations. Since there are m independent variables, $n \times m$ random numbers will be required to make n random evaluations of the function. This sampling is shown diagrammatically in Fig. 9.7.1 for a specific two dimensional function, i.e. a surface. The function is sampled using randomly chosen values of x and y. Now, from (9.7.2), in two dimensions,

$$\int_{a_1}^{b_1} dx \int_{a_2}^{b_2} f(x, y)\, dy = Af^*$$

where $A = r_1 r_2$, the base area.

Thus the value of the repeated integral is the average of the function values sampled, multiplied by the base area. It should be noted that in Fig. 9.7.1 some extra samples of the function are shown along the x and y boundaries; these samples have been included in the diagram to make it more easy to interpret. The extra samples would not be included in the estimate of the mean height of the surface because they were not chosen randomly.

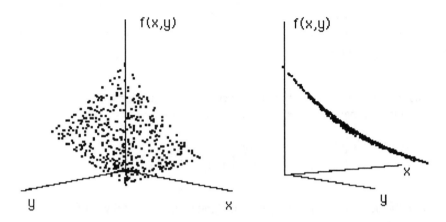

Fig. 9.7.1. The surface $f(x, y) = (2 - x)^2/2^y$ defined by 500 randomly chosen values of the function with
additional values along the boundaries to clarify the picture. In the right-hand diagram the
axes have been rotated so that the surface is viewed almost edge-on.

Having obtained an expression for the unbiased estimate of a repeated integral we
can estimate the variance, s^2, from

$$s^2 = [(r_1 r_2 \ldots r_m)^2 \sum_{i=1}^{n} g_i^2 - n I_e^2]/(n - 1)$$

From the foregoing we can see that there is no essential difference between   the
estimation of single or repeated integrals when using the Monte Carlo method. The
program 9.7.1 listed here is a relatively simple extension of program 9.4.1 to account
for m dimensions. A similar program, program 9.7.2, is also given: it is a simplification of program 9.7.1 in so far as the integral with respect to each variable must
be in the range 0 to 1.

```
100   REM --------------------------------------------------------
110   REM MULTIPLE INTEGRATION BY MONTECARLO METHOD   JP/GL (C)
120   REM --------------------------------------------------------
130   P=1:Q=1:S=0:T=0
140   PRINT "MONTE CARLO INTEGRATION "
150   PRINT "(MULTIDIMENSIONAL INTEGRALS)"
160   PRINT
170   PRINT "HOW MANY DIVISIONS OF INTERVAL REQUIRED";
180   INPUT D
190   PRINT "HOW MANY DIMENSIONS";
200   INPUT M
210   PRINT
```

                        Program 9.7.1.(continues).

```
220    DIM X(M),R(M),L(M)
230    FOR J=1 TO M
240    PRINT"ENTER UPPER LIMIT FOR VARIABLE X(";J;")";
250    INPUT U
260    PRINT "ENTER LOWER LIMIT FOR VARIABLE X(";J;")";
270    INPUT L(J)
280    R(J)=U-L(J): P=P*R(J): Q=Q*R(J)*R(J)
290    PRINT
300    NEXT J
310    PRINT "SAMPLE SIZE";TAB(13);"ESTIMATE";TAB(25);"ST ERROR";
320    PRINT TAB(36);"(EST-2*ST ERR)";TAB(49);"(EST+2*ST ERR)"
330    B=1
340    FOR L=1 TO D
350        N=4^L
360        FOR K=B TO N
365        REM J LOOP SAMPLES EACH DIMENSION
370            FOR J=1 TO M
380                X(J)=RND(1)*R(J)+L(J)
390            NEXT J
400            GOSUB 490
410            S=S+F: T=T+F*F
420        NEXT K
430        EST=S*P/N: T1=T*Q
440        S2=(T1-EST*EST*N)/(N-1): SE=SQR(S2/N)
450        PRINT N;TAB(12);EST;TAB(24);SE;TAB(38);EST-2*SE;TAB(50);EST+2*SE
460        B=N+1
470    NEXT L
480    END
485    REM DEFINE FUNCTION IN TERMS OF THE VARIABLES X(1)...X(N)
490    F=2*X(1)+3*X(2)+4*X(3)
500    RETURN
```

Program 9.7.1

```
100    REM --------------------------------------------------------
110    REM MULTIPLE INTEGRATION BY MONTE CARLO METHOD  JP/GL (C)
120    REM --------------------------------------------------------
130    PRINT"MONTE CARLO INTEGRATION "
140    PRINT"(MULTIDIMENSIONAL INTEGRALS, RANGE 0 TO 1 ONLY)"
150    PRINT
160    PRINT "HOW MANY DIVISIONS OF INTERVAL REQUIRED";
170    INPUT D
180    PRINT"HOW MANY DIMENSIONS";
190    INPUT M
200    PRINT
210    DIM X(M)
220    S=0:T=0
230    PRINT "SAMPLE SIZE";TAB(13);"ESTIMATE";TAB(25);"ST ERROR";
240    PRINT TAB(36);"(EST-2*ST ERR)";TAB(49);"(EST+2*ST ERR)"
250    B=1
260    FOR L=1 TO D
270        N=4^L
```

Program 9.7.2. (continues)

```
280     FOR K=B TO N
290         FOR J=1 TO M
300             X(J)= RND(1)
310         NEXT J
320         GOSUB 400
330         S=S+F: T=T+F*F:
340     NEXT K
350     EST=S/N:S2=(T-EST*EST*N)/(N-1): SE=SQR(S2/N)
360     PRINT N;TAB(12);EST;TAB(24);SE;TAB(38);EST-2*SE;TAB(50);EST+2*SE
370     B=N+1
380 NEXT L
390 END
395 REM DEFINE FUNCTION IN TERMS OF VARIABLES X(1)....X(N)
400 F=EXP(-X(1)*X(2)*X(3)*X(4)*X(5))
410 RETURN
```

Program 9.7.2.

Consider the repeated integral

$$\int_0^1 dx \int_0^1 e^{-xy}dy \qquad\qquad (9.7.3)$$

The value of this integral is 0.796599599 to 9 significant digits. Using program 9.7.2 the estimates given in Table 9.7.1 were obtained. If desired, the variance, and hence the standard error, can be reduced by creating an equivalent function with a smaller variance or by the removal of part of the function; techniques that have been shown above to be helpful when estimating single integrals. One problem that can arise with repeated integrals is the difficulty of visualising the shape of the function; this makes the choice of a similar but readily integrated function difficult.

Table 9.7.1. Estimates of the integral (9.7.3).

```
MONTE CARLO INTEGRATION
(MULTIDIMENSIONAL INTEGRALS, RANGE 0 TO 1 ONLY)

HOW MANY DIVISIONS OF INTERVAL REQUIRED? 6
HOW MANY DIMENSIONS? 2
```

SAMPLE SIZE	ESTIMATE	ST ERROR	(EST-2*ST ERR)	(EST+2*ST ERR)
4	.7993581	9.505539E-02	.6092473	.9894689
16	.759613	3.942686E-02	.6807593	.8384667
64	.7777122	1.890062E-02	.739911	.8155135
256	.777753	1.018316E-02	.7573867	.7981193
1024	.7943675	4.901909E-03	.7845637	.8041713
4096	.7954634	2.444171E-03	.790575	.8003517

Let (9.7.3) be replaced by

$$\int_0^1 dx \int_0^1 [\{e^{-xy} - e^{-(1-x)(1-y)}\}/2]dy \qquad (9.7.4)$$

The function $\{e^{-xy} - e^{-(1-x)(1-y)}\}/2$ is more symmetric in the ranges of integration than e^{-xy}, but (9.7.4) and (9.7.3) will have identical values. Estimates of (9.7.4) have been made using program 9.7.2 and these are shown in Table 9.7.2. It is seen that the variance of the original function of (9.7.3), and hence the standard error in the estimates of the integral, has been reduced approximately by a factor of three.

Table 9.7.2. Estimates of the integral (9.7.3) using the form (9.7.4).

```
MONTE CARLO INTEGRATION
(MULTIDIMENSIONAL INTEGRALS, RANGE 0 TO 1 ONLY)

HOW MANY DIVISIONS OF INTERVAL REQUIRED?  6
HOW MANY DIMENSIONS?  2
```

SAMPLE SIZE	ESTIMATE	ST ERROR	(EST−2*ST ERR)	(EST+2*ST ERR)
4	.8107954	3.039951E-02	.7499964	.8715944
16	.8051832	1.499283E-02	.7751976	.8351689
64	.7927161	5.820981E-03	.7810742	.8043581
256	.7906136	2.894854E-03	.7848239	.7964033
1024	.7961383	1.573016E-03	.7929923	.7992844
4096	.7971682	8.096779E-04	.7955489	.7987875

Alternatively (9.7.3) can be replaced by

$$\int_0^1 dx \int_0^1 [e^{-xy} - (1 - xy) + 0.75]dy \qquad (9.7.5)$$

The repeated integral of $1 - xy$ can be found analytically and in the ranges 0 to 1 it is equal to 0.75. Thus using program 9.7.2 to estimate the integral shown above gives an estimate of the integral defined by (9.7.3); the constant term has no effect on the variance. The estimates are given in Table 9.7.3; compared to the original function, the standard error in the estimates of the integral has been approximately halved.

An example of the use of program 9.7.1 for a three dimensional integral is in the

Table 9.7.3. Estimates of the integral (9.7.3) using the form (9.7.5).

```
MONTE CARLO INTEGRATION
(MULTIDIMENSIONAL INTEGRALS, RANGE 0 TO 1 ONLY)

HOW MANY DIVISIONS OF INTERVAL REQUIRED?  6
HOW MANY DIMENSIONS?  2

SAMPLE SIZE   ESTIMATE      ST ERROR      (EST-2*ST ERR) (EST+2*ST ERR)
    4         .7980821      3.911104E-02     .71986          .8763042
   16         .8080245      1.990817E-02     .7682082        .8478408
   64         .7999921      8.148941E-03     .7836942        .81629
  256         .8037546      4.40643E-03      .7949418        .8125675
 1024         .7968451      2.050314E-03     .7927445        .8009458
 4096         .7963086      1.007745E-03     .7942932        .7983241
```

estimation of the following integral:

$$\int_{-2}^{2} dx \int_{0}^{1} dy \int_{1}^{2} (2x + 3y + 4z) dz \tag{9.7.6}$$

The estimates for this integral are given in Table 9.7.4, these estimates may be compared with the exact value of 30.

Table 9.7.4. Estimates of the integral (9.7.6).

```
MONTE CARLO INTEGRATION
(MULTIDIMENSIONAL INTEGRALS)

HOW MANY DIVISIONS OF INTERVAL REQUIRED?  6
HOW MANY DIMENSIONS?  3

ENTER UPPER LIMIT FOR VARIABLE X( 1 )?  2
ENTER LOWER LIMIT FOR VARIABLE X( 1 )?  -2

ENTER UPPER LIMIT FOR VARIABLE X( 2 )?  1
ENTER LOWER LIMIT FOR VARIABLE X( 2 )?  0

ENTER UPPER LIMIT FOR VARIABLE X( 3 )?  2
ENTER LOWER LIMIT FOR VARIABLE X( 3 )?  1

SAMPLE SIZE   ESTIMATE      ST ERROR      (EST-2*ST ERR) (EST+2*ST ERR)
    4         32.48949      5.509421        21.47065        43.50833
   16         32.99706      2.541916        27.91322        38.08089
   64         30.48457      1.326448        27.83167        33.13747
  256         31.08455      .67264          29.73927        32.42983
 1024         30.59389      .3295738        29.93474        31.25304
 4096         30.03443      .1672693        29.69989        30.36897
```

9.8 REPEATED INTEGRALS WITH VARIABLE LIMITS

Integrals of this form cannot readily be estimated directly using Monte Carlo integration. Initially a series of transformations must be made in order to make the limits of integration constant, arbitrarily chosen here to be 0 and 1. Referring to (9.7.1), a_i and b_i are the limits of integration of x_i and all of these limits may be functions of the other variables except one pair that must be constant. Let the variable x_i be changed to s_i where x_i and s_i are related by

$$x_i = (b_i - a_i)s_i + a_i \qquad i = 1, 2, \ldots ,m \tag{9.8.1}$$

Note that in (9.8.1) when $x_i = b_i$, $s_i = 1$, and when $x_i = a_i$, $s_i = 0$. Thus replacing x_i by s_i changes the range of integration from a_i, b_i to 0, 1. From (9.8.1)

$$\mathrm{d}x_i = (b_i - a_i)\, \mathrm{d}s_i \qquad i = 1, 2, \ldots ,m \tag{9.8.2}$$

Using (9.8.1) we will begin by transforming any variable with constant limits of integration to the range 0 and 1. The order in which the remaining variables are transformed is governed by the fact that a variable cannot be transformed until all other variables on which its limits of integration depend have themselves been transformed. Once the transformation relationships have been established, all the functions of x_i can be replaced by functions of s_i in the integral and Monte Carlo integration can be carried out in the range 0 to 1 using program 9.7.2.

Three examples are given to illustrate this procedure for changing the range of integration.

Example 1: Evaluate

$$\int_0^2 \mathrm{d}x \int_0^2 \mathrm{d}y \int_0^{2-x} xyz \, \mathrm{d}z \tag{9.8.3}$$

Here the variables x, y and z will be replaced by the variables r, s and t respectively. Note z can only be transformed after x has been transformed. Using (9.8.1)

$$
\begin{aligned}
x &= (2-0)r + 0, & \text{hence } x = 2r & \quad \text{and } \mathrm{d}x = 2\mathrm{d}r \\
y &= (2-0)s + 0, & \text{hence } y = 2s & \quad \text{and } \mathrm{d}y = 2\mathrm{d}s \\
z &= [(2-x)-0]t + 0, & \text{hence } z = (2-2r)t & \quad \text{and } \mathrm{d}z = (2-2r)\mathrm{d}t
\end{aligned}
$$

Substituting these relationships in (9.8.3) gives

$$\int_0^1 \mathrm{d}r \int_0^1 \mathrm{d}s \int_0^1 64rs(1-r)^2 t \, \mathrm{d}t \tag{9.8.4}$$

Example 2: Evaluate

$$\int_0^z dx \int_{x-z}^{x+z} dy \int_1^3 (x+y+z)\,dz \tag{9.8.5}$$

Letting x, y, z be replaced by r, s, t respectively and using (9.8.1) gives

$$z = (3-1)t + 1, \qquad \text{hence } z = 2t + 1 \qquad \text{and } dz = 2dt$$
$$x = (z-0)r + 0, \qquad \text{hence } x = (2t+1)r \quad \text{and } dx = (2t+1)dr$$
$$y = [(x+z)-(x-z)]s + (x-z) = 2zs + x - z$$
$$= 2(2t+1)s + (2t+1)r - (2t+1)$$
$$= (2t+1)(2s+r-1), \quad \text{hence } dy = (2t+1)2ds$$

Substituting these relationships into (9.8.5) gives

$$\int_0^1 dr \int_0^1 ds \int_0^1 8(r+s)(2t+1)^3 dt \tag{9.8.6}$$

Example 3. Evaluate

$$\int_1^2 dx \int_x^{x^4} x^2 y\, dy \tag{9.8.7}$$

Letting x and y be replaced by s and t respectively, we have

$$x = (2-1)r + 1, \qquad \text{hence } x = r+1 \text{ and } dx = dr$$
$$y = (x^4 - x^2)s + x^2, \qquad \text{hence } dy = (x^4 - x^2)ds$$

The integral (9.8.7) becomes

$$\int_0^1 dr \int_0^1 x^2(x^4 - x^2)[(x^4 - x^2)s + x^2]\,ds \tag{9.8.8}$$

where $x = r + 1$

The integrals of Examples 1, 2 and 3 have been estimated and the results are shown in Tables 9.8.1, 9.8.2 and 9.8.3. Note that the exact values of the integrals are as follows:

Example 1: Integral = 4/3
Example 2: Integral = 80
Example 3: Integral = 6466/77 = 83.974026

Table 9.8.1. Estimates of the integral (9.8.3) using the form (9.8.4).

```
MONTE CARLO INTEGRATION
(MULTIDIMENSIONAL INTEGRALS, RANGE 0 TO 1 ONLY)

HOW MANY DIVISIONS OF INTERVAL REQUIRED? 6
HOW MANY DIMENSIONS? 3
```

SAMPLE SIZE	ESTIMATE	ST ERROR	(EST-2*ST ERR)	(EST+2*ST ERR)
4	1.012471	.7975596	-.5826485	2.60759
16	2.012874	.5714298	.8700141	3.155733
64	1.448016	.2051471	1.037722	1.85831
256	1.234781	9.550657E-02	1.043768	1.425794
1024	1.344354	4.996822E-02	1.244417	1.44429
4096	1.332001	2.501943E-02	1.281962	1.38204

Table 9.8.2. Estimates of the integral (9.8.5) using the form (9.8.6).

```
MONTE CARLO INTEGRATION
(MULTIDIMENSIONAL INTEGRALS, RANGE 0 TO 1 ONLY)

HOW MANY DIVISIONS OF INTERVAL REQUIRED? 6
HOW MANY DIMENSIONS? 3
```

SAMPLE SIZE	ESTIMATE	ST ERROR	(EST-2*ST ERR)	(EST+2*ST ERR)
4	100.8709	50.08831	.6943207	201.0475
16	94.91748	19.67004	55.57741	134.2576
64	81.77658	8.915792	63.945	99.60817
256	87.65714	4.680614	78.29591	97.01837
1024	83.06628	2.275774	78.51474	87.61783
4096	79.712	1.12377	77.46445	81.95954

Table 9.8.3. Estimates of the integral (9.8.7) using the form (9.8.8).

```
MONTE CARLO INTEGRATION
(MULTIDIMENSIONAL INTEGRALS, RANGE 0 TO 1 ONLY)

HOW MANY DIVISIONS OF INTERVAL REQUIRED? 6
HOW MANY DIMENSIONS? 2
```

SAMPLE SIZE	ESTIMATE	ST ERROR	(EST-2*ST ERR)	(EST+2*ST ERR)
4	102.255	68.32731	-34.39957	238.9097
16	108.1516	40.13004	27.89155	188.4117
64	88.54985	17.50762	53.53461	123.5651
256	84.44028	8.016206	68.40788	100.4727
1024	80.25941	3.89041	72.47859	88.04024
4096	82.45152	1.963229	78.52505	86.37798

9.9 A NOMINALLY MORE DIFFICULT EXAMPLE

To conclude these examples of the use of random numbers in the approximate evaluation of definite integrals the following integral with five dimensions will be estimated:

$$\int_0^1 dv \int_0^1 dw \int_0^1 dx \int_0^1 dy \int_0^1 \exp(-vwxyz)\, dz \qquad (9.9.1)$$

To reduce the variance the readily integrated function

$$p(v,w,x,y,z) = 1 - vwxyz$$

is subtracted from (9.9.1) and its mean value in the ranges of integration, 0.96875 is added to (9.9.1) to compensate for its removal. Thus the integral becomes

$$\int_0^1 dv \int_0^1 dw \int_0^1 dx \int_0^1 dy \int_0^1 [\exp(-vwxyz) - (1 - vwxyz) + 0.96875]dz$$

$$(9.9.2)$$

The value of this integral is 0.97065719 to 8 significant digits and estimates for this integral obtained using program 9.7.2. are shown in Table 9.9.1.

    This integral is described in the title of this section as being a *nominally* more difficult example. In reality it is not; the greater number of dimensions requires more random numbers to be generated and this in turn increases the time required to complete the computation but in all other respects the task of estimating the integral remains essentially simple.

    For further details on the use of Monte Carlo methods for evaluating integrals and other applications see Hammersley & Handscomb (1964) and Shreider (1967).

Table 9.9.1. Estimates of the integral (9.9.1) using the form (9.9.2).

```
MONTE CARLO INTEGRATION
(MULTIDIMENSIONAL INTEGRALS, RANGE 0 TO 1 ONLY)

HOW MANY DIVISIONS OF INTERVAL REQUIRED? 6
HOW MANY DIMENSIONS? 5
```

SAMPLE SIZE	ESTIMATE	ST ERROR	(EST-2*ST ERR)	(EST+2*ST ERR)
4	.9725659	2.047485E-03	.968471	.9766609
16	.9709511	7.378273E-04	.9694755	.9724268
64	.9702898	3.411319E-04	.9696075	.970972
256	.97034	2.700519E-04	.9697999	.9708801
1024	.9708276	2.493357E-04	.9703289	.9713262
4096	.9705384	1.126604E-04	.9703131	.9707637

9.10 CONCLUSIONS AND SUMMARY

In this chapter we have described Monte Carlo methods which are based on probability and provide, relatively efficiently, low accuracy answers to single variable and repeated integrals.

PROBLEMS

9.1   Use Monte Carlo integration, program 9.4.1, to evaluate the integral

$$\int_0^1 \frac{\log_e t}{(1-t)}\, dt$$

Use 4096 interval. Compare your answers with exact value, $-\pi^2/6$. Explain why Monte carlo methods can be used directly, whereas conventional methods cannot.

9.2   Use Monte Carlo integration, program 9.4.1, to evaluate the Fresnel integrals

$$C(1) = \int_0^1 \cos(\pi t^2/2)\, dt \text{ and } S(1) = \int_0^1 \sin(\pi t^2/2)\, dt$$

Using 4096 intervals. Compare your answers with exact values to three decimal places, $C(1) = 0.780$ and $S(1) = 0.438$.

9.3   The Elliptic integral of the second kind is given

$$\int_0^t \sqrt{(1 - \sin^2\beta \sin^2\theta)}\, d\theta$$

Taking β as $\pi/6$ and t as $\pi/2$ evaluate this integral using program 9.4.1 with 4096 intervals. Compare your answer with the exact value to three decimal places which is 1.211.

9.4   Solve the integral

$$\int_0^\pi \sin x \cos kx\ dx$$

for $k = 0$, 2 and 5 using program 9.4.1 with 4096 intervals. Compare your results with the exact answer, $2/(1 - k^2)$, $k \neq 1$. Why do you get such poor results when $k = 5$?

9.5 A definition of the Bessel function of the first kind of order zero, $J_0(z)$ is given by

$$J_0(z) = (1/\pi) \int_0^\pi \cos(z \sin q) \, dq$$

Evaluate this function using program 9.4.1 with 4096 samples for values of $z = 1, 3$ and 5.

9.6 A definition of the error function $\text{erf}(z)$ is given by

$$\text{erf}(z) = (2/\sqrt{\pi}) \int_0^z \exp(-t^2) \, dt$$

Evaluate this function using program 9.4.1 with 4096 samples for values of $z = 0.2, 0.76$ and 1.35

9.7 A definition of the sine integral, $\text{Si}(z)$ is given by

$$\text{Si}(z) = \int_0^z \frac{\sin t}{t} \, dt$$

Evaluate this function using program 9.4.1 with 4096 samples for values of $z = 0.5, 1$ and 2.

9.8 Evaluate the following repeated integrals using program 9.7.2 with 1024 and 4096 samples respectively.

(i) $\displaystyle\int_0^1 dx \int_x^{2-x} (x/y) \, dy$ (ii) $\displaystyle\int_0^2 dx \int_0^{2-x} dy \int_0^{2-x-y} (1 + x + y + z) \, dz$

10

The algebraic eigenvalue problem

10.1 INTRODUCTION

In this chapter we are concerned with the task of determining values of λ and $\mathbf{x}$ which satisfy a set of equations of the form

$$\mathbf{Ax} = \lambda\mathbf{x} \qquad (10.1.1)$$

This matrix equation describes an algebraic eigenvalue problem where $\mathbf{A}$ is a square matrix of coefficients called the system matrix, $\mathbf{x}$ is an unknown column vector and λ is an unknown scalar. The values of λ that satisfy (10.1.1) are called the eigenvalues, characteristic values or latent roots of $\mathbf{A}$ and the corresponding vector solutions $\mathbf{x}$ are called the eigenvectors, characteristic vectors or latent vectors of $\mathbf{A}$. The elements of $\mathbf{A}$ may be real or complex but in this text most emphasis will be placed on the situation where these elements are real.

A large number of algorithms have been developed to solve eigenvalue problems but in this chapter we will consider only some simple iterative methods. If a problem is ill-conditioned, very large or both then one of the more advanced algorithms must be used. In addition, double precision arithmetic may be required to obtain accurate solutions. Readers wishing to familiarise themselves with some of these more advanced methods should refer to the classic text by Wilkinson (1965). A more concise alternative is by Gourlay and Watson (1973).

Eigenvalue problems arise in many branches of science and engineering. For example, consider the torsional oscillations of the three rotor system shown in Fig. 10.1.1. Neglecting the inertia of the shaft, if the rotors are displaced by q_1, q_2 and q_3

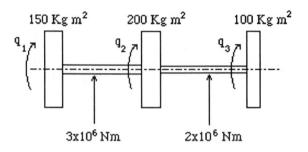

Fig 10.1.1. Three rotor system.

then the equations of motion for the system are

$$150\ddot{q}_1 + 3\times10^6 q_1 - 3\times10^6 q_2 = 0$$
$$200\ddot{q}_2 - 3\times10^6 q_1 + 5\times10^6 q_2 - 2\times10^6 q_3 = 0$$
$$100\ddot{q}_3 - 2\times10^6 q_2 + 2\times10^6 q_3 = 0$$

or in matrix notation

$$\mathbf{M\ddot{q} + Kq} = 0 \qquad\qquad (10.1.2)$$

where

$$\mathbf{M} = \begin{bmatrix} 150 & 0 & 0 \\ 0 & 200 & 0 \\ 0 & 0 & 100 \end{bmatrix} \text{kg m}^2 \text{ and } \mathbf{K} = \begin{bmatrix} 3 & -3 & 0 \\ -3 & 5 & -2 \\ 0 & -2 & 2 \end{bmatrix} 10^6 \text{N m}$$

For free torsional oscillations $\mathbf{q} = \mathbf{x} \sin \omega t$, where ω is the frequency of the oscillations. Differentiating twice and substituting in (10.1.2) gives the linear system of equations

$$-\omega^2 \mathbf{Mx + Kx} = 0$$

or

$$\mathbf{Ax} = \lambda \mathbf{x} \text{ where } \mathbf{A} = \mathbf{M^{-1}K} \text{ and } \lambda = \omega^2$$

The eigenvalues and eigenvectors for this system are

$$\lambda_1 = 0/\text{s}^2 \qquad \text{i.e. } \omega_1 = 0/\text{s or 0 Hz} \qquad \text{and } \mathbf{x}_1^T = [1 \quad 1 \quad 1]$$
$$\lambda_2 = 2\times10^4/\text{s}^2 \quad \text{i.e. } \omega_2 = 141/\text{s or 22.4 Hz and } \mathbf{x}_2^T = [2 \quad 0 \quad 3]$$
$$\lambda_3 = 4.5\times10^4/\text{s}^2 \text{ i.e. } \omega_3 = 212/\text{s or 33.7 Hz and } \mathbf{x}_3^T = [-8 \quad 10 \quad -8]$$

The values ω corresponding to the eigenvalues are the natural frequencies of torsional oscillations. Thus the system oscillates torsionally with natural frequencies 0, 22.4 and 33.7 Hz. The zero natural frequency can be explained as follows. The vector solution corresponding to the zero frequency tells us that each rotor is rotating by an equal amount and the system is behaving as a rigid body. Under these circumstances there are no elastic forces to cause the system to return to its initial position because the shaft is not distorted. If the shaft does not return to its starting position then its period of oscillation is infinitely long and its frequency, which is the reciprocal of its period, is zero. This zero frequency case is not, of course, a real oscillation. However, the non-zero natural frequencies are of considerable engineering significance because if the system is excited by a harmonic force having a frequency close to one of these frequencies then large torsional oscillations will develop in the shaft with potentially serious consequences.

10.2 BACKGROUND THEORY

Before developing computational algorithms we will now summarize some important theoretical properties of eigenvalues and eigenvectors. If (10.1.1) is rearranged into

$$(\mathbf{A} - \lambda \mathbf{I})\mathbf{x} = \mathbf{0} \tag{10.2.1}$$

where $\mathbf{0}$ is a vector of zeros, then the eigenvalues of $\mathbf{A}$ are those values of λ for which

$$\det (\mathbf{A} - \lambda \mathbf{I}) = 0 \tag{10.2.2}$$

Note that in this equation 0 is the scalar zero. If $\mathbf{A}$ is an $n \times n$ matrix then (10.2.2) represents an nth degree polynomial in λ called the characteristic polynomial. The characteristic polynomial has n roots, some of which may be repeated, so an $n \times n$ system has n eigenvalues, λ_i, where $i = 1, 2, \ldots, n$. However, it should be noted that forming and solving the characteristic polynomial explicitly is not a computationally efficient method for determining the eigenvalues of the system. Corresponding to each eigenvalue λ_i there will be an eigenvector solution $\mathbf{x}_i$ but since both sides of (10.1.1) are multiplied by $\mathbf{x}_i$ it follows that $\mathbf{x}_i$ multiplied by any scalar quantity is also a solution. Thus each eigenvector $\mathbf{x}_i$ may be scaled arbitrary.

We will now consider the properties of eigensolutions where the system matrix is (1) real and symmetric, (2) real and non-symmetric and (3) complex with symmetric real and skew-symmetric imaginary parts.

If $\mathbf{A}$ is a real symmetric matrix the eigenvalues of $\mathbf{A}$ are real, but not necessarily positive, and the corresponding eigenvectors are also real. In addition, if λ_i, $\mathbf{x}_i$ and λ_j, $\mathbf{x}_j$ satisfy the eigenvalue problem (10.1.1) and λ_i and λ_j are distinct then

$$x_i^T x_j = 0 \quad i \neq j \tag{10.2.3}$$

and

$$x_i^T A x_j = 0 \quad i \neq j \tag{10.2.4}$$

Equations (10.2.3) and (10.2.4) are called the orthogonality relationships. Note that if $i = j$, then in general $x_i^T x_i$ and $x_i^T A x_i$ are not zero. Recalling that x_i includes an arbitrary scaling factor, then the product $x_i^T x_i$ must also be arbitrary. However, if the arbitrary scaling factor is adjusted so that

$$x_i^T x_i = 1 \tag{10.2.5}$$

then

$$x_i^T A x_i = \lambda_i \tag{10.2.6}$$

and the eigenvectors are then said to be *normalised*. Sometimes the eigenvalues are not distinct and the eigenvectors associated with these equal or repeated eigenvalues are not, of necessity, orthogonal. If $\lambda_i = \lambda_j$ and the other eigenvalues, λ_k, are distinct then

$$x_i^T x_k = 0 \quad k = 1, 2, \dots, n \quad k \neq i, k \neq j.$$

$$x_j^T x_k = 0 \quad k = 1, 2, \dots, n \quad k \neq i, k \neq j. \tag{10.2.7}$$

and for consistancy we can choose to make

$$x_i^T x_j = 0$$

When $\lambda_i = \lambda_j$ the eigenvectors x_i and x_j are not unique and a linear combination of them, i.e. $\alpha x_i + \gamma x_j$ where α and γ are arbitrary constants, also satisfy the eigenvalue problem. One important result is that an order n symmetric matrix always has n distinct eigenvectors even if some of the eigenvalues are repeated.

Let us now consider the case where A is real but not symmetric. A pair of related eigenvalue problems can arise as follows

$$A x = \lambda x \tag{10.2.8}$$
$$A^T y = \beta y \tag{10.2.9}$$

and (10.2.9) can be transposed to give

$$y^T A = \beta y^T \tag{10.2.10}$$

The vectors x and y are called the right-hand and left-hand vectors of A respectively.

The eigenvalues of A and A^T are identical, i.e. $\lambda_i = \beta_i$ but the eigenvectors x and y will, in general, differ from each other. The eigenvalues and eigenvectors of a non-symmetric real matrix are either real or pairs of complex conjugates. If λ_i, x_i, y_i and λ_j, x_j, y_j are solutions that satisfy the eigenvalue problems of (10.2.8) and (10.2.9) and λ_i and λ_j are distinct then

$$y_j^T x_i = 0 \quad i \neq j \tag{10.2.11}$$

and

$$y_j^T A x_i = 0 \quad i \neq j \tag{10.2.12}$$

(10.2.11) and (10.2.12) are called the bi-orthogonal relationships. Again note that if, in these equations, $i = j$ then in general $y_i^T x_i$ and $y_i^T A x_i$ are not zero. The eigenvectors x_i and y_i include arbitrary scaling factors and so the product of these vectors will also be arbitrary. However, if the vectors are adjusted so that

$$y_i^T x_i = 1 \tag{10.2.13}$$

then

$$y_i^T A x_i = \lambda_i \tag{10.2.14}$$

We cannot, in these circumstances, describe neither x_i nor y_i as normalised; the vectors still include arbitrary scaling factor, only their product is uniquely chosen.

If, for a non-symmetric matrix $\lambda_i = \lambda_j$ and the remaining eigenvalues, λ_k, are distinct then

$$y_j^T x_k = 0, \quad y_i^T x_k = 0 \quad k = 1, 2, \dots, n \quad k \neq i \; k \neq j$$

$$y_k^T x_i = 0, \quad y_k^T x_j = 0 \quad k = 1, 2, \dots, n \quad k \neq i \; k \neq j$$

For certain matices with repeated eigenvalues the eigenvectors may also be repeated, consequently for an nth order matrix of this type we may have less than n distinct eigenvectors. This type of matrix is called deficient.

The properties of one particular complex matrix, the Hermitian matrix, will now be briefly discussed. The Hermitian matrix, H, is defined as

$$H = A + iB \tag{10.2.15}$$

Where A and B are real matrices such that $A = A^T$ and $B = -B^T$. Here A is symmetric and B is skew symmetric with zero terms on the leading diagonal. Thus by definition the Hermitian matrix, H, has a symmetric real part and a skew-symmetric imaginary part making H equal to the transpose of its complex conjugate, denoted by H^*. Considering now the eigenvalue problem

$$\mathbf{Hx} = \lambda \mathbf{x} \qquad (10.2.16)$$

If λ_i, $\mathbf{x}_i$, are solutions then $\mathbf{x}_i$ is complex but λ_i is *real*. In addition, if λ_i, $\mathbf{x}_i$ and λ_j, $\mathbf{x}_j$ satisfy the eigenvalue problem (10.2.16) and λ_i and λ_j are distinct then

$$\mathbf{x}_i^* \mathbf{x}_j = 0 \quad i \neq j \qquad (10.2.17)$$

and

$$\mathbf{x}_i^* \mathbf{H} \mathbf{x}_j = 0 \quad i \neq j \qquad (10.2.18)$$

where $\mathbf{x}_i^*$ is the transpose of the complex conjugate of $\mathbf{x}_i$. As before, $\mathbf{x}_i$ includes an arbitrary scaling factor and the product $\mathbf{x}_i^* \mathbf{x}_i$ must also be arbitrary. However, if the arbitrary scaling factor is adjusted so that

$$\mathbf{x}_i^* \mathbf{x}_i = 1 \qquad (10.2.19)$$

then

$$\mathbf{x}_i^* \mathbf{H} \mathbf{x}_i = \lambda_i \qquad (10.2.20)$$

and the eigenvectors are then said to be *normalised*.

10.3 FINDING THE DOMINANT EIGENVALUE BY ITERATION

It is the accepted convention that the n eigenvalues, λ_i, of a matrix are numbered such that the eigenvalue with largest magnitude or dominant eigenvalue is written as λ_1 and the smallest as λ_n. The simple iterative procedure described in this section determines the dominant eigenvalue λ_1. The method can be used on both symmetric and non-symmetric matrices. However, for a non-symmetric matrix the user must be alert to the possibility that there is not a single real dominant eigenvalue value but a complex conjugate pair. Under these conditions simple iteration does not converge and a modified form of iteration, described in section 10.10, must be used.

Consider the eigenvalue problem defined by (10.1.1) and let the vector $\mathbf{u}_0$ be an initial, trial, solution. The vector $\mathbf{u}_0$ is an unknown linear combination of all the eigenvectors of the system provided they are linearly independent. Thus

$$\mathbf{u}_0 = \sum_{i=1}^{n} \alpha_i \mathbf{x}_i \qquad (10.3.1)$$

where α_i are unknown coefficients and $\mathbf{x}_i$ are the unknown eigenvectors. Defining an iterative scheme

$$\mathbf{u}_1 = \mathbf{A}\mathbf{u}_0; \ \mathbf{u}_2 = \mathbf{A}\mathbf{u}_1; \ \dots \ ; \mathbf{u}_p = \mathbf{A}\mathbf{u}_{p-1} \qquad (10.3.2)$$

Substituting (10.3.1) into the sequence (10.3.2) we have

$$\mathbf{u}_1 = \sum_{i=1}^{n} \alpha_i \mathbf{Ax}_i \; = \sum_{i=1}^{n} \alpha_i \lambda_i \mathbf{x}_i \quad \text{since } \mathbf{Ax}_i = \lambda_i \mathbf{x}_i$$

$$\mathbf{u}_2 = \sum_{i=1}^{n} \alpha_i \lambda_i \mathbf{Ax}_i \; = \sum_{i=1}^{n} \alpha_i \lambda_i^2 \mathbf{x}_i$$

$$\cdots\cdots\cdots\cdots\cdots\cdots\cdots\cdots\cdots\cdots\cdots\cdots\cdots\cdots\cdots$$

$$\mathbf{u}_p = \sum_{i=1}^{n} \alpha_i \lambda_i^{p-1} \mathbf{Ax}_i \; = \sum_{i=1}^{n} \alpha_i \lambda_i^{p} \mathbf{x}_i \tag{10.3.3}$$

$$= \lambda_1^{p} \left[\alpha_1 \mathbf{x}_1 + \sum_{i=2}^{n} \alpha_i (\lambda_i/\lambda_1)^p \, \mathbf{x}_i \right] \tag{10.3.4}$$

Recalling that the eigenvalues are numbered such that $|\lambda_1| > |\lambda_2| > \ldots > |\lambda_n|$ then $(\lambda_i/\lambda_1)^p$, where $i = 2, 3, \ldots, n$, tends to zero as p tends to infinity. Thus as p becomes large

$$\mathbf{u}_p = \lambda_1^{p} \alpha_1 \mathbf{x}_1 + \delta$$

where δ is a vector of small components. As δ tends to a vector of zeros, $\mathbf{u}_p$ becomes proportional to $\mathbf{x}_1$ and the ratio between corresponding components of $\mathbf{u}_p$ and $\mathbf{u}_{p-1}$ tends to λ_1.

If the algorithm is used exactly as described above, problems could arise due to numeric overflows in the computer. For example, if in (10.3.4) the elements of $\alpha_1 \mathbf{x}_1$ are of the order of unity and the unknown value of λ_1 is 100, then after 19 iterations $\mathbf{u}_{19}$ is of the order 10^{38}; very close to the largest number allowed in many microcomputers. To overcome this problem the algorithm is modified as follows. After each iteration the resulting trial vector is normalised by dividing it by its largest element, thereby reducing the largest element in the vector to unity. This can be expressed mathematically as

$$\mathbf{v}_p = \mathbf{Au}_{p-1} \tag{10.3.5}$$

$$\mathbf{u}_p = \mathbf{v}_p / \max(\mathbf{v}_p) \tag{10.3.6}$$

where $\max(\mathbf{v}_p)$ is equal to the element of $\mathbf{v}_p$ with the maximum modulus. The pair of equations (10.3.5) and (10.3.6) are used iteratively for $p = 1, 2, 3 \ldots$ until convergence is achieved. This change in the algorithm does not effect the rate of convergence of the iteration. In addition to preventing the build up of very large numbers the modification described above has the added advantage that it is now much easier to decide at what stage the iteration should be terminated. Post multiplying the coefficient matrix A by one of its eigenvectors gives the eigenvector multiplied by the corresponding eigenvalue. Thus, when we stop the iteration because u_p is sufficiently close to u_{p-1} to ensure convergence, $\max(v_p)$ will be an estimate of the eigenvalue. If ε is a vector specifying the maximum change acceptable in the vector u from one iteration to the next then iteration can be stopped when

$$|u_p - u_{p-1}| < \varepsilon$$

Alternatively we could carry out a similar test on the change in the eigenvalue or, if required, both criteria could be used. It should be noted that this does not mean that the error in the eigenvector is less than ε, only that the estimated eigenvector is changing by less than ε per iteration.

10.4 RATE OF CONVERGENCE OF THE ITERATION

The rate of convergence of the iteration is primarily dependent on the distribution of the eigenvalues; the smaller the ratios $|\lambda_i/\lambda_1|$, where $i = 2, 3, \ldots ,n$, the faster the convergence. To a lesser extent the rate of convergence will also be determined by the starting vector. Thus when carrying out an iteration by hand it is sensible to use judgement and any available previous experience of similar problems to try to choose an initial approximation as close as possible to the first eigenvector, thereby making α_1 of (10.3.1) large relative to $\alpha_2, \alpha_3, \ldots ,\alpha_n$. When using a computer this particular consideration is of little importance and any starting vector can be used provided that it is not totally deficient in the first eigenvector, making $\alpha_1 = 0$. In practice this is not usually a problem since with limited precision arithmetic and rounding errors it is unlikely that α_1 could be exactly zero and remain so during the iteration. A program to determine the dominant eigenvalue of a system by iteration is not presented here since this basic iterative scheme will be incorporated in several programs later in this chapter. The dominant eigenvalue of the matrix

$$A = \begin{bmatrix} 5 & 4 & 1 & 1 \\ 4 & 5 & 1 & 1 \\ 1 & 1 & 4 & 2 \\ 1 & 1 & 2 & 4 \end{bmatrix} \qquad (10.4.1)$$

will now be determined by iteration.

Table 10.4.1 shows the progress of the iteration using the starting vector $[1, 1, 1, 1]^T$. Note that convergence requires less than 30 iterations to determine the eigenvalue and eigenvector exactly. Table 10.4.2 repeats this iteration with a

Table 10.4.1. Iteration of the matrix (10.4.1) to determine
the dominant eigensolution.

ITER'N	EIGENVALUE	X(1)	X(2)	X(3)	X(4)
STARTING VECTOR >		1	1	1	1
2	10.45455	1	1	.6086956	.6086956
4	10.10638	1	1	.5263158	.5263158
6	10.02618	1	1	.5065274	.5065274
8	10.00652	1	1	.5016286	.5016287
10	10.00163	1	1	.5004069	.5004069
12	10.00041	1	1	.5001017	.5001017
14	10.0001	1	1	.5000254	.5000255
16	10.00002	1	1	.5000063	.5000064
18	10.00001	1	1	.5000015	.5000016
20	10	1	1	.5000004	.5000004
22	10	1	1	.5000001	.5000001

Table 10.4.2. Iteration of the matrix (10.4.1), using a different starting vector
from that used in Table 10.4.1, to find the dominant eigensolution.

ITER'N	EIGENVALUE	X(1)	X(2)	X(3)	X(4)
STARTING VECTOR >		1	.5	.3333333	.25
2	9.593407	1	.9931272	.4719358	.467354
4	9.967988	1	.9999303	.4928292	.4926437
6	9.992711	1	.9999993	.498188	.4981806
8	9.998184	1	1	.499546	.4995457
10	9.999546	1	1	.4998864	.4998864
12	9.999886	1	1	.4999716	.4999717
14	9.999971	1	1	.4999929	.4999929
16	9.999992	1	1	.4999982	.4999982
18	9.999998	1	1	.4999996	.4999996
20	10	1	1	.4999999	.4999998

different starting vector, $[1, 1/2, 1/3, 1/4]^T$. It is seen that the iteration is equally fast; the different starting vector having little effect except in the early stages of the iteration. In the test matrix (10.4.1) the eigenvalues are known to be 10, 5, 2 and 1 so that the ratios $|\lambda_i/\lambda_1|$ are 0.5, 0.2 and 0.1. Hence, we would expect convergence to be fast.

Since the eigenvectors are known we can, for test purposes, construct a trial vector which is totally deficient in $\mathbf{x}_1$. For example, if the vector is constructed from $\mathbf{x}_2 + \mathbf{x}_3 + \mathbf{x}_4$ then theoretically the iteration will converge to λ_2 rather than to λ_1. Table 10.4.3 shows that from the starting vector $[-2, 0, 1, 3]^T$ (i.e. $\mathbf{x}_2 + \mathbf{x}_3 + \mathbf{x}_4$) rounding errors cause the iteration to converge to λ_1 although, because of the very poor starting vector, convergence is initially very slow.

Table 10.4.3. Iteration of the matrix (10.4.1) to determine the dominant eigensolution. The starting vector is deficient in the dominant eigenvector but rounding errors cause convergence.

ITER'N	EIGENVALUE	X(1)	X(2)	X(3)	X(4)
STARTING VECTOR >		-2	0	1	3
5	4.962085	-.4976125	-.4972942	.9898122	1
10	4.999597	-.4999842	-.4999841	.9998951	1
15	4.999665	-.5003303	-.5003303	.9999989	1
20	4.989399	-.5106232	-.5106232	1	1
25	4.636915	-.8915157	-.8915157	1	1
30	9.482706	1	1	.363622	.363622
35	9.982035	1	1	.4955004	.4955003
40	9.999437	1	1	.4998591	.4998591
45	9.999983	1	1	.4999956	.4999955

Let us now determine the dominant eigenvalue of

$$A = \begin{bmatrix} 2 & 1 & 3 & 4 \\ 1 & -3 & 1 & 5 \\ 3 & 1 & 6 & -2 \\ 4 & 5 & -2 & -1 \end{bmatrix} \tag{10.4.2}$$

Table 10.4.4 shows that the convergence of the iteration is extremely slow, requiring more than 800 iterations to obtain 3 significant digit accuracy and approximately 1500 iterations to obtain 7 significant digit accuracy. The reason for this very slow convergence can be found by examining the distribution of the eigenvalues. The eigenvalues of the matrix (10.4.2) are approximately –8.0286, 7.9329, 5.6689 and –1.5732. Thus $|\lambda_2/\lambda_1| = 0.9881$. This ratio is close to unity and only when this quantity is raised to a very large power does it become negligible compared to one.

Iteration can also be used to determine the smallest eigenvalue of a system. The eigenvalue problem $Ax = \lambda x$ can be rearranged to give

$$A^{-1}x = (1/\lambda)x$$

Here iteration will converge to the largest value of $1/\lambda$, that is the smallest value of λ. However, as a general rule, matrix inversion should be avoided, particularly in large systems and the methods of sections 10.7, 10.8 and 10.9 should be used in preference.

Table 10.4.4. Iteration of matrix (10.4.2) converges very slowly
to the dominant eigensolution.

ITER'N	EIGENVALUE	X(1)	X(2)	X(3)	X(4)
STARTING VECTOR >		1	1	1	1
100	7.576461	.7694737	.357043	1	.1774226
200	6.808997	.8906255	.5691854	1	-8.528865E-03
300	7.048663	.9504864	1	.6708257	-.5475776
400	7.894418	.6001356	1	.0513622	-.8513534
500	7.852035	.4642009	1	-.1889874	-.9692173
600	-7.900104	-.4163072	-.9921674	.2657602	1
700	-7.989618	-.3978767	-.9805174	.286584	1
800	-8.016809	-.3923596	-.97703	.2928177	1
900	-8.025028	-.3906995	-.9759806	.2946932	1
1000	-8.027507	-.3901994	-.9756646	.2952584	1
1100	-8.028255	-.3900486	-.9755692	.2954287	1
1200	-8.028481	-.3900031	-.9755405	.2954801	1
1300	-8.028549	-.3899893	-.9755318	.2954957	1
1400	-8.028569	-.3899852	-.9755291	.2955004	1
1500	-8.028574	-.3899843	-.9755285	.2955014	1
1600	-8.028574	-.3899843	-.9755285	.2955014	1

10.5 SHIFTING THE ORIGIN OF EIGENVALUES

We have seen that the rate of convergence of the iteration is controlled by the ratio $|\lambda_i/\lambda_1|$ where $i = 2, 3, \ldots, n$. The smaller these ratios, the faster is the convergence. If the origin of the spectrum of eigenvalues is moved, these ratios would change and under suitable conditions the change could be beneficial. This change of origin can be achieved by subtracting μI from both sides of the (10.1.1), giving

$$(A - \mu I)x = (\lambda - \mu)x \qquad (10.5.1)$$

The eigenvalues of this system are now $\lambda - \mu$: the origin of the spectrum of eigenvalues has been moved in a positive sense by μ and iteration will lead to the largest value of $\lambda - \mu$.

The shifting process can be used to advantage in many problems but in order to use the technique as a systematic procedure, rather than by trial and error, some knowledge of the distribution of the eigenvalues is required. We will now give examples of the use of this method in three situations.

(1) To determine the smallest eigenvalue of a system: By shifting the origin we can turn the smallest eigenvalue into the dominant eigenvalue of the new system. For example, consider the matrix

$$A = \begin{bmatrix} 8.8 & -2.6 & -1.2 \\ -0.8 & 6.6 & -0.8 \\ -1.4 & -2.6 & 8.6 \end{bmatrix} \qquad (10.5.2)$$

This system has eigenvalues equal to 10 and close to 9.08 and 4.92. If we shift the origin by 9 the new eigenvalues are 1 and close to 0.08 and −5.08. Thus −5.08 is the eigenvalue of largest magnitude and the iteration will converge to it. Having determined this value the shift value can be added to it to provide the smallest eigenvalue. Recalling that λ_1 is the largest eigenvalue and λ_n the smallest eigenvalue, then theoretically if

$\mu > (\lambda_n + \lambda_1)/2$ where $(\lambda_n + \lambda_1)$ is positive

or

$\mu < (\lambda_n + \lambda_1)/2$ where $(\lambda_n + \lambda_1)$ is negative

the iteration converges to the smallest eigenvalue. In practice we cannot use these relationships directly to help us determine λ_n since the relationships require a knowledge of λ_n. The only certain course of action is to shift the origin by an amount greater than λ_1. Table 10.5.1 shows the number of iterations required to reach 5 digit accuracy in the determination of the smallest eigenvalue for this system. The optimum rate of convergence is when the shift is in the region of 9.5 but shifts greater than 12 converge satisfactorily.

Table 10.5.1. The effect of the shift of origin on convergence. When the shift is 9.5 only six iterations are required to obtain the eigenvalue of the matrix (10.5.2) to 4 decimal places.

```
ERROR =  .0001
STARTING VECTOR = 1   1   1

SHIFT      ITER'NS    EIGENVALUE
 7.5         264        4.921587
 8            21        4.921511
 8.5          11        4.921512
 9             8        4.921548
 9.5           6        4.92154
10             7        4.921535
11             8        4.921495
12            10        4.921489
15            15        4.92145
```

(2) To speed up convergence: If an eigensystem has a second eigenvalue close
     to the dominant eigenvalue then shifting the origin will speed the
     convergence. For example, consider the matrix (10.5.2). With the given
     distribution of eigenvalues the rate of convergence is controlled by the ratios
     0.908 and 0.492 and the first value is too close to one to give fast convergence.
     If the origin is shifted by 6.9 the eigenvalues for the shifted system become
     3.1 and close to 2.18 and −1.98. Thus the ratios controlling the rate of
     convergence are now 0.703 and 0.639; a worth while improvement. Table
     10.5.2 shows the approximate number of iterations required to reach 4
     significant digit accuracy. If the shift is greater than about 6.9 the
     convergence begins to slow again. Shifts greater than about 7.5 converge to
     λ_3 as previously described.

Table 10.5.2. The effect of shifting the origin on the rate of convergence to find
the dominant eigenvalue of (10.5.2). Results given single and double precision.

```
ERROR =   .0001
STARTING VECTOR = 1   1   1

  0            78        9.99944      78      9.9994395179455
  2            63        9.999515     63      9.9995143937793
  4            49        9.99968      49      9.9996796572778
  6            33        9.999781     33      9.9997799417909
  6.5          29        9.999816     29      9.9998155366236
  6.7          28        9.999853     28      9.9998523833745
  6.9          39        10.00001     39      10.000004718926
```

A second example of an eigensystem that has eigenvalues of similar magnitude is the matrix of (10.4.2). Table 10.4.4 shows that iteration converges
very slowly because $\lambda_1 = -8.029$ and $\lambda_2 = 7.933$, a ratio of approximatey
0.988. However, if the origin is shifted by 7, the eigenvalues of the shifted
system become −15.029 and 0.933 approximately; a ratio of 0.062. This
significant change in the critical ratio has a dramatic effect on the rate of
convergence as a comparison of tables 10.4.4 and 10.5.3 illustrate.

(3) To achieve convergence to a dominant eigenvalue where the system has a
     pair of dominant eigenvalues that are identical in magnitude but of opposite
     sign. Under these conditions iteration does not converge since $|\lambda_1/\lambda_2| = 1$.
     If the origin of the system is shifted then the eigenvalues can be separated.
     For example, the matrix

$$\begin{bmatrix} 3 & 5 \\ -1 & -3 \end{bmatrix}$$

has eigenvalues equal to 2 and –2. Iteration of this matrix fails to converge but a shift of 1 gives rapid convergence.

In conclusion, we see that shifting the origin is a simple and powerful tool to accelerate convergence but to use the technique to greatest effect requires some knowledge of the distribution of the eigenvalues; information that may not be available.

Table 10.5.3. Shifting the origin by 7 produces a rapid convergence to the dominant eigenvalue for matrix (10.4.2).

ITER'N	EIGENVALUE	X(1)	X(2)	X(3)	X(4)
STARTING VECTOR >		1	1	1	1
2	16.66667	-.6551725	1	.2413793	.1034483
4	-3.94425	-.3409743	1	5.189433E-02	-.309774
6	-5.721647	3.805472E-02	1	-.1297447	-.6764849
8	-7.039551	.259227	1	-.2356314	-.8896462
10	-7.669629	.3509508	1	-.2795446	-.9780416
12	22.04784	-.3799405	-.9906547	.2923813	1
14	-8.045215	-.3867252	-.9804356	.2944898	1
16	-8.033982	-.3889242	-.9771234	.2951732	1
18	-8.030336	-.3896388	-.976047	.2953953	1
20	-8.02915	-.3898713	-.9756969	.2954676	1
22	-8.028764	-.3899469	-.975583	.295491	1
24	-8.028639	-.3899716	-.9755459	.2954987	1
26	-8.028599	-.3899795	-.9755338	.2955012	1
28	-8.028584	-.3899822	-.9755299	.295502	1
30	-8.028581	-.389983	-.9755286	.2955022	1

10.6 RAYLEIGH'S QUOTIENT

Consider the eigenvalue problem (10.1.1). If x_i and λ_i are solutions then $Ax_i = \lambda_i x_i$ where in this case we take A as a symmetric matrix. Premultiplying this expression by x_i^T and rearranging gives

$$\lambda_i = x_i^T A x_i / x_i^T x_i$$

This expression in itself is not useful since if an eigenvector x_i is known then the eigenvalue would also be known. However, suppose we consider an arbitrary vector **u**. Then we may define $Q(\mathbf{u})$ as

$$Q(\mathbf{u}) = \mathbf{u}^T A \mathbf{u}/\mathbf{u}^T \mathbf{u}$$

The scalar quantity $Q(\mathbf{u})$ is called Rayleigh's quotient and if $\mathbf{u} = \mathbf{x}_i$ then $Q(\mathbf{u}) = \lambda_i$. We will now show how this ratio may be used to provide approximations for eigenvalues.

An arbitrary vector may be expressed as a linear combination of the system eigenvectors providing they are linearly independent. Thus

$$\mathbf{u} = \sum_{i=1}^{n} \alpha_i \mathbf{x}_i \qquad (10.6.1)$$

where, without loss of generality, we will assume that the $\mathbf{x}_i$ are normalised such that $\mathbf{x}_i^T \mathbf{x}_i = 1$. Thus

$$A\mathbf{u} = \sum_{i=1}^{n} \alpha_i A\mathbf{x}_i = \sum_{i=1}^{n} \alpha_i \lambda_i \mathbf{x}_i$$

Hence

$$Q(\mathbf{u}) = (\sum_{i=1}^{n} \sum_{j=1}^{n} \alpha_i \alpha_j \lambda_i \mathbf{x}_i^T \mathbf{x}_j)/(\sum_{i=1}^{n} \sum_{j=1}^{n} \alpha_i \alpha_j \mathbf{x}_i^T \mathbf{x}_j) \qquad (10.6.2)$$

In these double summations terms of the form

$$\mathbf{x}_i^T \mathbf{x}_j = 0, \text{ if } i \neq j, \text{ due to orthogonality.}$$
$$= 1, \text{ if } i = j, \text{ due to normalisation.}$$

Hence (10.6.2) becomes

$$Q(\mathbf{u}) = (\sum_{i=1}^{n} \alpha_i^2 \lambda_i)/(\sum_{i=1}^{n} \alpha_i^2) \qquad (10.6.3)$$

Suppose the choice of $\mathbf{u}$ is not completely arbitrary but instead $\mathbf{u}$ is chosen to be a reasonable estimate of $\mathbf{x}_k$. Then we can write

$$|\alpha_i/\alpha_k| = \varepsilon_i \ll 1 \quad i = 1, 2, \dots ,n ; i \neq k$$

where $\ll$ means much less than. Dividing (10.6.3) by α_k we obtain

$$Q(\mathbf{u}) = (\lambda_k + \sum_{\substack{i=1 \\ i \neq k}}^{n} \varepsilon_i^2 \lambda_i)/(1 + \sum_{\substack{i=1 \\ i \neq k}}^{n} \varepsilon_i^2)$$

$$Q(\mathbf{u}) = \lambda_k(1 + O(\varepsilon_i^2))$$

here $O(\varepsilon_i^2)$ denotes an expression involving terms in ε_i of the second or higher degree. Thus we may conclude that if $\mathbf{u}$ differs from an eigenvector $\mathbf{x}_k$ by a small quantity in the first degree then $Q(\mathbf{u})$ differs from λ_k by a small quantity in the second degree. Further more, if $k = 1$ then

$$Q(\mathbf{u}) = (\lambda_1 + \sum_{i=2}^{n} \varepsilon_i^2 \lambda_i)/(1 + \sum_{i=2}^{n} \varepsilon_i^2)$$

$$= (\lambda_1 + \sum_{i=2}^{n} \varepsilon_i^2 \lambda_i)(1 + \sum_{i=2}^{n} \varepsilon_i^2)^{-1}$$

$$= \lambda_1 + \sum_{i=2}^{n} \varepsilon_i^2 \lambda_i - \lambda_1 \sum_{i=2}^{n} \varepsilon_i^2$$

where terms ε_i^4 and higher powers have been neglected in the above equation. Since

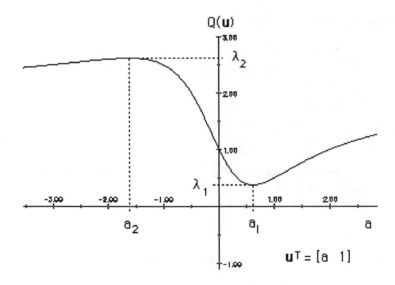

Fig. 10.6.1. Variation of Rayleigh's quotient $Q(\mathbf{u})$ with trial vector for a second order eigensystem. The extreme values of $Q(\mathbf{u})$ are equal to the two eigenvalues of the system. $Q(\mathbf{u})$ is insensitive to change in the vector in the region of the extreme values.

$$\lambda_1 > \lambda_i \quad i = 2, 3, \dots, n$$

$$Q(\mathbf{u}) \le \lambda_1$$

It can be shown by an analogous argument that $Q(\mathbf{u}) \ge \lambda_n$. Fig. 10.6.1 illustrates the variation of $Q(\mathbf{u})$ with $\mathbf{u}$ for a second order eigensystem. Note that this simple graphical presentation is only possible for a second order system.

The Rayleigh quotient can be used in two ways. Firstly, if a reasonable estimate of an eigenvector can be made, perhaps based on previous experience of a similar problem, then a useful estimate can be made of the eigenvalue. This is particularly true of the dominant eigenvalue since in some physical systems it is relatively easy

```
100   REM -------------------------------------
110   REM ITERATION & RAYLEIGH QUOTIENT JP/GL (C)
120   REM -------------------------------------
130   PRINT "SYMMETRIC EIGENVALUE PROBLEM"
140   PRINT "DOMINANT EIGENSOLUTION"
150   PRINT
160   It=1:V0=0
170   PRINT "MAX CHANGE IN EIGENVECTORS";
180   INPUT Ep
190   READ N
200   DIM A(N,N),Z(N),V(N)
205   REM READ UPPER TRIANGLE OF MATRIX
210   FOR I=1 TO N
220      FOR J=I TO N
230           READ A(I,J)
240            A(J,I)=A(I,J)
250      NEXT J
260   NEXT I
270   PRINT "SHIFT IN ORIGIN";
280   INPUT P
290   FOR I=1 TO N
300      A(I,I)=A(I,I)-P
310   NEXT I
320   FOR I=1 TO N
330      Z(I)=1
340   NEXT I
345   REM BEGIN ITERATION
350   MAX=0
360   FOR I=1 TO N
370      V(I)=0
380      FOR J=1 TO N
390           V(I)=V(I)+A(I,J)*Z(J)
400      NEXT J
410      IF ABS(V(I))>MAX THEN MAX=ABS(V(I)):IR=I
420   NEXT I
430   Flg=0
440   FOR I=1 TO N
450      Zt=V(I)/V(IR)
```

Program 10.6.1.(continues).

```
460      IF ABS(Zt-Z(I))>Ep THEN Flg=1
470      Z(I)=Zt
480  NEXT I
490  IF ABS(V(IR)-V0)>Ep THEN Flg=1
500  V0=V(IR):It=It+1
510  IF Flg=1 THEN 350
520  PRINT:PRINT "NO OF ITERATIONS = ";It
530  PRINT:PRINT "EIGENVECTOR"
540  De=0:Nu=0
545  REM COMPUTE NUMERATOR AND DENOMINATOR FOR RAYLEIGH QUOTIENT
550  FOR I=1 TO N
560      PRINT Z(I)
570      V(I)=0
580      FOR J=1 TO N
590          V(I)=V(I)+A(I,J)*Z(J)
600      NEXT J
610      De=De+Z(I)*Z(I)
620      Nu=Nu+Z(I)*V(I)
630  NEXT I
640  PRINT
650  PRINT "EIGENVALUE = ";V0+P;" BY ITERATION"
660  PRINT "                  = ";Nu/De+P;" RAYLEIGH QUOTIENT"
670  END
680  DATA 4: REM ORDER OF EIGENVALUE PROBLEM
690  DATA 2,1,3,4,-3,1,5,6,-2,-1:REM UPPER TRIANGLE OF SYMMETRIC MATRIX
```

Program 10.6.1.

Table 10.6.1. Iteration of matrix (10.4.1) to find the dominant eigenvalue.
Rayleigh's quotient gives a better estimate than does iteration alone and always
underestimates the dominant eigenvalue.

STARTING VECTOR = 1 1 1 1

| ITER'N | EIGENVALUE FROM | |
	ITERATION	RAYLEIGH QUOTIENT
2	10.45455	9.965517
4	10.10638	9.997831
6	10.02618	9.999863
8	10.00652	9.999992
10	10.00163	10
12	10.00041	10
14	10.0001	10
16	10.00002	10
18	10.00001	10
20	10	9.999999
22	10	10

to make an approximation to the dominant eigenvector. If several guesses for the dominant eigenvector are made then the highest value $Q(\mathbf{u})$ is chosen since it must be closest to the dominant eigenvalue. Secondly, and more importantly in the context of this work, the Rayleigh quotient can be used to refine the estimate of the dominant eigenvalue that has been computed by iteration. Program 10.6.1 implements direct iteration of a symmetric matrix and then refines the eigenvalue using the Rayleigh quotient.

Table 10.6.1 shows the Rayleigh quotient computed from the estimated eigenvector and also the estimated eigenvalue after every second iteration of the eigensystem of (10.4.1). At every stage the Rayleigh quotient is a more accurate approximation to the dominant eigenvalue (equal to 10) than the estimate given directly from iteration. Furthermore, whilst we know that the Rayleigh quotient always underestimates the dominant eigenvalue the estimate deduced directly from the iteration may be an over or under estimation of the dominant eigenvalue, depending on the starting vector.

Table 10.6.2. Estimate of the dominant eigenvalue for matrix (10.4.2).

```
SYMMETRIC EIGENVALUE PROBLEM
DOMINANT EIGENSOLUTION

MAX CHANGE IN EIGENVECTORS?  1E-3
SHIFT IN ORIGIN?  7

NO OF ITERATIONS =  20

EIGENVECTOR
-.3897868
-.9758241
 .2954413
 1


EIGENVALUE = -8.02958   BY ITERATION
           = -8.028579  RAYLEIGH QUOT
```

Program 10.6.1 has also been used to estimate the dominant eigenvalue of the matrix of (10.4.2) and the output from the program is shown in Table 10.6.2. Note that the origin of the eigenvalues has been shifted by 7 and that iteration stops when the change in the eigenvector is less than 10^{-3}. The error in the dominant eigenvalue deduced directly from iteration is approximately 1.75×10^{-3} whilst that from the Rayleigh quotient is 3×10^{-6}.

10.7 SUBDOMINANT EIGENVALUES BY HOTELLING'S METHOD

In section 10.3 it was shown that, starting with an arbitrary trial vector, iteration of the eigenvalue problem converges to the dominant eigenvalue. If, however, an initial trial vector is chosen that is deficient in x_1 then, theoretically, iteration converges to λ_2 and x_2. In practice, due to rounding errors, this seldom occurs and convergence is still to λ_1 and x_1: a phenomena that has already been illustrated in Table 10.4.3. Thus, if convergence to the dominant eigenvalue is to be avoided we must choose a trial vector that is deficient in x_1 *and remains so during the iteration.*

Procedures that implement these conditions are usually called 'deflation' procedures and one of the simplest is Hotelling's Algorithm. Let us assume the dominant eigensolution for the symmetric matrix A has been determined, presumably by iteration. We can then form a new matrix A_1 such that

$$A_1 = A - \lambda_1 x_1 x_1^T \tag{10.7.1}$$

where x_1 has been normalised according to (10.2.9). Now any arbitrary trial vector u can be expressed as

$$u = \sum_{i=1}^{n} \alpha_i x_i \tag{10.7.2}$$

Thus

$$A_1 u = A u - \lambda_1 x_1 x_1^T u$$

$$= A \sum_{i=1}^{n} \alpha_i x_i - \sum_{i=1}^{n} \alpha_i \lambda_1 x_1 x_1^T x_i \tag{10.7.3}$$

Recalling that, due to the orthogonality relationship, (10.2.3), $x_1^T x_i = 0$ if $i \neq 1$ and that due to normalisation, (10.2.5), $x_1^T x_1 = 1$, we have

$$A_1 u = \sum_{i=1}^{n} \alpha_i \lambda_i x_i - \alpha_1 \lambda_1 x_1 = \sum_{i=2}^{n} \alpha_i \lambda_i x_i \tag{10.7.4}$$

Thus iteration of A_1 will converge to λ_2 and x_2 since u is deficient in x_1. Note that after each iteration the necessary conditions to make the next trial vector deficient in x_1 are reimposed so that rounding errors cannot cause iteration to drift back to the dominant eigensolution. Once x_2 and λ_2 have been determined a new matrix,

A_2, can be formed to ensure that the trial vectors will be deficient in both x_1 and x_2. Thus

$$A_2 = A_1 - \lambda_2 x_2\, x_2^T$$

Iteration of A_2 will give x_3 and λ_3 etc.

A similar method can be developed to find the subdominant eigenvalues and eigenvectors for non-symmetric matrices. If λ_1, x_1 and y_1, the first eigenvalue and corresponding right-hand and left-hand eigenvectors of matrix A have been determined then A_1 can be formed from

$$A_1 = A - \lambda_1 x_1 y_1^T \tag{10.7.5}$$

where x_1 and y_1 have been scaled such that $y_1^T x_1 = 1$. Now the trial vector for the right-hand vector, u, can be expressed as

$$u = \sum_{i=1}^{n} \alpha_i\, x_i \tag{10.7.6}$$

Thus

$$A_1 u = A \sum_{i=1}^{n} \alpha_i\, x_i - \sum_{i=1}^{n} \alpha_i\, \lambda_1 x_1 y_1^T\, x_i \tag{10.7.7}$$

$$= \sum_{i=1}^{n} \alpha_i\, \lambda_i\, x_i - \alpha_1 \lambda_1\, x_1 = \sum_{i=2}^{n} \alpha_i\, \lambda_i\, x_i \tag{10.7.8}$$

Iterating A_1 produces a vector deficient in x_1 and thus convergence is to λ_2. If we wish to deflate further then y_2 is required and this vector can also be determined from A_1. A trial left-hand vector v can be expressed as

$$v^T = \sum_{i=1}^{n} \gamma_i\, y_i^T$$

and thus

$$v^T A_1 = \sum_{i=1}^{n} \gamma_i\, y_i^T\, A - \sum_{i=1}^{n} \gamma_i\, \lambda_1 y_1^T x_1 y_1^T$$

$$= \sum_{i=2}^{n} \gamma_i\, \lambda_i\, y_i^T$$

Thus convergence is to λ_2 and y_2.

```
100   REM -------------------------------------
110   REM DEFLATION OF SYMMETRIC EVP. JP/GL (C)
120   REM -------------------------------------
130   PRINT "SOLUTION OF SYMMETRIC EIGENVALUE PROBLEM"
140   C=1
150   PRINT:PRINT "MAX CHANGE IN EIGENVECTOR";
160   INPUT Ep
170   READ N
180   DIM A(N,N),Z(N),V(N)
185   REM READ UPPER TRIANGLE OF MATRIX
190   FOR I=1 TO N
200       FOR J=I TO N
210           READ A(I,J)
220           A(J,I)=A(I,J)
230       NEXT J
240   NEXT I
250   PRINT "SHIFT IN ORIGIN";
260   INPUT P
270   FOR I=1 TO N
280       A(I,I)=A(I,I)-P
290   NEXT I
300   It=0
310   FOR I=1 TO N
320       Z(I)=1/I
330   NEXT I
340   It=It+1:MAX=0
345   REM BEGIN ITERATION
350   FOR I=1 TO N
360       V(I)=0
370       FOR J=1 TO N
380           V(I)=V(I)+A(I,J)*Z(J)
390       NEXT J
400       IF ABS(V(I))>MAX THEN MAX=ABS(V(I)):IR=I
410   NEXT I
420   Flg=0
430   FOR I=1 TO N
440       Zt=V(I)/V(IR)
450       IF ABS(Zt-Z(I))>Ep THEN Flg=1
460       Z(I)=Zt
470   NEXT I
480   IF Flg=1 THEN 340
490   PRINT:PRINT "AFTER ";It;" ITERATIONS"
500   PRINT "EIGENVALUE(";C;") = ";V(IR)+P
510   PRINT "EIGENVECTOR"
520   FOR I=1 TO N
530       PRINT Z(I)
540   NEXT I
550   IF C=N THEN END
560   C=C+1:S=0
565   REM DEFLATION PROCEDURE
570   FOR I=1 TO N
580       S=S+Z(I)*Z(I)
590   NEXT I
```

Program 10.7.1.(continues)

```
600   FOR I=1 TO N
610      FOR J=1 TO N
620         A(I,J)=A(I,J)-V(IR)*Z(I)*Z(J)/S
630      NEXT J
640   NEXT I
650   GOTO 300
660   DATA  4: REM ORDER OF EIGENVALUE PROBLEM
670   DATA 6,4,4,1,6,1,4,6,4,6: REM COEFFS OF UPPER TRIANGLE OF MATRIX
```

Program 10.7.1.

Table 10.7.1. Complete eigensolution for matrix (10.4.1) using program 10.7.1.

```
SOLUTION OF SYMMETRIC EIGENVALUE PROBLEM

MAX CHANGE IN EIGENVECTOR?  1E-7
SHIFT IN ORIGIN?  0

AFTER  21  ITERATIONS        │  AFTER  26  ITERATIONS
EIGENVALUE( 1 ) = 10         │  EIGENVALUE( 3 ) = 2
EIGENVECTOR                  │  EIGENVECTOR
   1                         │      2.980219E-08
   1                         │     -1.490115E-07
    .4999999                 │      1
    .4999999                 │     -.9999998

AFTER  19  ITERATIONS        │  AFTER  3  ITERATIONS
EIGENVALUE( 2 ) = 5          │  EIGENVALUE( 4 ) = 1
EIGENVECTOR                  │  EIGENVECTOR
  -.4999998                  │      1
  -.4999998                  │     -1
   .9999999                  │     -1.788137E-07
   1                         │      1.788137E-07
```

```
100   REM ---------------------------------------
110   REM DEFLATION OF NONSYMMETRIC EVP JP/GL (C)
120   REM ---------------------------------------
130   PRINT "SOLUTION OF NONSYMMETRIC EIGENVALUE PROBLEM"
140   It=0
150   C=1
160   PRINT:PRINT"MAX CHANGE IN EIGENVECTOR";
170   INPUT Ep
180   READ N
190   DIM A(N,N),Z(N),V(N),P(N),Q(N)
195   REM READ COMPLETE MATRIX
200   FOR I=1 TO N
210      FOR J=1 TO N
```

Program 10.7.2.(continues).

```
220          READ A(I,J)
230       NEXT J
240    NEXT I
250    PRINT "SHIFT IN ORIGIN";
260    INPUT P
270    FOR I=1 TO N
280       A(I,I)=A(I,I)-P
290    NEXT I
300    FOR I=1 TO N
310       Z(I)=1/I:P(I)=1/I
320    NEXT I
325    REM ITERATION BEGINS
330    It=It+1:MAX=0:MAX1=0
340    FOR I=1 TO N
350       V(I)=0:Q(I)=0
360       FOR J=1 TO N
370          V(I)=V(I)+A(I,J)*Z(J)
380          Q(I)=Q(I)+P(J)*A(J,I)
390       NEXT J
400       IF ABS(V(I))>MAX THEN MAX=ABS(V(I)):IR=I
410       IF ABS(Q(I))>MAX1 THEN MAX1=ABS(Q(I)):IS=I
420    NEXT I
430    Flg=0
440    FOR I=1 TO N
450       Zt=V(I)/V(IR): Pt=Q(I)/Q(IS)
460       IF ABS(Zt-Z(I))>Ep THEN Flg=1
470       IF ABS(Pt-P(I))>Ep THEN Flg=1
480       Z(I)=Zt: P(I)=Pt
490    NEXT I
500    IF Flg=1 THEN 330
510    PRINT:PRINT "AFTER ";It;" ITERATIONS"
520    LA=(V(IR)+Q(IS))/2
530    PRINT "EIGENVALUE(";C;") = ";LA+P
540    PRINT "R H EIGENVECTOR      L H EIGENVECTOR"
550    It=0
560    FOR I=1 TO N
570       PRINT Z(I);TAB(20);P(I)
580    NEXT I
590    IF C=N THEN END
595    REM DEFLATION PROCEDURE
600    C=C+1: S=0
610    FOR I=1 TO N
620       S=S+Z(I)*P(I)
630    NEXT I
640    FOR I=1 TO N
650       FOR J=1 TO N
660          A(I,J)=A(I,J)-LA*Z(I)*P(J)/S
670       NEXT J
680    NEXT I
690    GOTO 300
700    DATA 3:REM ORDER OF EIGENVALUE PROBLEM
710    DATA 33 ,16,72,-24,-10,-57,-8,-4,-17: REM COEFFICIENTS OF MATRIX
```

Program 10.7.2.

Programs for these two deflation procedures are given: program 10.7.1 for a symmetric matrix and program 10.7.2 for a non-symmetric matrix. For symmetric matrices iteration stops when the change in the vector per iteration is less than a specified quantity. In the case of non-symmetric matrices this test is applied to both the left-hand and right-hand vector and, in addition, a check is made to ensure that the difference between the eigenvalue computed from the left and right-hand vectors is also less than the same specified quantity.

The Hotelling algorithm may not be numerically stable. However, for well conditioned problems no difficulty should be experienced. For example, Table 10.7.1 shows the output from program 10.7.1 when this program is used to determine all the eigensolutions for the matrix defined in (10.4.1). The eigenvalues are found after a relatively small number of iterations because they are well separated. The accuracy is high; the computed eigenvalues are close to the exact values of 10, 5, 2 and 1.

The same program has been used to determine the eigensolution of the system given by

$$A = \begin{bmatrix} 6 & 4 & 4 & 1 \\ 4 & 6 & 1 & 4 \\ 4 & 1 & 6 & 4 \\ 1 & 4 & 4 & 6 \end{bmatrix} \tag{10.7.9}$$

Table 10.7.2. Complete eigensolution for matrix (10.7.9) using program 10.7.1.

```
SOLUTION OF SYMMETRIC EIGENVALUE PROBLEM

MAX CHANGE IN EIGENVECTOR?  1E-6
SHIFT IN ORIGIN?  0

AFTER  14  ITERATIONS        AFTER  22  ITERATIONS
EIGENVALUE( 1 ) =  15          EIGENVALUE( 3 ) =  5
EIGENVECTOR                    EIGENVECTOR
 1                              -.2222222
 .9999999                        1
 .9999998                       -1
 .9999997                        .2222222


AFTER  10  ITERATIONS        AFTER  3  ITERATIONS
EIGENVALUE( 2 ) =  4.999999    EIGENVALUE( 4 ) = -.9999999
EIGENVECTOR                    EIGENVECTOR
 -.9999995                      -.9999999
 -.2222219                        .9999997
  .2222224                       1
 1                              -.9999993
```

It is seen from Table 10.7.2 that for this system λ_2 and λ_3 are very close to each other and one may suspect that they are, in fact, repeated eigenvalues. This latter condition would imply that eigenvectors corresponding to these eigenvalues are not unique. The exact solutions for this problem are $\lambda = 15, 5, 5, -1$ and the eigenvector corresponding to the repeated eigenvalues is $[-\alpha, -1, 1, \alpha]$ where α is arbitrary. A second vector, orthogonal to this one, is $[1, -\alpha, \alpha, -1]$. At first sight the eigenvectors given in Table 10.7.2 do not appear to be in agreement with these solutions. However, recalling that all eigenvectors include an arbitrary scaling factor, we can divide the vectors given here by α and, writing $\beta = 1/\alpha$, we have the vectors $[-1, -\beta, \beta, 1]$ and $[\beta, -1, 1, -\beta]$. The correspondence between these vectors and those in Table 10.7.2 is now apparent.

We will now use program 10.7.2 to determine the eigensolutions of the non-symmetric matrix

$$\mathbf{A} = \begin{bmatrix} 33 & 16 & 72 \\ -24 & -10 & -57 \\ -8 & -4 & -17 \end{bmatrix} \qquad (10.7.10)$$

Table 10.7.3. The results of using program 10.7.2. to find the complete eigensolution for matrix (10.7.10).

```
SOLUTION OF NONSYMMETRIC EIGENVALUE PROBLEM

MAX CHANGE IN EIGENVECTOR?  1E-6
SHIFT IN ORIGIN?  0

AFTER  34  ITERATIONS
EIGENVALUE( 1 ) =  2.999998
R H EIGENVECTOR    L H EIGENVECTOR
   1                 .9999993
-.7499997            1
-.25                 .7499991

AFTER  19  ITERATIONS
EIGENVALUE( 2 ) =  2.000023
R H EIGENVECTOR    L H EIGENVECTOR
   1                 3.099443E-06
-.8124999           -.3333293
-.2499999            1

AFTER  3  ITERATIONS
EIGENVALUE( 3 ) =  1.000007
R H EIGENVECTOR    L H EIGENVECTOR
   1                 .2500029
-.8                  3.814635E-06
-.2666662            1
```

The estimates obtained from this program for the right and left-hand eigenvectors and the eigenvalues are shown in Table 10.7.3. The exact eigenvalues are 3, 2 and 1.

10.8 DEFLATION: AN ALTERNATIVE PROCEDURE

The procedure described in this section will be used to determine the subdominant eigenvalues of a symmetric or non-symmetric system. The procedure can be adapted to determine the complete eigensolution but the recovery of the system eigenvectors is more difficult to program. Furthermore, inverse iteration, described in the next section, provides a powerful technique to determine eigenvectors once a reasonable estimate of an eigenvalue has been made.

Once again we will assume that the the dominant eigensolution λ_1, x_1 has been determined for A and that x_1 is scaled *so that its largest element equals unity*. Assuming that this largest element is the sth element then we can form A_1 such that

$$A_1 = A - x_1 r_s^T \qquad (10.8.1)$$

where r_s^T is the sth row of the array A. An arbitrary trial vector u can be expressed as

$$u = \sum_{i=1}^{n} \alpha_i x_i \qquad (10.8.2)$$

Note that in (10.8.2) the vectors x_i are normalised so that their sth element is unity. This equation is identical to (10.7.2) except that in (10.7.2) the vectors x_i are normalised according to (10.2.5). Substituting (10.8.2) into (10.8.1) gives

$$A_1 u = A \sum_{i=1}^{n} \alpha_i x_i - x_1 r_s^T \sum_{i=1}^{n} \alpha_i x_i \qquad (10.8.3)$$

Now the sth row of A multiplied by x_i is equal to λ_i times the sth element of x_i which is equal to λ_i since the sth element of x_i is one. Thus

$$A_1 u = \sum_{i=1}^{n} \lambda_i \alpha_i x_i - x_1 \sum_{i=1}^{n} \alpha_i \lambda_i$$

$$= \sum_{i=2}^{n} \lambda_i \alpha_i (x_i - x_1)$$

Since $A_1 u$ is always deficient in x_1 convergence is to λ_2. Repeated use of this algorithm allows all the subdominant eigenvalues to be determined. Note that in the first deflation, the resulting vector z is equivalent to $x_2 - x_1$ and is not an eigenvector of the original system. Recovery of x_2 is not quite so straight forward

```
100    REM --------------------------------------
110    REM  DEFLATION, EIGENVALUES ONLY  JP/GL (C)
120    REM --------------------------------------
130    PRINT "SOLUTION OF EIGENVALUE PROBLEM: EIGENVALUES ONLY"
140    It=0:C=1
150    PRINT:PRINT "MAX CHANGE IN EIGENVECTOR";
160    INPUT Ep
170    READ N
180    DIM A(N,N),Z(N),V(N),R(N)
185    REM READ COMPLETE MATRIX
190    FOR I=1 TO N
200       FOR J=1 TO N
210          READ A(I,J)
220       NEXT J
230    NEXT I
240    PRINT "SHIFT IN ORIGIN";
250    INPUT P
260    FOR I=1 TO N
270       A(I,I)=A(I,I)-P
280    NEXT I
290    FOR I=1 TO N
300       Z(I)=1/I
310    NEXT I
315    REM ITERATION BEGINS
320    It=It+1:MAX=0
330    FOR I=1 TO N
340       V(I)=0
350       FOR J=1 TO N
360          V(I)=V(I)+A(I,J)*Z(J)
370       NEXT J
380       IF ABS(V(I))>MAX THEN MAX=ABS(V(I)):IR=I
390    NEXT I
400    Flg=0
410    FOR I=1 TO N
420       Zt=V(I)/V(IR)
430       IF ABS(Zt-Z(I))>Ep THEN Flg=1
440       Z(I)=Zt
450    NEXT I
460    IF Flg=1 THEN 320
470    PRINT:PRINT "AFTER ";It;" ITERATIONS"
480    PRINT "EIGENVALUE(";C;") = ";V(IR)+P
490    IF C=N THEN END
500    C=C+1:It=0
505    REM DEFLATION PROCEDURE
510    FOR J=1 TO N
520       R(J)=A(IR,J)
530       FOR I=1 TO N
540          A(I,J)=A(I,J)-Z(I)*R(J)
550       NEXT I
560    NEXT J
570    GOTO 290
580    DATA  3: REM ORDER OF EIGENVALUE PROBLEM
590    DATA -2,-1,4,2,1,-2,-1,-1,3: REM COEFFICIENTS OF MATRIX
```

Program 10.8.1.

as might at first appear since z includes an unknown scaling factor. An explanation of the procedure required to determine x_2 is given in Goodwin (1961).

Program 10.8.1 uses this algorithm to determine all the eigenvalues of a system. The eigenvalue problems defined by (10.4.5), (10.7.9) and (10.7.10) have been solved using this program and the resulting eigenvalues are essentialy the same as those given in Tables 10.7.1, 2 and 3. Thus, in these examples there is no particular advantage to be gained from using this deflation procedure in preference to the ones given in section 10.7.

As a further example of the use of this program, consider the eigenvalue problem defined by

$$A = \begin{bmatrix} -2 & -1 & 4 \\ 2 & 1 & -2 \\ -1 & -1 & 3 \end{bmatrix} \tag{10.8.4}$$

Using program 10.8.1 we find that, with a maximum change in the vector of 10^{-6}, $\lambda_1 = 2$. After 1400 iterations no further eigenvalues were found, suggesting very close pairs of eigenvalues. The program was then re-run with the origin of the eigenvalue problem shifted by 3. The output is shown in Table 10.8.1. and it is seen that the original difficulty in obtaining convergence was caused by two eigenvalues having opposite sign but identical absolute values. Shifting the origin allowed the three eigenvalues to be found easily. The exact eigenvalues for this problem are 2, 1 and –1.

Table 10.8.1. The eigenvalues of the matrix (10.8.4) determined
using program 10.8.1.

```
SOLUTION OF EIGENVALUE PROBLEM: EIGENVALUES ONLY

MAX CHANGE IN EIGENVECTOR?  1E-7
SHIFT IN ORIGIN?  3

AFTER  26  ITERATIONS
EIGENVALUE( 1 ) = -.9999998

AFTER  21  ITERATIONS
EIGENVALUE( 2 ) =  .9999995

AFTER  2  ITERATIONS
EIGENVALUE( 3 ) =  2
```

10.9 INVERSE ITERATION

In section 10.3 we have seen that direct iteration of $Ax = \lambda x$ leads to the largest value of λ and iteration of $A^{-1}x = (1/\lambda)x$ leads to the largest value of $1/\lambda$; i.e. the smallest value of λ. It has also been shown, in section 10.5, that by shifting the origin of the spectrum of eigenvalues by forming $A - \mu I$, direct iteration can be made to converge to the smallest or largest value of λ but not to any intermediate values. Inverse iteration combines these two notions to give a powerful method of determining subdominant eigensolutions. Consider now the eigensystem

$$Ax = \lambda x$$

then subtracting μx from both sides of this equation gives

$$(A - \mu I)x = (\lambda - \mu)x \tag{10.9.1}$$

$$(A - \mu I)^{-1} x = [1/(\lambda - \mu)]x \tag{10.9.2}$$

Thus iteration of $(A - \mu I)^{-1}$ leads to the largest value of $1/(\lambda - \mu)$, that is the smallest value of $\lambda - \mu$. The smallest value of $\lambda - \mu$ implies that the value of λ determined will be the value closest to μ. Thus by a suitable choice of μ we have a procedure for finding subdominant eigensolutions.

In practice $(A - \mu I)^{-1}$ is not formed explicitly; instead $A - \mu I$ is decomposed into the product of a lower triangular matrix, L, and an upper triangular matrix, U. If u_s is a trial vector for the solution of (10.9.2) then

and

$$(A - \mu I)^{-1} u_s = v_s, \text{ say,} \tag{10.9.3}$$

$$u_{s+1} = v_s / \max(v_s) \tag{10.9.4}$$

Rearranging (10.9.3) we have

$$u_s = (A - \mu I) v_s$$

$$= LUv_s \tag{10.9.5}$$

Let

$$Uv_s = z \tag{10.9.6}$$

then

$$Lz = u_s \tag{10.9.7}$$

Assuming a starting value for the trial vector u_s we can find z from (10.9.7) by forward-substitution and, knowing z, we can find v_s from (10.9.6) by back-substitution. The new estimate for the trial vector, u_{s+1}, can then be found from (10.9.4). Explicit matrix inversion is avoided; it is replaced by two efficient substitution procedures which are described in section 3.12.

Iteration is terminated when u_{s+1} is sufficiently close to u_s and it can easily be shown that when convergence is complete

$$1/(\lambda - \mu) = \max(v_s)$$

Thus the value of λ nearest to μ is given by

$$\lambda_\mu = \mu + 1/\max(v_s).$$

```
100  REM --------------------------
110  REM INVERSE ITERATION JP/GL (C)
120  REM --------------------------
130  PRINT "INVERSE ITERATION OF EIGENVALUE PROBLEM"
140  PRINT:PRINT "MAXIMUM CHANGE IN EIGENVECTOR";
150  INPUT Ep
160  READ N
170  DIM A(N,N),Z0(N),Z(N),Y(N),U(N,N),L(N,N)
175  REM READ COMPLETE MATRIX
180  FOR I=1 TO N
190      FOR J=1 TO N
200          READ A(I,J)
210      NEXT J
220  NEXT I
230  PRINT "ENTER ESTIMATE OF EIGENVALUE";
240  INPUT Q
250  FOR I=1 TO N
260      A(I,I)=A(I,I)-Q
270  NEXT I
275  REM LU DECOMPOSITION OF MATRIX (A-I*LAMBDA)
280  FOR I=1 TO N
290      FOR J=I TO N
300          F=0:G=0
310          IF I=1 THEN 350
320          FOR K=1 TO I-1
330              F=F+L(I,K)*U(K,J):G=G+L(J,K)*U(K,I)
340          NEXT K
350          U(I,J)=A(I,J)-F
360          IF U(I,I)=0 THEN PRINT "ESTIMATE MAY BE EXACT: PERTURB AND
                 TRY AGAIN": END
```

Program 10.9.1 (continues).

```
370          L(J,I)=(A(J,I)-G)/U(I,I)
380      NEXT J
390  NEXT I
400  FOR I=1 TO N
410      Z0(I)=1/I
420  NEXT I
425  REM BEGIN INVERSE ITERATION USING L AND U MATRICES
430  It=1
440  GOTO 530
445  REM FORWARD SUBSTITUTION
450  FOR I=1 TO N
460      F=0
470      IF I=1 THEN 510
480      FOR K=1 TO I-1
490          F=F+L(I,K)*Z0(K)
500      NEXT K
510      Z0(I)=Z(I)-F
520  NEXT I
525  REM BACKWARD SUBSTITUTION
530  FOR I=N TO 1 STEP -1
540      G=0
550      IF I=N THEN 590
560      FOR K=N TO I+1 STEP -1
570          G=G+U(I,K)*Y(K)
580      NEXT K
590      Y(I)=(Z0(I)-G)/U(I,I)
600  NEXT I
610  MAX=0
620  FOR I=1 TO N
630      IF ABS(Y(I))>MAX THEN MAX=ABS(Y(I)):IR=I
640  NEXT I
650  CR=1/Y(IR)
660  Flg=0
670  FOR I=1 TO N
680      Zt=Y(I)/Y(IR)
690      IF ABS(Zt-Z(I))>Ep THEN Flg=1
700      Z(I)=Zt
710  NEXT I
720  It=It+1
730  IF Flg=1 THEN 450
740  PRINT:PRINT "AFTER ";It-1;" ITERATIONS"
750  PRINT "EIGENVALUE = ";CR+Q
760  PRINT
770  PRINT "EIGENVECTOR"
780  FOR I=1 TO N
790      PRINT Z(I)
800  NEXT I
810  END
820  DATA 4: REM ORDER OF MATRIX
830  DATA 6,4,4,1,4,6,1,4,4,1,6,4,1,4,4,6: REM COEFFICIENTS OF MATRIX
```

Program 10.9.1.

The rate of convergence is fast, particularly if the chosen value of μ is close to an eigenvalue. If μ is equal to an eigenvalue then $A - \mu I$ is singular but in practice this seldom presents difficulties because it is unlikely that μ would be chosen, by chance, to equal an eigenvalue exactly. However, if $A - \mu I$ is singular then we have confirmation that the eigenvalue is known to a high precision. The corresponding eigenvector can then be obtained by changing μ by a small quantity and iterating to determine the eigenvector.

Although inverse iteration can be used to find the eigensolutions of a system about which we have no previous knowledge it is more usual to use inverse iteration to refine the approximate eigensolution obtained by some other technique.

Program 10.9.1 determines eigensolutions by inverse iteration. The system matrix is decomposed into the product of a lower and upper triangular matrix and and then, for a given starting value, inverse iteration finds the eigenvalue closest to the starting value and the corresponding eigenvector.

Table 10.9.1. Using program 10.9.1 to determine two eigenvalues and their eigenvectors for matrix (10.4.2).

```
INVERSE ITERATION OF EIGENVALUE PROBLEM
MAXIMUM CHANGE IN EIGENVECTOR?  1E-6

ENTER ESTIMATE OF EIGENVALUE?  8

AFTER  6  ITERATIONS
EIGENVALUE =  7.932905

EIGENVECTOR
 .7211775
 .272474
 1
 .2515509

INVERSE ITERATION OF EIGENVALUE PROBLEM

MAXIMUM CHANGE IN EIGENVECTOR?  1E-6
ENTER ESTIMATE OF EIGENVALUE?  -8

AFTER  4  ITERATIONS
EIGENVALUE = -8.028578

EIGENVECTOR
 -.3899834
 -.9755281
  .2955024
  1
```

In Table 10.9.1 two eigenvalues and their associated eigenvectors for the matrix (10.4.2) have been found using program 10.9.1. One of these eigenvalues is close to –8 and direct iteration of the matrix has previouly been shown to converge very slowly to this (dominant) eigenvalue. A second eigenvalue close to 7.9 has also been found. We see that the technique of inverse iteration is much faster than direct iteration.

A second example of the use of program 10.9.1 is given in Table 10.9.2. Using direct iteration and deflation it has been shown that the matrix of (10.7.9) has two eigenvalues close to 5. This is confirmed by inverse iteration; the two eigenvalues are repeated. Note that the eigenvectors found using the two iterative schemes are different. This difference might be expected since we have repeated eigenvalues and linear combinations of one pair of eigenvectors will reproduce the other pair. The table also shows that for this problem double precision working provides no significant improvement in accuracy.

Table 10.9.2. Use of program 10.9.1 working in single and double precision to find a pair of repeated eigenvalues and associated eigenvectors.

```
INVERSE ITERATION OF EIGENVALUE PROBLEM

MAXIMUM CHANGE IN EIGENVECTOR?  1E-6    MAXIMUM CHANGE IN EIGENVECTOR?  1E-8
ENTER ESTIMATE OF EIGENVALUE?  4.8      ENTER ESTIMATE OF EIGENVALUE?  4.8

AFTER  7  ITERATIONS                    AFTER  12  ITERATIONS
EIGENVALUE =  5                         EIGENVALUE =  4.9999999999997

EIGENVECTOR                             EIGENVECTOR
-.1369527                                .13694837498155
-.9999999                               1
 1                                      -.99999999999998
 .1369524                               -.13694837498148

INVERSE ITERATION OF EIGENVALUE PROBLEM

MAXIMUM CHANGE IN EIGENVECTOR?  1E-5    MAXIMUM CHANGE IN EIGENVECTOR?  1E-8
ENTER ESTIMATE OF EIGENVALUE?  5.1      ENTER ESTIMATE OF EIGENVALUE?  5.1

AFTER  18  ITERATIONS                   AFTER  7  ITERATIONS
EIGENVALUE =  5                         EIGENVALUE =  4.999999999982

EIGENVECTOR                             EIGENVECTOR
-.9999997                               1
-.8723024                                .87242854423915
 .8723023                               -.8724285442454
 1                                      -.999999999993
```

10.10 COMPLEX EIGENSOLUTIONS

So far we have tacitly assumed that the eigensolutions being sought are real. If the eigenvalue problem is not symmetric then this may not be the case; pairs of complex eigensolutions can exist. Under these conditions direct iteration from a real trial vector will oscillate rather than converge and inverse iteration will converge to the nearest real solution. If dominant eigenvalues are a complex conjugate pair they can be obtained using a modified form of the method of iteration decribed in section 10.3. The procedure, which is described in detail by Wilkinson (1965) is as follows: if

$$\mathbf{A}\mathbf{u}_s = \mathbf{v}_{s+1} = k_{s+1}\mathbf{u}_{s+1}$$

where $k_{s+1} = \max(\mathbf{v}_{s+1})$ and $\mathbf{u}$ and $\mathbf{v}$ are real vectors, then after the second and each subsequent iteration we must solve

$$\begin{bmatrix} \mathbf{u}_s^T\mathbf{u}_s & \mathbf{u}_s^T\mathbf{u}_{s+1} \\ \mathbf{u}_s^T\mathbf{u}_{s+1} & \mathbf{u}_{s+1}^T\mathbf{u}_{s+1} \end{bmatrix} \begin{bmatrix} q_s \\ k_{s+1}p_s \end{bmatrix} = k_{s+1}k_{s+2} \begin{bmatrix} \mathbf{u}_s^T\mathbf{u}_{s+2} \\ \mathbf{u}_{s+1}^T\mathbf{u}_{s+2} \end{bmatrix}$$

to determine the scalars p_s and q_s. Iteration is continued until, within a specified tolerance, both p_s and q_s converge. Letting the values of p_s and q_s, at convergence, be p and q then

$$\lambda_1 = (p/2) \pm i[\sqrt{(p^2 + 4q)}]/2.$$

It can also be proved that if $\lambda_1 = \xi_1 \pm i\eta_1$ then

$$\mathbf{x}_1 = \eta_1\mathbf{u}_s \pm i(\xi_1\mathbf{u}_s - k_{s+1}\mathbf{u}_{s+1})$$

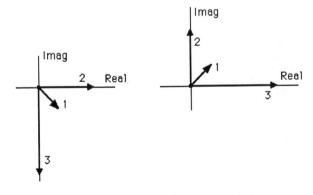

Fig. 10.10.1. Identical complex vector solutions of the matrix (10.10.1). The $\pi/2$ phase shift is of no importance since the reference phase and amplitude of an eigenvector is arbitrary.

```
100   REM -----------------------------
110   REM COMPLEX EIGENSYSTEM  JP/GL (C)
120   REM -----------------------------
130   PRINT "COMPLEX EIGENVALUE PROBLEM"
140   PRINT "PROGRAM GIVES THE REAL AND IMAGINARY PARTS OF THE"
150   PRINT "COMPLEX CONJUGATE PAIRS FOR THE DOMINANT EIGENSOLUTION"
160   PI=4*ATN(1):It=0
170   PRINT:PRINT "MAX CHANGE IN EIGENVALUES";
180   INPUT Ep
190   READ N
200   DIM A(N,N),Z(N),V(N),G(N,3),Ip(N)
210   DIM Rp(N),An(N),Md(N),K(3),C(2,2),B(2)
215   REM READ COMPLETE MATRIX
220   FOR I=1 TO N
230      FOR J=1 TO N
240         READ A(I,J)
250      NEXT J
260   NEXT I
270   FOR I=1 TO N
280      Z(I)=1/I
290   NEXT I
300   Nu=0: Ea=0
305   REM BEGIN ITERATION
310   Nw=Nu: Eb=Ea: It=It+1: MAX=0
320   FOR I=1 TO N
330      V(I)=0
340      FOR J=1 TO N
350         V(I)=V(I)+A(I,J)*Z(J)
360      NEXT J
370      IF ABS(V(I))>MAX THEN MAX=ABS(V(I)):IR=I
380   NEXT I
390   FOR I=1 TO N
400      Z(I)=V(I)/V(IR)
410   NEXT I
420   FOR I=1 TO N
430      G(I,1)=G(I,2): G(I,2)=G(I,3): G(I,3)=Z(I)
440   NEXT I
450   K(1)=K(2): K(2)=K(3): K(3)=V(IR)
460   IF It<=3 THEN 310
470   FOR I=1 TO 2
480      FOR J=1 TO 2
490         C(I,J)=0
500         FOR K=1 TO N
510            C(I,J)=C(I,J)+G(K,I)*G(K,J)
520         NEXT K
530      NEXT J
540      B(I)=0
550      FOR K=1 TO N
560      B(I)=B(I)+G(K,I)*G(K,3)
570      NEXT K
580   NEXT I
```

Program 10.10.1. (continues).

```
585   REM SOLVE PAIR OF EQUATIONS TO FIND P AND Q
590   DET=C(1,1)*C(2,2)-C(1,2)*C(2,1)
600   Q=K(2)*K(3)*(C(2,2)*B(1)-C(1,2)*B(2))/DET
610   P=K(3)*(C(1,1)*B(2)-C(1,2)*B(1))/DET
615   REM SOLVE QUADRATIC
620   Nu=SQR(ABS(P*P+4*Q))/2
630   Ea=P/2
640   IF ABS(Nu-Nw)>Ep THEN 310
650   IF ABS(Ea-Eb)>Ep THEN 310
660   PRINT "AFTER ";It;" ITERATIONS"
670   PRINT "REAL PART OF EIGENVALUE = ";Ea
680   PRINT "IMAGINARY PART OF EIGENVALUE = ";Nu
690   PRINT: PRINT "VECTOR:"
700   PRINT "REAL PART        IMAGINARY PART"
705   REM COMPUTE REAL AND IMAGINARY PARTS OF VECTOR
706   REM VECTOR SCALED AND ROTATED SO THAT LARGEST ELEMENT =1.
710   MAX=0
720   FOR I=1 TO N
730       Rp=Nu*G(I,1)
740       Ip=Ea*G(I,1)-K(2)*G(I,2)
750       Md(I)=SQR(Rp*Rp+Ip*Ip)
760       An(I)=0
770       IF Md(I)=0 THEN 860
780       IF ABS(Rp)<.000001 THEN 850
790       An(I)=ATN(Ip/Rp)
800       IF Rp>0 THEN 860
810       Pc=PI
820       IF Ip<0 THEN Pc=-PI
830       An(I)=An(I)+Pc
840       GOTO 860
850       An(I)=ABS(Ip)/Ip*PI/2
860       IF Md(I)>MAX THEN MAX=Md(I):IR=I
870   NEXT I
880   Rot=An(IR)
890   FOR I=1 TO N
900       Md(I)=Md(I)/MAX: An(I)=An(I)-Rot
910       PRINT Md(I)*COS(An(I));TAB(16);Md(I)*SIN(An(I))
920   NEXT I
930   END
940   DATA 3: REM ORDER OF MATRIX
950   DATA 8 ,-1,-5,-4,4,-2,18,-5,-7: REM COEFFICIENTS OF MATRIX
```

Program 10.10.1

This procedure is used in program 10.10.1 to determine the dominant complex conjugate eigensolutions of a system. It should be noted that the complex elements of $\mathbf{x}_1$ may be expressed in polar coordinate form and thus the jth element of $\mathbf{x}_1$ may be written as

$$x_j = (r_j, \theta_j) \quad j = 1, 2, \dots, n$$

It can be proved that x_j can be written equivalently as

$$x_j = (\alpha r_j, \theta_j + \beta) \quad j = 1, 2, \dots, n$$

Thus the amplitude of each element of x_1 may be multiplied by the same arbitrary scaling factor α and the polar angle of each element may be rotated by the same angle β. This implies that for each eigenvector there are an infinite number of alternative but equivalent orientations and scales in the complex plane. Fig. 10.10.1 shows a plot of two orientations of an eigenvector in the complex plane.

Table 10.10.1. Use of program 10.10.1 to find the dominant complex conjugate pair of eigenvalues and associated eigenvectors.

```
COMPLEX EIGENVALUE PROBLEM
PROGRAM GIVES REAL AND IMAGINARY PARTS OF THE
COMPLEX CONJUGATE PAIRS FOR THE DOMINANT EIGENSOLUTION

MAX CHANGE IN EIGENVALUES? 1E-6
AFTER  14  ITERATIONS
REAL PART OF EIGENVALUE =  2
IMAGINARY PART OF EIGENVALUE =  4.000001

VECTOR:
REAL PART       IMAGINARY PART
 .4999999         .4999999
-2.384185E-07     .9999998
 1                0
```

Program 10.10.1 normalises the eigenvector solution by choosing an orientation and scale so that the largest element of the eigenvector is real and equal to one. For example, consider the eigensystem given by

$$A = \begin{bmatrix} 8 & -1 & -5 \\ -4 & 4 & -2 \\ 18 & -5 & -7 \end{bmatrix} \tag{10.10.1}$$

Direct iteration using program 10.6.1 fails to converge because the dominant eigenvalues form a complex conjugate pair. This is confirmed by using program 10.10.1 where a complex eigensolution is found after 14 iterations, see Table 10.10.1. The matrix (10.10.1) is not symmetric and a left hand complex conjugate pair of eigenvectors can be found by iterating $A^T y = \lambda y$ rather than $Ax = \lambda x$.

Once a dominant pair of complex eigenvalues have been found it is not possible to find any subdominant eigensolutions by the simple deflation and iteration procedure described in section 10.7. However, suppose, for example that two real eigenvalues have been found using iteration and deflation and we suspect that the third eigenvalue is complex. Complex iteration can then be applied to the deflated matrix to reveal the complex conjugate pair of eigenvalues.

For example, direct iteration using program 10.7.2 of the matrix

$$A = \begin{bmatrix} 4 & -5 & 0 & 3 \\ 0 & 4 & -3 & -5 \\ 5 & -3 & 4 & 0 \\ 3 & 0 & 5 & 4 \end{bmatrix}$$

gives $\lambda_1 = 12$. If program 10.7.2 is modified to output the deflated matrix it is found to be

$$\begin{bmatrix} 1 & -2 & -3 & 0 \\ 3 & 1 & 0 & -2 \\ 2 & 0 & 1 & -3 \\ 0 & 3 & 2 & 1 \end{bmatrix}$$

Further simple iteration fails to converge, but complex iteration of the deflated matrix (using program 10.10.1) converges and $\lambda_2 = 1 + i5$ and $\lambda_3 = 1 - i5$.

10.11 EIGENSOLUTIONS OF THE HERMITIAN MATRIX

In section 10.2 it was shown that Hermitian matrices have real eigenvalues but complex eigenvectors. Consider

$$Hx = \lambda x \tag{10.11.1}$$

where H is an $n \times n$ Hermitian matrix. We may replace H by $A + iB$ and x by $u + iv$, where A, B, u and v are real. Thus (10.11.1) becomes

$$(A + iB)(u + iv) = \lambda(u + iv) \tag{10.11.2}$$

$$(Au - Bv) + i(Bu + Av) = \lambda u + i\lambda v$$

Separating real and imaginary parts gives

$$Bu + Av = \lambda v \quad \text{(imaginary part)}$$

$$Au - Bv = \lambda u \quad \text{(real part)}$$

or in matrix notation

$$\begin{bmatrix} A & B \\ -B & A \end{bmatrix} \begin{bmatrix} v \\ u \end{bmatrix} = \lambda \begin{bmatrix} v \\ u \end{bmatrix} \tag{10.11.3}$$

Table 10.11.1. Use of program 10.7.1 to find the eigensolution
of the Hermitian matrix (10.11.4).

```
SOLUTION OF SYMMETRIC EIGENVALUE PROBLEM

MAX CHANGE IN EIGENVECTOR?  1E-7
SHIFT IN ORIGIN?  0

AFTER  11  ITERATIONS      |  AFTER  3  ITERATIONS
EIGENVALUE( 1 ) =  2.414214 |  EIGENVALUE( 3 ) = -.4142135
EIGENVECTOR                 |  EIGENVECTOR
 .815615                    |   1
 1                          |  -.5378963
 .5985985                   |  -.2392996
-.1534538                   |   .8763176


AFTER  20  ITERATIONS      |  AFTER  4  ITERATIONS
EIGENVALUE( 2 ) =  2.414214 |  EIGENVALUE( 4 ) = -.414213E
EIGENVECTOR                 |  EIGENVECTOR
-.5985985                   |   .2392997
 .1534538                   |  -.8763179
 .8156149                   |   1
 1                          |  -.5378965
```

Thus we have replaced an nth order Hermitian system by a $2n$th order system with
real coefficients. Note also that since $\mathbf{A} = \mathbf{A}^T$ and $\mathbf{B} = -\mathbf{B}^T$ the matrix of (10.11.3)
is symmetric. Thus any procedure suitable for find the eigensolution for a system
matrix with real coefficients may be used to solve (10.11.3).

Consider the eigensolution of the Hermitian eigensystem given by

$$\mathbf{H} = \begin{bmatrix} 1 & 1-i \\ 1+i & 1 \end{bmatrix} \qquad (10.11.4)$$

The matrix (10.11.4) may be rearranged into a 4 x 4 matrix of real elements, using
(10.11.3) to give

$$\begin{bmatrix} 1 & 1 & 0 & -1 \\ 1 & 1 & 1 & 0 \\ 0 & 1 & 1 & 1 \\ -1 & 0 & 1 & 1 \end{bmatrix}$$

This real eigensystem is solved using program 10.7.1. and the results shown in Table 10.11.1. The eigenvalues, 2.41421356 and –0.41421352 are close to the exact solution, $1 \pm \sqrt{2}$. However, the eigenvectors are more difficult to interpret. Recalling from (10.11.3) that the upper half of the vector is the imaginary part of the solution and the lower part is the real part of the solution of the original Hermitian matrix we have

$$x_1 = \begin{bmatrix} 0.59859 + i0.81562 \\ -0.15345 + i1 \end{bmatrix} = \begin{bmatrix} (1.0117, 53.7^\circ) \\ (1.0117, 98.7^\circ) \end{bmatrix} \equiv \begin{bmatrix} (1, 0^\circ) \\ (1, 45^\circ) \end{bmatrix}$$

$$x_2 = \begin{bmatrix} 0.81562 - i0.59859 \\ 1 + i0.15345 \end{bmatrix} = \begin{bmatrix} (1.0117, -36.3^\circ) \\ (1.0117, 8.7^\circ) \end{bmatrix} \equiv \begin{bmatrix} (1, 0^\circ) \\ (1, 45^\circ) \end{bmatrix}$$

Thus we see that x_1 and x_2 represent the same vector, once an adjustment has been made to account for arbitrary scaling and the arbtrary orientation of the vector. Similarly

$$x_3 = \begin{bmatrix} -0.23930 + i1 \\ 0.87632 - i0.53790 \end{bmatrix} = \begin{bmatrix} (1.028, 103.5^\circ) \\ (1.028, -31.5^\circ) \end{bmatrix} \equiv \begin{bmatrix} (1, 135^\circ) \\ (1, 0^\circ) \end{bmatrix}$$

$$x_4 = \begin{bmatrix} 1 + i0.23930 \\ -0.53790 - i0.87632 \end{bmatrix} = \begin{bmatrix} (1.028, 13.5^\circ) \\ (1.028, -121.5^\circ) \end{bmatrix} \equiv \begin{bmatrix} (1, 135^\circ) \\ (1, 0^\circ) \end{bmatrix}$$

Thus the two eigenvectors for the original Hermitian matrix consist of two components of the same amplitude but with phase difference of 45° in one solution and 135° in the other.

10.12 THE PERTURBED SYMMETRIC EIGENVALUE PROBLEM

It is sometimes desirable to discover the effect on the eigenvalues of making some small change in the elements of a symmetric system matrix. Let us assume that the eigenvalue problem for the original system matrix has been solved. We now wish to solve a new eigenvalue problem that is similar to the original one but in which A is replaced by $A + \delta A$ such that the symmetry of the problem is maintained. We could solve this new eigenvalue problem directly but providing the changes in A are relatively small, we can avoid the need to solve the new eigenvalue problem by adopting the following procedure.

Symmetric changes in the symmetric matrix A will cause λ and x to become $x + \delta x$ and $\lambda + \delta \lambda$ respectively. Hence

$$(A + \delta A)(x + \delta x) = (\lambda + \delta \lambda)(x + \delta x)$$

Assuming the changes are small, terms involving the product of these small changes can be neglected. Thus

$$Ax + \delta Ax + A\delta x = \lambda x + \delta \lambda x + \lambda \delta x$$

Now $Ax = \lambda x$ and so

$$\delta Ax + A\delta x = \delta \lambda x + \lambda \delta x \qquad (10.12.1)$$

Premultiplying by x^T gives

$$x^T\delta Ax + x^TA\delta x = \delta \lambda\, x^Tx + \lambda\, x^T\delta x$$

Now $x^TA = x^T \lambda$ so that

$$\delta \lambda = x^T\delta Ax/x^Tx \qquad (10.12.2)$$

Thus the change in the eigenvalue caused by the change in the system matrix can easily be computed from (10.12.2). It is also possible to develop an expression for the change in eigenvectors due to small changes in the system matrix. However, the expression requires a knowledge of the complete eigensolution for the original system. If the eigenvectors of the modified system are required then it is probably easier to estimate the new eigenvalues from the perturbed eigenvalue problem and then, by inverse iteration, refine these eigenvalues and compute the new eigenvectors.

10.13 REDUCING EIGENVALUE PROBLEMS TO STANDARD FORM

In many physical systems the eigenvalue problems may be more general than the form presented here and methods to reduce eigenvalue problems to the standard form will now be briefly examined.

(1)　$\lambda Kx = Mx$ where K and M are real symmetric matrices. This formulation can occur, for example, when finding the natural frequencies of vibration of an elastic system. One approach is to premultiply this equation by the inverse of K or M, providing that the chosen matrix is non-singular, to give

$$K^{-1}Mx = \lambda x \text{ or } M^{-1}Kx = (1/\lambda)x$$

Whilst these equations are of the form $Ax = \lambda x$ the products $K^{-1}M$ and $M^{-1}K$ are, in general, non-symmetric; thus the original symmetry of the equation has been lost. Furthermore, inverting matrices, particularly large matrices, should be avoided since it is a time consuming process that can be prone to inaccuracy. An alternative procedure is to decompose K into the product a lower triangular matrix L and L^T. Thus if $\lambda Kx = Mx$ then

$$\lambda LL^Tx = Mx$$

Premultiplying by L^{-1} we have

$$\lambda L^{-1}LL^Tx = L^{-1}M(L^{-T}L^T)x$$

$$\lambda y = (L^{-1}ML^{-T})y$$

where $y = L^Tx$.

The product $L^{-1}ML^{-T}$ is symmetric and the inversion of L and L^T is relatively fast and straightforward. Solving these equations gives eigenvalues that are identical to those of the original system but the eigenvectors for the original system must be determined from $x = L^{-T}y$.

(2) $STx = \lambda x$ where S and T are symmetric matrices. Using reasoning similar to that above, if T is decomposed so that $T = LL^T$, then

$$L^TSLy = \lambda y$$

where
$$y = L^Tx.$$

which reduces it to standard form.

(3) $(A_p\lambda^p + A_{p-1}\lambda^{p-1} + ... + A_1\lambda + A_0)x = 0$ (10.13.1)

Equations of this form are called generalised eigenvalue problems. We illustrate how this can be reduced to standard form by considering the case $p = 2$. Let

$$x_1 = \lambda x_0 \quad \text{and} \quad x_0 = x$$

Note that here the subscripts 0 and 1 to do not indicate that x_0 and x_1 are eigenvector solutions. Substituting for x_0 and x_1 in (10.13.1) with $p = 2$ we have

$$A_2 \lambda x_1 + A_1 x_1 + A_0 x_0 = 0 \tag{10.13.2}$$

Provided A_2 is non–singular, let $Z = A_2^{-1}$. Premultiplying (10.13.2) by Z gives

$$\lambda x_1 + Z A_1 x_1 + Z A_0 x_0 = 0$$

letting

$$B_i = -Z A_i \quad i = 0, 1$$

then

$$B_1 x_1 + B_0 x_0 = \lambda x_1$$

By combining this equation with the equation

$$x_1 = \lambda x_0$$

we can express these equations in matrix form thus

$$\begin{bmatrix} 0 & I \\ B_0 & B_1 \end{bmatrix} \begin{bmatrix} x_0 \\ x_1 \end{bmatrix} = \lambda \begin{bmatrix} x_0 \\ x_1 \end{bmatrix}$$

If the original matrices A_0, A_1, A_2 are of order n this eigenvalue problem is order $2n$. A problem can arise in the inversion of A_2 if it is ill-conditioned. In the general case, given in (10.13.1), the number of equations to be solved will equal pn and this can be large and may present computational difficulties on a microcomputer.

10.14 SOME EXAMPLES

The examples which follow have been chosen to test the programs presented in this chapter using relatively large matrices. Some may be ill-conditioned and might be expected to cause numerical difficulties. These examples are taken from Gregory and Karney (1969) which contains many more interesting test problems. The accurate values quoted in the tables are taken from Gregory and Karney.

Example 1 : Consider the symmetric matrix

$$\begin{bmatrix} 1 & 0 & 0 & 0 & ... & 0 & 1 \\ 0 & 1 & 0 & 0 & ... & 0 & 2 \\ 0 & 0 & 1 & 0 & ... & 0 & 3 \\ \multicolumn{7}{c}{\dotfill} \\ 0 & 0 & 0 & 0 & ... & 1 & n{-}1 \\ 1 & 2 & 3 & 4 & ... & n{-}1 & n \end{bmatrix}$$

This matrix has two well separated eigenvalues given by

where

$$\lambda_1, \lambda_2 = (n + 1)[1 \pm Q]/2$$

$$Q = 2\sqrt{\{(n - 1)(n - 3/4)/(3(n + 1))\}}$$

and $n - 2$ repeated eigenvalues equal to unity. Using programs 10.7.1 and 10.8.1 to solve this problem with $n = 10$, the first two eigenvalues ($\lambda_1 \approx 23$ and $\lambda_2 \approx -12$) were determined to an accuracy of 6 significant digits and the equal eigenvalues were obtained exactly. This test was repeated with $n = 20$ and again both programs were successful. When the maximum change in any element of the vector was less than 10^{-6} iterations were terminated.

Example 2 : The next symmetric eigenvalue problem to be examined is where **A** is defined by

$$a_{ij} = 1/(i + j - 1) \quad i,j = 1, 2, ... ,n$$

This is the Hilbert matrix and it was shown in Chapter 3 that this matrix is very ill-conditioned. For this reason, we may anticipate difficulties in determining the system eigenvalues. Table 10.14.1 shows a comparison of the exact eigenvalues and their estimates obtained using programs 10.7.1 and 10.8.1 for the case when $n = 10$. Using these programs with 7 digit precision, only the first 6 eigenvalues could be found with reasonable accuracy. To obtain improved results we have used double precision working and it can be seen that the accuracy of the estimated eigenvalues decrease as they become smaller and the estimates for the 9th and 10th eigenvalue are still grossly inaccurate.

Table 10.14.2 shows a comparison between the values quoted in Gregory and Karney for the 6th eigenvector and those estimated by programs 10.7.1 and 10.9.1 working in double precision. It is seen that inverse iteration gives a superior estimate to that given by direct iteration and deflation and, furthermore, it is much faster. Direct iteration and deflation required 46 iterations to determine the 6th eigenvector; inverse iteration is dependent on the initial estimate of the eigenvalue and in this example only 11 iterations were required using a starting value 4×10^{-6}.

Table 10.14.1 A comparison of the results of two programs used to find all the eigenvalues of the 10th order Hilbert matrix. Convergence criteria = 1E–6.

Eigenvalue			
	Using Prog 10.7.1	Using 10.8.1	Exact
1	1.7519191763156	1.7519191763156	1.7519196702652
2	.34292945933258	.34292946011993	.34292954848351
3	.035741796508306	.035741852689279	.03574181627164
4	.0025308906412991	.0025308893792434	.00253089076867
5	.0001287495953974	1.2874961530977D-04	1.28749614276D-04
6	4.7296873974648D-06	4.7296892297864D-06	4.7296892931 8D-06
7	4.9394929363924D-07	1.2289678117618D-07	1.22896773875D-07
8	1.2289682112468D-07	2.1474372853827D-09	2.14743881735D-09
9	8.9150909027378D-08	2.2666356438488D-11	2.26674674776D-11
10	1.9763348621795D-08	1.1034905334160-13	1.09315381938D-13

Table 10.14.2. Comparison between the results of two programs working in double precision to find the 6th eigenvector of the 10th order Hilbert matrix.

MAX CHANGE IN EIGENVECTOR? 1e-6		
Prog 10.7.1	Prog 10.9.1	
EIGENVECTOR	EIGENVECTOR	EXACT EIGENVECTOR
.0134198097412	.013419707183732	.013419707196314
-.25634892541384	-.25634900036434	-.256349000536856
1	1	1
-.82625441592081	-.82625449618943	-.826254493098565
-.73039881330578	-.73039887054279	-.730398877200419
.097979994452187	.097979968464468	.097979970771839
.69003518892259	.69003519207004	.690035194494965
.70955949171078	.70955950026691	.709559501029808
.15232811029263	.15232809798995	.152328096630799
-.86461805559242	-.86461811216497	-.864618112593268

Example 3 : We will now determine the eigenvalues of the following non-symmetric matrix

$$
\begin{bmatrix}
12 & 11 & 0 & 0 & 0 & \dots & 0 & 0 \\
11 & 11 & 10 & 0 & 0 & \dots & 0 & 0 \\
10 & 10 & 10 & 9 & 0 & \dots & 0 & 0 \\
\hdotsfor{8} \\
2 & 2 & 2 & 2 & 2 & \dots & 2 & 1 \\
1 & 1 & 1 & 1 & 1 & \dots & 1 & 1
\end{bmatrix}
$$

Using programs 10.7.2 and 10.8.1, estimates of the eigenvalues were obtained and these are compared with the exact results as shown in Table 10.14.3. Both deflation procedures fail to locate the smaller eigenvalues. If program 10.9.1, which implements inverse iteration, is used with double precision and a starting value of 8×10^{-2} it is found that the estimate for λ_{10} is 0.081227961784603 which is accurate to 6 decimal places. Using inverse iteration to estimate the eigenvector gives the result shown in Table 10.14.4 where each of the eigenvectors have been normalised so that the second element in each is identical.

Table 10.14.3. Use of programs 10.7.2 and 10.8.1 to find the eigenvalues of *Example* 3 matrix.

SOLUTION OF NONSYMMETRIC EIGENVALUE PROBLEM

PROGRAM 10.7.2			PROGRAM 10.8.1		
MAX CHANGE IN EIGENVECTOR? 1e-6			MAX CHANGE IN EIGENVECTOR? 1e-6		
SHIFT IN ORIGIN? 0			SHIFT IN ORIGIN? 0		
NUMBER.	EIGENVALUE	ITER'NS	NUMBER	EIGENVALUE	ITER'NS
1	32.22894	28	1	32.22895	28
2	20.19901	25	2	20.19897	26
3	12.31108	21	3	12.31106	24
4	6.961534	26	4	6.961531	20
5	3.511863	22	5	3.511861	22
6	1.554155	21	6	1.554051	19
7	.6369171	25	7	.6448993	21
8	.3488688	25			

Table 10.14.4. Comparison between results obtained by inverse iteration and accurate results given by Gregory & Karney for *Example* 3 matrix.

AFTER 4 ITERATIONS
EIGENVALUE = .081227961784603

EIGENVECTOR	Accurate values
-.67714107409239	-0.6771411
.7336991	0.7336991
-.056254140252775	-0.0562542
-8.4536290388988D-04	-0.0008453
6.0057855406197D-04	0.0006006
-6.0575245200879D-05	-0.0000606
8.995301122306D-07	0.0000009
7.0390557210221D-07	0.0000007
-1.4757518871985D-07	-0.0000001
2.0110548920153D-08	0.0000000
-2.2295105514029D-09	0.0000000
2.1621895159652D-10	0.0000000

Example 4 : Consider the eigenvalue problem defined by the matrix

$$A = \begin{bmatrix} 5C & -C & 10C & -2C \\ 5C & C & 10C & 2C \\ 20C & -4C & 15C & -3C \\ 20C & 4C & 15C & 3C \end{bmatrix}$$

where the sub-matrix C is

$$C = \begin{bmatrix} -2 & 2 & 2 & 2 \\ -3 & 3 & 3 & 3 \\ -2 & 0 & 4 & 2 \\ -1 & 0 & 0 & 5 \end{bmatrix}$$

Direct iteration fails to converge. This causes us to suspect that either a pair of dominant eigenvalues are complex conjugates or they are equal in magnitude but opposite in sign. Using complex iteration, implemented in program 10.10.1, the following results are obtained setting the maximum change in the eigenvector as 10^{-4}:

	Exact	After 69 iterations.
Real part	60	59.99995
Imaginary part	20	20.00002

Example 5 : We will now consider the Hermitian matrix $H = A + iB$ where

$$A = \begin{bmatrix} -0.845 & 5.2 & 0.301 & -9.6 & 0.0734 \\ & -6.2 & -3.39 & 0.122 & 4.19 \\ & & 0.019 & 0.935 & -0.0572 \\ & \text{symmetric} & & 7.21 & 0.337 \\ & & & & -1.23 \end{bmatrix}$$

$$B = \begin{bmatrix} 0 & 0.103 & -0.0454 & 0.936 & 7.26 \\ & 0 & -.407 & 0.91 & -3.66 \\ & & 0 & -0.271 & 2.82 \\ & \text{skew-symmetric} & & 0 & 0.0603 \\ & & & & 0 \end{bmatrix}$$

Rearranging H into the form of (10.11.3) we may solve the resulting 10th order real symmetric eigenvalue problem by any of the methods of sections 10.7, 10.8 and 10.9. Using inverse iteration the results of Table 10.14.5 were obtained. It is seen that the five unique eigenvalues are accurately computed.

Table 10.14.5. A comparison of the exact and estimated eigenvalues
for *Example* 5.

```
INVERSE ITERATION OF EIGENVALUE PROBLEM

MAXIMUM CHANGE IN EIGENVECTOR? 1E-6

ENTER ESTIMATE OF EIGENVALUE? 15
AFTER  6  ITERATIONS
EIGENVALUE =  15.180165223827  EXACT=15.1801652

ENTER ESTIMATE OF EIGENVALUE? 5
AFTER  9  ITERATIONS
EIGENVALUE =  5.6787292250848  EXACT=5.67872935

ENTER ESTIMATE OF EIGENVALUE? 0
AFTER  10  ITERATIONS
EIGENVALUE = -.8339874622953  EXACT=-0.8339868

ENTER ESTIMATE OF EIGENVALUE? -5
AFTER  6  ITERATIONS
EIGENVALUE = -5.1498456221786    EXACT=-5.14984563

ENTER ESTIMATE OF EIGENVALUE? -15
AFTER  8  ITERATIONS
EIGENVALUE = -15.92106198852  EXACT=-15.9210621
```

10.15 CONCLUSIONS AND SUMMARY

In this chapter we have described a number of techniques based on iteration to
determine the eigensolutions for real symmetric, non-symmetric and Hermitian
matrices. The methods cannot be guaranteed to work efficiently for all problems and
in some cases may fail. The user must be alert to particular problems that may arise
when dealing with non-symmetric matrices. For example, an apparant failure to
converge during iteration may be caused by a dominant eigensolution which is
complex or by a repeated eigenvalue. Table 10.15.1 summarises the methods
presented in this Chapter and indicates the type of problem for which they can be most
efficiently used.

Table 10.15.1. Methods of solving eigenvalue problems.

Method	Program	Nature of matrix	Eigenvalues	Eigenvectors
Simple iter'n Iteration &	None	Real	Dominant only§	Dominant only§
Rayleigh quot.	10.6.1	Symm	Dominant only	Dominant only
Hotelling	10.7.1	Symm	All	All
Hotelling	10.7.2	Non-symm	All§	All§
Deflation	10.8.1	Real	All§	None
Inverse iter'n	10.9.1	Real	Any§	Any§
Complex iter'n	10.10.1	Real	Dominant only¶	Dominant only¶

§ Convergence to real eigensolutions only.
¶ Determines dominant complex conjugate pair.

PROBLEMS

10.1. Consider the matrix

$$\begin{bmatrix} 10 & 1 & 2 & 3 & 4 \\ 1 & 9 & -1 & 2 & -3 \\ 2 & -1 & 7 & 3 & -5 \\ 3 & 2 & 3 & 12 & -1 \\ 4 & -3 & -5 & -1 & 15 \end{bmatrix}$$

Determine the dominant eigenvalue of this matrix using program 10.6.1. Use a zero shift of origin and allow a maximum change in the eigenvector of
(i) 1E–3 and (ii) 1E–5.
Note that the Rayleigh quotient gives a better estimate of the dominant eigenvalue than does iteration alone. The value of the dominant eigenvalue is 19.1754203.

10.2 The nth order triple-diagonal matrix **A** where

$$A = \begin{bmatrix} a & b & 0 & ... & & & \\ b & a & b & 0 & ... & & \\ 0 & b & a & b & 0 & ... & \\ & & .... & & & & \\ & & ... & 0 & b & a & b \\ & & & ... & 0 & b & a \end{bmatrix}$$

can be shown to have eigenvalues

$$\lambda_k = a + 2b \cos(k\pi/(n + 1)) \quad k = 1, 2, \dots ,n$$

Use program 10.7.1 to find the eigenvalues of this matrix when $a = 2$, $b = 5$
   (i) when $n = 3$ and shift $= 0$
   (ii) when $n = 5$ and shift $= 3$. Repeat with shift $= 0$.

In all cases use a maximum change of vector of 1E–5. Check your results using the formula which gives the exact eigenvalues and explain why a shift of zero in case (ii) gives such poor results.

10.3 Find the eigenvalues of the matrix. Use inverse iteration, program 10.9.1 with a trial vector [1, –1, 0.1] and a maximum change in the eigenvector of 1E–5.

$$\begin{bmatrix} 0 & 1 & 0 \\ 1 & 0.1 & 0.1 \\ 0 & 0.1 & 0.1 \end{bmatrix}$$

10.4 Find all the eigenvalues of the matrix using program 10.7.2 with a maximum change in the eigenvector of 1E–5 and a shift of zero.

$$\begin{bmatrix} 1/2 & 0 & 1/8 & 1/8 \\ 0 & 1/2 & 1/8 & 1/8 \\ 1/16 & 1/16 & 17/32 & 1/32 \\ 1/16 & 1/16 & 1/32 & 17/32 \end{bmatrix}$$

10.5 The matrix

$$A = \begin{bmatrix} 33 & 16 & 72 \\ -24 & -10 & -57 \\ -8 & -4 & -17 \end{bmatrix}$$

has eigenvalues 1, 2 and 3. Form the product A times A and verify that the eigenvalues of this matrix are the squares of the original ones.

10.6 The matrix below is known to have eigenvalues in the regions of 23, 7, 4.5, 1 and –1 refine these eigenvalues using inverse iteration program 10.9.1 and for the eigenvalue in the region of 23 find the eigenvector.

$$\begin{bmatrix} 5 & 4 & 3 & 2 & 1 \\ 4 & 6 & 0 & 4 & 3 \\ 3 & 0 & 7 & 6 & 5 \\ 2 & 4 & 6 & 8 & 7 \\ 1 & 3 & 5 & 7 & 9 \end{bmatrix}$$

10.7 Illustrate the effect of inverse iteration on a matrix with complex conjugate pair of eigenvalues by considering the matrix given by (10.10.1) in the text.

10.8 Use program 10.7.1 to find the eigenvalues of the matrix

$$\begin{bmatrix} 2 & -i & 0 \\ i & 2 & 0 \\ 0 & 0 & 3 \end{bmatrix}$$

10.9 The following matrix has a dominant pair of complex eigenvalues use program 10.10.1, with a maximum change in the eigenvector of 1E–5, to find this dominant eigenvalue.

$$\begin{bmatrix} -13.25 & 91.25 & 0 & -237.5 \\ -28.75 & -143.25 & 0 & -112.5 \\ 8.375 & 65.125 & 0 & 13.25 \\ -16.25 & 91.25 & 0 & -234.5 \end{bmatrix}$$

11

Testing microcomputers

11.1 INTRODUCTION

Most microcomputers can be used for numerical analysis and in this chapter we compare the performance of a selection of them to examine their suitability for numerical analysis. The features most relevant for this comparison are accuracy and computing speed. Memory capacity is not a major limitation for the algorithms considered except in solving linear equation systems and the eigenvalue problem. We do not consider factors such as graphics capabilities and the disc operating system since they are not directly relevant to numerical analysis. The microcomputers we have chosen for our comparisons are listed here and the user may apply the tests we describe to computers not included in the list and to others which may be developed in the future.

Acorn Archimedes 310
Amstrad PC1512 using Loco and Microsoft BASIC
Apple II+
Apple Macintosh 512k/400k using Microsoft and Z BASIC
Apricot portable using GW BASIC
Atari 512ST
BBC Model B
CBM 64
HP9845
IBM PC-AT with a maths coprocessor, GW BASIC and TurboBASIC
Sinclair QL

These machines have been chosen because they are current models or, if they are no longer in production, they are still widely used.

Most modern microcomputers have large random access memories (RAM) and for this reason a variety of languages and versions of these languages may be loaded into the system from disc. This is in contrast to almost all of the older microcomputers which had BASIC in read only memory (ROM) and provided the user with a relatively small amount RAM. Some of these disc based versions of BASIC cannot access all the memory of the system even though some software such as a word processors may access all available memory.

11.2 SIMPLE TESTS OF ACCURACY AND COMPUTING SPEED

An important feature of a computer language is the working precision used in calculations. Sometimes this precision can be controlled by the user who may specify single or double precision. Single precision varies with the language implementation and can be as few as 6 digits or as many as 12 digits. Double precision is not available in all versions of BASIC but when it is, it allows an increase in precision upto twice that offered in single precision. When using double precision a difficulty may arise because functions may not be evaluated in double precision although the standard arithmetic operations are. If the manual is not clear on this point then the user would be wise to evaluate some standard functions in double precision and compare the results with those given in Table 11.2.1. When converting a program to work in double precision it is important to follow carefully the rules for this process given with the particular dialect of BASIC. If the rules for conversion are not correctly implemented the program may appear to be working in double precision when in fact it is not.

An important accuracy parameter is the machine precision. This is defined as the smallest number E which the machine can recognise such that the comparison $1 + E > 1$ is true. It is conventional for an initial value of E to be taken as 1 and its value approximated by successively halving E while $1 + E > 1$ is true. Program 11.2.1 estimates machine precision using this procedure. The results are compared for selected machines in Table 11.2.2. The method used to calculate machine precision

Table 11.2.1. Values of some mathematical constants and functions.

π	3.1415 92653 58979 32384
$\sqrt{2}$	1.4142 13562 37309 50488
e	2.7182 81828 45904 52353
$\log_e 2$	0.6931 47180 55994 53094
$\sin 0.5$	0.4794 25538 60420 30003
$\cos 0.5$	0.8775 82561 89037 27161

```
10 E=1
20 E=E/2
30 IF E+1>1 THEN 20
40   PRINT "MACHINE PRECISION ="; 2*E
50 END
```

Program 11.2.1.

Table 11.2.2. Machine precision for some microcomputers.
DP indicates double precision.

Computer	
Acorn Archimedes 310	2.32830644E–10
Amstrad PC1512DD, LocoBASIC	4.65661287E–10
Apple II+	2.32830644E–10
Apple Macintosh, MS BASIC	1.192093E–7
Apple Macintosh, MS BASIC (DP)	2.220446... D–16
Atari 520ST	1.19209E–7
BBC Model B	2.32830644E–10
Casio BP700	1.164153218E–10
CBM 64	2.328 .... E–10
HP 9845	7.27595761414E–12
IBM PC-AT, GW BASIC	1.192093E–7
IBM PC-AT, TurboBASIC	1.084202172485504E–19

does not give its exact value and the reader should not be surprised that some of the accuracies given for different machines are identical. The results show that there is a wide variation in machine precision. Numerical difficulties may arise in dealing with ill-conditioned problems when the machine precision is as large as 1E–7.

An interesting and dramatic test of computer accuracy was given by Gruenberger (1984). In this test the number 1.0000001 is squared and the result repeatedly squared until a total of 27 squaring operations have been made. This is equivalent to raising 1.0000001 to the power 2^{27} (i.e to the power 134217728). The correct result, accurate to 10 digits, is 674530.4707. Table 11.2.3 shows the results derived from program

```
10 A=1.0000001
20 FOR I=1 TO 27
30     A=A*A
40 NEXT I
50 PRINT A
60 END
```

Program 11.2.2.

Table 11.2.3. Squaring 1.0000001 by itself twenty seven times.
DP indicates double precision.

Computer	By using A^2	By using A*A
Acorn Archimedes 310	685090.596	685090.596
Amstrad PC1512DD, LocoBASIC	685090.596	685090.596
Apple II+	728339.418	22723.9709
Apple Macintosh, MS BASIC	8850398	8850397
Apple Macintosh, MS BASIC (DP)	674530.4755	674530.4755
Atari 520ST	1	7.89733E13
BBC Model B	685090.513	685090.513
Casio BP700	674475.396	674475.396
CBM 64	728339.418	22723.9709
HP 9845	674514.8688	674514.8688
IBM PC-AT, GW BASIC	8850273	8850273
IBM PC-AT, TurboBASIC	8850397	8850397
Sinclair QL	643391.2	643391.2
EXACT	674530.4707	674530.4707

11.2.2 for a variety of microcomputers. The results of Table 11.2.3 for Gruenberger's problem illustrate the limitation of many computers, even when working in double precision. The results shown here are very dependent on the accuracy of the system and strikingly bad results are obtained when the number of significant digits used in the calculation is limited.

A further interesting test can be devised, based on the result $(1 + 1/x)^x$ tends to e as x tends to infinity. The use of formulae such as this in computers working with finite precision arithmetic will always lead to problems. In practice microcomputers will behave differently as x becomes large, ultimately making $1/x$ smaller than the machine precision, see Table 11.2.2. Then the term in the bracket becomes unity and the expression evaluates to one. The results shown in Table 11.2.4 where produced by program 11.2.3. This test gives similar results to the machine precision test. If we had set $X = 2$ and used a multiplier of 2 in program 11.2.3 then the best result for e coincides with $1/X$ taking the value of machine precision.

```
10 X=10
20 FOR I=1 TO 15
30    X=X*10
40    PRINT X, (1+1/X)^X
50 NEXT I
60 END
```

Program 11.2.3.

Table 11.2.4. Calculating e from $(1 + 1/x)^x$.

Computer	Best approx to e	Value of x for best approx
Acorn Archimedes 310	2.71828894	1E5
Amstrad PC1512DD, LocoBASIC	2.7182901	1E5
Apricot 256K port. MS BASIC	2.718023	1E4
Apple II+	2.71814671	1E4
Apple Macintosh, MS BASIC	2.718597	1E4
Apple Macintosh, MS BASIC (DP)	2.718281798347358	1E8
Atari 520ST	2.71549	1E4
BBC Model B	2.71828893	1E5
Casio BP700	2.7182818	1E10
CBM 64	2.71814671	1E4
HP 9845	2.71828182844	1E6
IBM PC-AT, GW BASIC	2.718023	1E4
IBM PC-AT, TurboBASIC	2.718281827142704	1E9
EXACT	2.718281828459045	

The following mathematical identities, which are given in BASIC form, allow us to test the accuracy of individual arithmetic operations and functions.

$$SQR(X^2) = X$$
$$LOG(X^N) = N*LOG(X)$$
$$EXP(-X/2)*EXP(X/2) = 1$$
$$X^{P^{(1/P)}} = X$$
$$SIN(X) = SIN(X+2*PI)$$
$$COS(2*X) = 2*COS(X)*COS(X) + 1$$

Note that on some computers it is necessary to define PI, either by using the value given in Table 11.2.1 or by letting PI = 4*ATN(1). The values on both sides of the identity should be equal and any differences will indicate inaccuracies in the machine computation. These and other similar tests may be used by the reader to investigate the accuracy of any microcomputer and by varying the values of X a great number of results can be produced. Tests must be devised with some care since in some compiled versions of BASIC the nature of the test may be changed. For example, the compiler may simplify X+X–X to X and Y*Y/Y to Y so that no real arithmetic test is involved.

To avoid having to assess a large amount of numerical data we obtain an overall measure of the accuracy by using a test due to Savage, described in Fried (1985). This test is implemented in program 11.2.4 and it can be seen that since the expression on

```
10 A=1
20 FOR I=1 TO 2499
30   A = TAN(ATN(EXP(LOG(SQR(A*A))))) + 1
40 NEXT I
50 PRINT A
60 END
```

Program 11.2.4.

Table 11.2.4. Results from Savage test.

Computer	Value	Time (secs)
Acorn Archimedes 310	2498.5004	5.2
Amstrad PC1512DD, LocoBASIC	2499.6831	28
Amstrad PC1512DD, MS BASIC	2716.961	70
Apricot 256K port. MS BASIC(SP)	2716.961	128
Apple II+	2500.0001	474
Apple Macintosh, MS BASIC (SP)	2769.853	42
Apple Macintosh, MS BASIC (DP)	2500	85
Apple Macintosh II MS BASIC(SP)	2769.853	12.1
Atari 520ST	1452.94	28
BBC	2499.6860	311
Casio BP700	2499.9991	875
HP 9845	2499.9993	22
IBM PC-AT, GW BASIC	2716.961	57
IBM PC-AT, TurboBASIC	2500	4.9

the righthand side of line 30 in the program evaluates to A + 1 then the final answer should be 2500. The results of running this program are compared for several micro-computers in Table 11.2.5. This table shows the widely varying speed for this particular problem. Three machines, the Macintosh II, the IBM PC-AT running TurboBASIC and the Archimedes are strikingly faster for varying reasons. The Macintosh II runs at 15.6 MHz and has a 68881 maths coprocessor, the Archimedes uses RISC technology and the IBM/TurboBASIC combination uses a maths coprocessor and a compiled BASIC which uses an efficient compiler which simplifies arithmetic expressions. There is also a very significant variation in accuracy of the result and it would appear that at least 8 digit precision is required to obtain a reasonable result. The Savage test not only allows us to test accuracy but also speed of computation. The designer of an interpreter must balance the requirement for speed with the need for accuracy. This balance in relation to the Savage test are considered in Sinnott (1987a). The Savage test has been criticised in Sinnott (1987b).

Table 11.2.6. Time taken to duplicate data between arrays.

Computer	Time (secs) for arrays ... 25x25	50x50	75x75
Acorn Archimedes 310	<1	<1	1.7
Amstrad PC1512DD, LocoBASIC	1.4	4.5	10
Amstrad PC1512DD, MS BASIC	1.9	7.5	17
Apple II+	6.3	25	*
Apple Macintosh, MS BASIC	2.1	8.5	19
Apple Macintosh II MS BASIC	<1	2.5	5.5
Atari 520ST	2.3	9.5	*
BBC Model B	3	*	*
HP 9845	<1	<1	1.9
IBM PC-AT, GW BASIC	1.5	6	14
IBM PC-AT, TurboBASIC	<1	1	2.4

* Indicates insufficient memory for test.

Other types of operations are important in numerical analysis and one important example is duplicating an array. Table 11.2.6 illustrates this point by showing the time taken by a variety of microcomputers to duplicate an array of data. This test illustrates that a significant time is required to duplicate arrays.

To further emphasise the influence of precision on calculations we ran Gruenberger's problem, program 11.2.2, using ZBASIC on the Macintosh computer. This version of BASIC has the unusual facility of allowing the user to specify a wide range of precision for the computation. The results of these tests are given in Table 11.2.7. There is no simple relation between the precision and the number of correct significant figures.

Table 11.2.7. Effect of precision in calculating Gruenberger's problem.

Precision	Result	Correct significant figures
6digits	1	0
10digits	674494.0561	4
14digits	674529.4131	5
18digits	674530.4707	10
EXACT	674530.4707	

We have considered above some simple tests which are useful for giving a basic comparison between machines and which have the advantage that they can be rapidly implemented on any computer. In the following sections we compare the performance of several microcomputers performing practical tasks in numerical analysis. In section 11.3 we consider algorithms which require a fixed number of operations to achieve the solution of a given problem. In section 11.4 we will examine problems where the solution is found by iteration and a specified accuracy must be given.

11.3. TESTS USING PROCEDURES WITH A FIXED NUMBER OF OPERATIONS

Examples of procedures involving a fixed number of operations are matrix inversion, the fast Fourier transform, integration and differentiation. Matrix inversion and the Fourier transform are examples were a finite number of operations are required, dependent only on the size of the problem. Integration and differentiation again require only a finite number of operations once the user has specified an interval size.

The fast Fourier transform provides a useful way of testing speed of computation Table 11.3.1 gives the results of running program 8.21.1 with a relatively large amount of data on the range of microcomputers. The table illustrates the advantages computers with maths coprocessors or RISC processor. The only compiled BASIC used in these tests is ZBASIC and the advantage this provides is only marginal.

Table 11.3.1. Time taken to perform a fast Fourier transform using program 8.21.1. Number of data values = 256.

Computer	Time (sec)
Acorn Archimedes 310	3
Apple Macintosh Plus, MS BASIC	32
Apple Macintosh Plus, ZBASIC	22
Apple Macintosh II	10
BBC Model B	38
IBM PC-AT, GW BASIC	30

The result of integrating $y = x^9 - x^{10}$ in the range 0 to 1 using Simpson's rule is shown in Table 11.3.2. The striking feature here is that there is very little difference between machines. This is because integration is a smoothing process and errors caused by machine accuracy limitations tend to even out. The major element of the error is due to the limitations of the algorithm and if we take sufficiently small intervals most microcomputers give satisfactory results.

Table 11.3.2. Numerical integration of $y = x^9 - x^{10}$ in the range 0 to 1 using Simpson's rule, program 4.3.1.

Computer	Result
Acorn Archimedes 310	9.09083771E–3
Apple Macintosh, MS BASIC	9.090836E–3
Atari 520ST	9.09084E–3
BBC	9.09083771E–3
IBM PC-AT, GW BASIC	9.090837E–3
EXACT	9.09090909E–3

Table 11. 3.3 Numerical differentiation of $y = \exp(\sin x)$ using extrapolation method, program 5.5.1.

Computer	1st derivative	2nd derivative
Acorn Archimedes 310	1.25338087	–1.27479657
Apople II+	1.25338087	–1.27479472
Apple Macintosh, MS BASIC	1.253386	–1.267493
Atari 520ST	1.25338	–1.29929
BBC	1.25338087	–1.27486578
HP9845	1.25338076388	–1.27481823047
IBM PC-AT, GW BASIC	1.25335	–1.276322
EXACT	1.2533807	–1.27482037

In contrast with integration, differentiation is machine sensitive. The first derivatives are quite accurate but significant errors arise in some of the second derivatives. Machines working to higher accuracy give much better results. For example the HP9845 gives the second derivative accurate to five decimal places while the Apple Macintosh gives only two decimal places and the Atari is worse.

We have made reference in Chapter 3 to the difficulties of inverting accurately the Hilbert matrix. We now examine in detail the effect of varying machine precision on the accuracy of the determinant of a 10 x 10 Hilbert matrix, calculated using program 3.9.1. The results are given in Table 11.3.4. These results show that a high precision of fourteen significant figures is required to obtain even one significant figure accuracy in the determinant. Above this precision there is an almost linear relationship between machine precision and the accuracy of the result.

Table 11.3.4. Determinant of a 10 x 10 Hilbert matrix using ZBASIC.

Precision	Significant digits
10	*
14	1
18	7
22	10
26	13
30	18
34	22
38	26

* The computed value was in error by a factor of about 100.

11.4. TESTS USING PROCEDURES WHERE THE NUMBER OF OPERATIONS DEPEND ON THE CONVERGENCE CRITERIA

The nature of the iterative process itself compensates for errors made during the calculations and consequently one would expect the influence of machine precision to be less significant in this type of algorithm. Machine precision is, however, important in relation to the criteria used to terminate the iteration and if an accuracy is requested which is beyond the machines ability to detect then the program will continue indefinitely. Thus it is very important for the user to specify an accuracy taking into consideration the machine precision. The methods considered here are Brent's method for finding the roots of equations and the Gauss-Seidel iteration to solve a set of linear equations. The results from the Brent test are shown in Table 11.4.1 and from the Gauss-Seidel test in Table 11.4.2. In both cases there is little difference in the machines tested. This illustrates the points described above. The problem chosen for the Gauss-Seidel method does not satisfy the sufficient conditions for convergence given in Chapter 3. However, it is seen that covergence is achieved after about 38 iterations.

Table 11.4.1. Brent's method to find a root of $\exp(\sin x) - 0.5 = 0$ searching in the range $x = 2$ to $x = 5$ with accuracy 1E–5.

Computer	No of iterations	Result
Acorn Archimedes 310	9	3.90743883
Apple Macintosh, MS BASIC	9	3.907439
Atari 520ST	8	3.90743
BBC	9	3.90743883
IBM PC-AT, GW BASIC	9	3.90744

Table 11.4.2. Gauss-Seidel method used to solve

$$\begin{bmatrix} 1.5 & 1 & 1 & 1 & 1 \\ 1 & 2 & 1 & 1 & 1 \\ 1 & 1 & 1 & 1 & 1 \\ 1 & 1 & 1 & 2 & 1 \\ 1 & 1 & 1 & 1 & 2 \end{bmatrix} \begin{bmatrix} x_1 \\ x_2 \\ x_3 \\ x_4 \\ x_5 \end{bmatrix} = \begin{bmatrix} 1 \\ 2 \\ 3 \\ 4 \\ 5 \end{bmatrix}$$

Computer	No of iterations	Solution vector
Acorn		
Archimedes	33	[–4.00000772, –1.00000768, 5.0000154, .999999998, 2]
Apple Macintosh		
MS BASIC	38	[–4.000001, –1.000001, 5.000002, 1, 2]
Atari 520ST	38	[–4, –1, 5, .999999, 1.99999]
BBC	33	[–4.00000772 –1.00000768 5.0000154, 1, 2]
IBM PC-AT		
GW BASIC	33	[–4.000008, –1.000008, 5.000016, .9999996, 2]

11.5 SOFTWARE LIBRARIES AND SPREADSHEETS

It has become possible to implement libraries of subroutines for numerical methods on modern, powerful microcomputers such as the IBM PC-AT and the Macintosh II. For example, a subset of subroutines from the the Numerical Algorithms Group (NAG) library is available for the IBM PC and other microcomputers. The NAG library has been widely available in universities and industry for many years but only on mainframe computers and is constantly updated to accommodate improvements and developments. The library of routines now available for microcomputers is called "The NAG FORTRAN PC50 Library" and contains the fifty most popular subroutines from the original NAG library. These routines are directly compatible with the mainframe NAG library subroutines. This selection of subroutines includes algorithms for solving systems of ordinary differential, linear equations, eigenvalue problems and quadrature. Another library available for microcomputers is the IMSL library. The IMSL library is divided into three parts for microcomputers: the MATH/ PC- LIBRARY which includes a range of numerical subroutines and printing, plotting and sorting subroutines; the STAT/PC- LIBRARY which contains statistical subroutines; the special functions library called SFUN- LIBRARY.

Another type of software that may be of value to the numerical analyst is the spreadsheet. This provides, by means of computer monitor screen, a window onto a spreadsheet which may be many times larger than the screen itself. The software allows the user to scroll rapidly to any portion of the spreadsheet and the package includes the usual mathematical functions and operators which allow a wide range of calculations to be performed. The spreadsheet is divided into a large number of cells which may contain formulas, titles or numerical values. This allows the user to set up a series of complex calculations. A key feature of spreadsheets is that they allow immediate recalculation of all related values throughout the sheet when a particular value is changed. The spreadsheet may thus be used in instructive and experimental way were the user may set out the formulas and data for the problem as if using pen and paper but use the calculation facilities of the spreadsheet to avoid arduous hand computation. The effects of any experimental adjustments to variables can be made to ripple throughout the whole table by using the recalculation facility. An interesting example of the application of this is the solution of partial differential equations by using the standard finite difference formula these formula can be set up in appropriate cells and duplicated throughout the spreadsheet in this way the cells of the spreadsheet become the nodes of the network.

An interesting recent development is a package called MATLAB produced by The MathWorks Inc. This software is a combination of a script language and a collection of powerful mathematical subroutines. A fundamental feature of this package is that a variable is in general an array and vectors and scalars are special cases of this. Examples of the subroutines available are matrix inversion, eigen solutions, roots of polynomials. Two and three dimensional graphical output routines are provided. This system is easier to use than a software library such as NAG because (i) the script language provided is specifically designed to meet the needs of the numerical analyst, it is less general purpose than FORTRAN but easier to use and (ii) it is easier to use the subroutines.

11.6 PROGRAM PORTABILITY

We have indicated earlier that an important feature of the programs presented in this book is that they are portable because they are written in a subset of BASIC which is common to the versions of BASIC implemented on most microcomputers. It is important however for the user to be aware of some of the differences between versions of BASIC.

The majority of differences in dialects of BASIC occur in the areas of string handling, graphics and disc input/output, features which are not usually important in numerical analysis. In the areas that relate directly to numerical analysis the differences between dialects are relatively small but may, none the less, be critical.

First of all the user should be mindful of the possibility of relatively large and unexpected errors arising in simple calculations. As an example of this, MS BASIC Version 2.10.00 running on an Apple Macintosh gives an absolute error of 0.0078125 when evaluating 8.5^2. This is a very large error and would lead to spurious results if this particular evaluation was required in some general calculation. Software manufacturers do their best to remove such errors in later versions of the software but undetected errors of this sort could exist for a long time. Another interesting problem concerns arithmetic hierarchy. The expected interpretation of -4^2 is -16 since according to the rules of both arithmetic hierarchy and BASIC language hierarchy, raising to a power has a higher precedence for evaluation than subtraction. However, in Applesoft BASIC used on the Apple II+ computer this is interpreted as $(-4)^2$ and evaluated to 16. Evaluation of $-4^{1.5}$ fails because it is interpreted as $(-4)^{1.5}$. This problem arises with the Apple II+ but not with the Apple Macintosh or indeed any other machine that the author have used.

Care must be taken when using different dialects of BASIC if the dialect uses a statement or function in a manner that is different from the norm. For example, in BBC BASIC logarithms to the base e are called using the function LN and and logarithms to the base 10 are called by LOG. Since LOG is normally used to call logarithms to be base e then a BBC microcomputer user must remember to change LOG to LN, failure to do so would be disastrous. Slightly less serious is when a version of BASIC uses a different, but not confusing, function name. For example the Sinclair QL uses ATAN instead of ATN to call the inverse tangent. Further problems may arise in using random numbers. Machine generation of random numbers involves two processes, resetting the random number generator and generating the sequence of numbers. The statements which are used to reset the random number generator vary considerably from machine to machine. Sometime the statement RANDOMISE is used, sometimes RAND( ) function is used with a particular parameter, such as time and a minus sign. Furthermore, the significance of this parameter varies from machine to machine.

The facilities offered in some dialects of BASIC are more generous than others. For example, whilst most versions of BASIC allow the user to define functions of one variable only, some allow functions in two variables. Array specifications usually allow a subscript value of zero but this is not a universal practice and in ZBASIC it is user definable.

As a final example of the pit-falls which face the user of BASIC, care must be taken when performing FOR-NEXT loops. Consider the loop:

```
FOR I=1 TO N
    PRINT I
NEXT I
```

If $N = 2$, the numbers 1 and 2 are printed, if $N = 1$ the number 1 is printed and if $N = 0$ we would logically expect that no values would be printed but most microcomputers print the number 1. To further emphasise the problems with FOR-NEXT loops we consider the case when the final value of the loop variable does not coincide with the final loop value is given by loop. This is illustrated by the following program segment:

```
FOR X=1 TO PI STEP PI/3420
    PRINT X
NEXT X
```

Due to rounding errors the last value of X may not be printed because the penultimate value of X exceeds the value PI by some small amount.

In conclusion we must reiterate *the need for care when ranslating programs from one system to another*. Readers must be aware of the details of the BASIC they are using and able to recognise and surmount any difficulties which may arise in the translation process. It is also important to note that even if the features of the BASICS on different systens are identical, programs may fuction differently because of different levels of machine precision. For eample, a convergence criteria may work satisfactorily with one system and not with another.

References

Abramowitz, M. & Stegun, I.A. (1964). *The Handbook of Mathematical Functions*. National Bureau of Standards.

Brent, R.P. (1971). "An algorithm with guaranteed convergence for finding a zero of a function". *Comp. J.* **14** 422-425.

Brigham, E.O. (1974). *The Fast Fourier Transform*. Prentice Hall, Englewood Cliffs. N. J.

Butcher, J.C. (1964). "On Runge Kutta processes of higher order". *J. Austral. Math. Soc.* **4** 179-194.

Carre, B.A. (1961). "The determination of the optimum accelerating factor for successive over-relaxation.". *Comp. J.* **4** 73-78.

Clenshaw, C.W. & Curtis, A.R. (1960). "A method for numerical integration on an automatic computer". *Numerische Math.* **2** 197-205.

Cooley, P.M. & Tukey, J.W. (1965). "An algorithm for the machine calculation of complex Fourier series". *Math. Comp.* **19** 297-301.

Davis, P. & Rabinowitz, P. (1956). "Abscissas and weights for Gaussian quadratures of high order". *J. Res. Nat. Bur. Stand.* **56** 35-37.

Dekker, T.J. (1969). "Finding a zero by means of successive linear interpolation". in Dejon, B & Henrici, P. (Eds). *Constructive Aspects of the Fundamental Theorem of Algebra*. Wiley-Interscience. London.

Dowell, M. & Jarrett, P. (1971). "A modified *regula falsi* method for computing the root of an equation". *BIT* 168-174.

Fox, L. (1964). *An Introduction to Numerical Linear Algebra*. OUP.

Fox, L & Mayers, D.F. (1977). *Computing Methods for Scientists and Engineers*. OUP.

Frankel, S.P. (1950). "Convergence rates of iterative treatments of partial differential equations". *Math. Tables Aids Comput*. **4** 65-75.

Fried, S.S. (1985). "The 8087/80287 performance curve". Byte **10** 11 67-88.

Gear, C.W. (1971). *Numerical Initial Value Problems in Ordinary Differential Equations*. Prentice Hall, Englewood Cliffs. N. J.

Gere, J.M. & Weaver, W. (1965). *Matrix Algebra for Engineers*. Van Nostrand, Princeton. N.J.

Gill, S. (1951). "Process for the step by step integration of differential equations in an automatic digital computing machine". *Proc. Cambridge Philos. Soc*. **47**, 96-108.

Gill, P.E. *et al*. (1986). "On projected Newton methods for linear programming and an equivalence to Karmarkar's projective method". *Math. Programming* **36**, 183-209.

Goldstine, H.H. (1977). *A history of numerical analysis from the 16th through the 19th century*. Springer Verlag.

Goodwin, E.T. (1961). *Modern Computing Methods*. HMSO., London.

Gourlay, A.R. & Watson, G.A. (1973). *Computational Methods for Matrix Eigenproblems*. John Wiley.

Gragg, W.B. (1965). "On extraplolation algorithms for ordinary initial value problems". *SIAM. J. Numer. Anal*. **2** 384-403.

Gregory, R.T. & Karney, D.L. (1969). *A Collection of Matrices for Testing Computational Algorithms*. Wiley-Interscience.

Greville, T.N.E. (ed.) (1969). *Theory and Application of Spline Functions*. Academic Press, New York.

Gruenberger, F. (1984) "Computer recreations". *Scientific American* **250**, 4 10-14

Hammersley, J.M. & Handscomb, D.C. (1964). *Monte Carlo Methods*. Methuen & Co.

Hamming, R.W. (1959). "Stable predictor-corrector methods for ordinary differential equations". *J. of the ACM*. **6** 37-47.

Hamming, R.W. (1973). *Numerical Methods for Scientists and Engineers*. 2nd Ed. International Student Edition, McGraw-Hill Kogakusha.

Hildebrand, F.B. (1974). *Intrduction to Numerical Analysis*. McGraw-Hill.

Hollingdale, S.H. (1972). "The history of numerical analysis". *Bul. of The Inst. of Mat. and its Application*. September/October.

Issacson, E. & Keller, H. (1966). *Analysis of Numerical Methods*. Wiley, New York.

Jeffrey, A. (1979). *Mathematics for Engineers and Scientists*. Nelson.

Jennings, A. (1966). "A compact storage scheme for the solution of symmetric linear simultaneous equations". *Comp. J.* **9** 281-285.

Karmarkar, N. (1984). "A new polynomial-time algorithm for linear programming". *A T & T Bell Laboratories, Murray Hill, New Jersey. U.S.A.* Undated.

Khachiyan, L.G. (1979). "A polynomial algorithm in linear programming". *Doklady Akademiia, Novaia seviia*. 244 **5** Nauk SSSR 1093-1096.

Kreyzig, E. (1967). *Advanced Engineering Maths*. John Wiley.

Lewis, H.R. & Papadimitriou, C.H. (1977). "The efficiency of algorithms". *Scientific American* **239**, 1 96-109.

Merson, R.H. (1957). "An operational method for the study of integration processes". *Proc. Conf. on Data Processing and Automatic Computing Machines. Weapons Research Establishment, Salisbury, South Australia.*

Milne, W.E. (1949). "A note on the numerical integration of differential equations". *J. Res. NBS* **43** 537-542 RP2046.

Modienis, D.T., Scott, R.C. & Cornwell, L.W. (1987) "Testing intrinsic random-number generators". *Byte* **12**, No1, 175-178.

Moore, J.B. (1967). "A convergent algorithm for solving polynomial equations". *J. of ACM*. **14** 311-315.

Naylor, T.H. et al, (1966). *Computer Simulation Techniques*. Wiley, New York.

Newmark, N.M. (1959). "A method of computation for structural dynamics". *ASCE* **85** EM3 67-94

Patterson, T.N.L. (1973). "Algorithm for automatic numerical integration over a finite interval". *CACM* **16** 694-699.

Powell, M.J.D. (1981). *Approximation Theory and Methods*. CUP.

Ralston, A. (1962). "Runge Kutta methods with minimum error bounds". *Math. Comp.* **16** 431-437.

Ralston, A. & Rabinowitz, P. (1978). *A First Course in Numerical Analysis*. McGraw-Hill.

Ramirez, R.W. (1986). *The FFT, Fundamentals and Concepts*. Prentice Hall, Englewood Cliffs. N. J.

Redish, K.A. (1961). *An Introduction to Computational Methods*. EUP.

Ripley, B.D. (1983). "Computer generation of random variables, a tutorial". *Int. Statistical Review*, **51**, 301-319

Ripley, B.D. (1987). *Stochastic Simulation*. Wiley, New York.

Salvadori, M.G. & Baron, M.L. (1961). *Numerical Methods in Engineering*. Prentice Hall, Englewood Cliffs. N. J.

Salzer, H.E., Zucker, R. & Capuano, R. (1952). "Table of the zeros and weight factors of the first twenty Hermite polynomials". *J. Res. Nat. Bur. Stand.* **48** 111-116.

Sawitzki, G. (1985). "Another random number generator which should be avoided". *Statistical Software Newsletter*. **11**, 2 81-82.

Scheid, F. (1968). *Numerical Analysis*. Schaum's Outline Series. McGraw-Hill.

Shreider, Y.A. (1967). *The Monte Carlo Method*. Pergamon.

Simmons, G.F. (1972). *Differential equations with application and historical notes.* McGraw-Hill.

Sinnott, R.W. (1984). "TRS-80 versus a giant brain of yesteryear". *Sky & Telescope* **67** 358-359.

Sinnott, R.W. (1987a). "Astronomical Computing". *Sky & Telescope.* **73** 309-310.

Sinnott, R.W. (1987b). "Astronomical Computing". *Sky & Telescope.* **73** 646-647.

Wilkinson, J.H. (1963). *Rounding Errors in Algebraic Processes.* HMSO. London.

Wilkinson, J.H. (1965). *The Algebraic Eigenvalue Problem.* OUP.

Wilkinson, J.H. (1967). In *Mathematical Methods for Digital Computers.* Vol 2. Ralston, A. & Wilf H. S. John Wiley & Sons, Inc.

Solutions to problems

Chapter 2

2.1   Root = 27.8235, 4 iterations required.

2.2   Root = 27.8235, bisection requires 22 iterations, secant requires 5 iterations.

2.3   Roots = –2 and 1.6344. Note when $x < 0$, $df/dx = -3x^2 + 1$, when $x \geq 0$
       $df/dx = 3x^2 + 1$. Insert `DEF FND(X)=SGN(X)*3*X*X+1` in program.

2.4 $c = 5$, root = 1.3734, 4 iterations required.
       $c = 10$, root = 1.4711, 15 iterations required.
       As c increases the root becomes closer to a discontinuity.

2.5 $c = 5$, root = 1.3734, 7 iterations required.
       $c = 10$, root = 1.4711, 9 iterations required.
       $c = 500$, root =1.5688, 10 iterations required.

2.6   Using program 2.9.2, root = 1 (which is exact), 4 iterations required.
       Using program 2.6.1, root = 0.9444, 13 iterations required. This is a very poor
       result because the root is repeated at $x = 1$.

2.7   Using program 2.6.1, root = 1.1183 after 5 iterations.
       Iterating $x = e^{x/10}$ gives a similar result. Iterating $x = 10 \log_e x$ and $x = e^x/x^9$ fail
       to converge.

2.8 $E = 0.1280231$ after 7 iterations.

2.9   By the secant method, root = 8.1739 x 10^{-4} after 109 iterations.
       By Brent's method, root = 2.019 x 10^{-4} after 25 iterations. Exact result is, of
       course, zero.

2.10 Using 4 terms in series, root = 1.429968.
Using 5 terms in series, root = 1.446744.
Using 6 terms in series, root = 1.445725.

2.11 Eliminating y from the pair of equations gives $x/5 - \cos x - 2 = 0$ and solving this quation using program 2.6.1 gives $x = 8.2183$ after 7 iterations. Solving the equations directly using 2.18.1 gives $x = 8.2183$ and $y = 2.2747$ after 9 iterations.

2.12 $x = 0.1605$, $y = 0.4931$.

2.13 $x = 1.1132$, $y = 0.2366$ after 9 iterations.

2.14/15

Method	ε	Root1	Root2	Root3	Root4
Bairstow	0.1	1	2	4.999948	5.100046
Bairstow	0.01	1	2	4.999965	5.010036
Bairstow	0.001	0.9999998	2.000001	4.997687	5.003312
Laguerre	0.1	1.000001	1.999997	4.999997	5.100005
Laguerre	0.01	1	2	4.999897	5.010103
Laguerre	0.001	1	2	4.998726	5.002275
Moore	0.1	0.9999998	2.000001	4.999972	5.100027
Moore	0.01	0.9999999	2	4.999766	5.010234
Moore	0.001	0.9999999	2.000001	4.997477	5.003522
QD	1	0.9999998	2.000001	4.999542	6.00046
QD	0.1	1	2.000001	4.9§	5.17§

Note: roots denoted by § were not output by program 2.14.1 because the required convergence criteria could not be met.

2.16 Bairstow and Moore's method gave roots close to the exact answers which are $\sqrt{2}, -\sqrt{2}, i, -i$ and 1.

2.17 Moore's method gives the roots
$-0.7660525 - i0.642781$
$0.9396955 - i0.342024$
$-0.1736431 + i0.984805$

2.18 27.83349 after 8 iterations.

2.19 (a) 10.0056 after 10 iterations (b) 10.0056 after 7 iterations.

2.20   For complex $z = x + iy$, Newton's method is $z_{n+1} = z_n - f(z_n)/f'(z_n)$. Letting
       $f(z) = f_R + if_I$ gives
       $$x_{n+1} = x_n - (f_R f'_R + f_I f'_I)/D$$
       $$y_{n+1} = y_n - (f_I f'_R - f_R f'_I)/D \text{ where } D = f'^2_R + f'^2_I$$

       (a) $4.12563427 \pm i20.2190687$, $3.19096274 \pm i7.44926919$
       (b) $1.57079633 \pm i1.3169579$, $7.85398163 \pm i1.3169579$
       (c) $\pm 0.7271361 \pm i0.4300143$

Chapter 3

3.1 $x_1 = 1.022761$, $x_2 = 1.138032$, $x_3 = -0.5374449$, determinant $= -1362$.

3.2   The first two equations are placed after the third.

ε	Jacobi iterations	Gauss Seidel iterations
0.0005	10	7
0.0000005	18	12

3.3   Solution is $x_1 = 2.4545\ldots$, $x_2 = 1.4545\ldots$, $x_3 = -0.2727\ldots$ Using $\omega = 1$ requires
       17 iterations; $\omega = 1.05$ requires 13 iterations; $\omega = 1.1$ requires 7 iterations;
       $\omega = 1.15$ requires 31 iterations.

3.4   The exact solutions is $x_1 = -12.5$, $x_2 = -24$, $x_3 = -34$, $x_4 = -42$, $x_5 = -47.5$,
       $x_6 = -50$, $x_7 = -49$, $x_8 = -44$, $x_9 = -34.5$, $x_{10} = -20$. Iterations required, 167.

3.5   The optimum value of ω is 1.55, 40 iterations required. For $\omega > 1.66$ the
       number of iteration tends to increase very rapidly.

3.6   See 3.4.

3.7   (i)
       $$A^{-1} = (1/6)\begin{bmatrix} 24 & -36 & 24 & -6 \\ -26 & 57 & -42 & 11 \\ 9 & -24 & 21 & -6 \\ -1 & 3 & -3 & 11 \end{bmatrix}$$

       (ii) $A^{-1} = A$.
       (iii) If $B = A^{-1}$ then $B_{ij} = (-1)^{i+j}A_{ij}$

3.8   Exact solution for $n = 10$ is $x_1 = 181.3$ and $x_2 = -180.1$. When $n = 100$,
       $x_1 = 18011.2$ and $x_2 = -18010$. Using program 3.6.1, for $n = 10$, $x_1 = 181.3002$
       and $x_2 = -180.1002$. When $n = 100$, $x_1 = 18008.21$ and $x_2 = -18007.01$. When

$n = 10$, condition number is 400 (approx.). When $n = 100$, condition number is 4E4 (approx.). Answers are less accurate because condition number is much larger.

3.9　Using inverse matrix, determinant = 3.462979E–11 and solution is
$z^T = [-1787.837, 23433.8, -64678.41, 73892.41, -38364.29, 7505.637]$,
Using Gaussian elimination to solve the system directly solution is
$z^T = [-1834.16, 23625.02, -64992.35, 74148.34, -38468.26, 7522.418]$,
and the determinant = 3.452945E–11. This is a more accurate result. The poor result obtained by inversion is due to the system being ill-conditioned.

3.10　$A^{-1} = \begin{bmatrix} 9 & -36 & 30 \\ -36 & 192 & -180 \\ 30 & -180 & 180 \end{bmatrix}$　Determinant = 4.629636E–4.

Now $A^TA = \begin{bmatrix} 1.3611111 & 0.75 & 0.525 \\ 0.75 & 0.4236111 & 0.3 \\ 0.525 & 0.3 & 0.2136111 \end{bmatrix}$

and its determinant = 2.134385E–7. The inverse of this matrix is

$$\begin{bmatrix} 2277 & -12636 & 12150 \\ -12636 & 70560 & -68040 \\ 12150 & -68040 & 65700 \end{bmatrix}$$

However, calculating the inverse using the formula $A^{-1}(A^{-1})^T$ gives

$$\begin{bmatrix} 2286.469 & -12688.96 & 12201.12 \\ -12688.97 & 70856.27 & -68325.91 \\ 12201.12 & -68325.91 & 65975.94 \end{bmatrix}$$

The matrix A^TA is more ill-conditioned than A.

3.11　$x = 2.049978, y = 0.5618268$

3.12　If the equations are rearranged into the form $x = \sqrt{(4 - y^2)}$, $y = 1/x$ then the convergence criteria is met in the region $x = 2, y = 0.25$. Gauss-Seidel iteration leads to $x = 1.931852$ and $y = 0.5176381$.

3.13　The determinant of the coefficient matrix is zero but, dependant on the computer used, the determinant will be small, but not necessarily zero due to rounding errors. This will allow a solution to be found which will satisfy the equations. These equations are consistent and have an infinity of solutions where $x_1 = 4 + 11x_3$, $x_2 = 3 + 6x_3$ and x_3 can take any value.

3.14 $A = 100$, 3 iterations; $A = 10$, 7 iterations; $A = 4.5$, 37 iterations; $A = 1$, fails
       to converge.

3.15   The determinant of the coefficient matrix is zero but, dependant on the
       computer used, the determinant will be small, but not necessarily zero due to
       rounding errors. Theoretically cases (i) and (ii) are different. In case (i) the
       solutions are indeterminant, in case (ii) the values are infinite.

3.16   (i) $\mathbf{x}^T = [3.0000001, 1.916667, 1.433334]$
           (ii) $\mathbf{x}^T = [-2.333334, -1.083334, -0.7000002]$

Chapter 4

4.1   Estimate equals 0.2250709 when $\varepsilon = 1E{-}2$, 0.2222466 when $\varepsilon = 1E{-}4$,
       0.2222701 when $\varepsilon = 1E{-}8$.

4.2   Estimate equals 0.2250699 when $\varepsilon = 1E{-}2$, 0.2222330 when $\varepsilon = 1E{-}4$,
       0.2222448 when $\varepsilon = 1E{-}8$.

4.3   The theoretical error is $(b - a)h^4/180 = 5E{-}9$ and so we would expect our
       solutions to be accurate to 8 decimal places.

4.4   Estimate equals 1.467463.

4.5   By Romberg's method integral = 170. There are 168 prime numbers below
       1000 and this solution appears quite satisfactory. The exact value of the
       integral is given by $\mathrm{Ei}(\log_e x) = \mathrm{Ei}(\log_e 1000) = \mathrm{Ei}(6.907755)$. $\mathrm{Ei}(x)$ is not
       tabulated but tabulated values of $x\exp(-x)\,\mathrm{Ei}(x)$ are available, see   Abra-
       mowitz and Stegun (1964), p242. From this table we can deduce that
       $\mathrm{Ei}(6.907755) = 177.6$.

4.6
k	Filon	Simpson	Exact
0	2	2	2
4	−0.1333335	1.547493E−8	−0.1333333
100	−2.325092E−3	−2.193768E−6	−2.000200E−4

       Note the very poor performance of Simpson's method when $k = 4$ and 100.
       If 4096 divisions are used when $k = 100$ then Simpson's rule gives
       −2.000515E−4.

4.7   Estimate = 9.901942E−3. Even the 2-point rule gives excellent results.

4.8 Using 16 point rule, estimate = 1.380388. This is an excellent agreement with the exact result.

4.9 For even n

$$\int_{-1}^{1} dy \int_{-\pi}^{\pi} x^n y^n dx = \frac{4\pi^{n+1}}{(n+1)^2}.$$

When $n = 4$, approximation = 48.963; $n = 10$, approximation = 9726.557.

4.10 (i) −1.718758, (ii) 0.2222223

4.11 (i) 9725.754, (ii) 0.222222. Both results are excellent.

4.12

	Exact.	64 point Simpson.
$J_0(1)$	0.7651976865	0.7651976
$J_0(3)$	−0.2600519549	−0.2600518
$J_0(5)$	−0.1775967713	−0.1775967

4.13

	Exact	64 point Simpson	4 point Gauss
erf(0.20)	0.2227025892	0.2227026	0.2227026
erf(0.76)	0.1715367528	0.7175367	0.7175367
erf(1.35)	0.9437621961	0.9437621	0.9437718

4.14

	Exact	4 point Gauss
Si(0.5)	0.493107418	0.4931074
Si(1)	0.9460830704	0.9460831
Si(2)	1.6054129768	1.605413

The Romberg and Simpson methods fail, (Division by zero error).

4.15 The 8-point Gauss-Laguerre method gives 0.7284324.
The functions $1/(1 + x^3)$ and $1/(1 + x^3 + x^{-x})$ are shown and it may be noted that they functions are virtually identical for $x > 5$. This fact provides an alternative method of evaluating the required integral.

By Simpson, $\displaystyle\int_0^5 \frac{dx}{(1+x^3)} = 1.189264;$ $\displaystyle\int_0^5 \frac{dx}{(1+x^3+x^{-x})} = 0.7047915$

Now $\displaystyle\int_0^\infty \frac{dx}{(1+x^3)} = \frac{\pi}{3\sin(\pi/3)} = 1.2091995$

Thus $\int_s^\infty \dfrac{dx}{(1+x^3)} = 1.2091995 - 1.189264 = 0.0199355$

Hence $\int_0^\infty \dfrac{dx}{(1+x^3+x^{-x})} \approx \int_0^s \dfrac{dx}{(1+x^3+x^{-x})} + \int_s^\infty \dfrac{dx}{(1+x^3)}$

$= 0.7047915 + 0.0199355 = 0.724727$

The difference between the two methods is about 0.5%.

Chapter 5

5.1   Using $h = 0.1, y' = 0.2267364$; $h = 0.01, y' = 0.2390087$. (Exact $= 0.2391336$).
       Using $h = 0.1, y'' = -2.824932$; $h = 0.01, y'' = -2.82526$. (Exact $= -2.8255816$).

5.2 $y' = 0.2391289$, $y'' = -2.82406$. (Exact $= 0.2391336$, -2.8255816).

5.3 $x = 1$: $h = 0.1$ (Richardson), $y' = -5.048839$; $h = 0.001, y' = -5.048931$.
       (Exact $= -5.04882591$).
       $x = 2, h = 0.1$ (Richardson), $y' = -174.2738$; $h = 0.001, y' = -175.5747$.
       (Exact $= -176.644999$).

5.4 | Func'n | h | y' | Exact y' | y'' | Exact y'' |
       |---|---|---|---|---|---|
       | $x^2\cos x$ | 0.1 | 0.2390627 | 0.2391336 | -2.825602 | -2.825581633 |
       | $x^2\cos x$ | 0.01 | 0.2391329 | 0.2391336 | -2.825161 | -2.825581633 |
       | $\cos x^6$ | 0.1 | -5.448516 | -5.0488259 | -46.53707 | -44.6950126 |
       | $\cos x^6$ | 0.01 | -5.048857 | -5.0488259 | -44.69415 | -44.6950126 |

5.5 $x = 3.1$ $y' = -1707.538$ (Exact $= -1717.7369$).
       $x = 3.01$ $y' = -1113.842$ (Exact $= -1119.02927$).
       $x = 3.001$ $y' = -1452.777$ (Exact $= -1459.32545$).
       $x = 3$ $y' = -217.582$ (Exact $= -218.6078794$).

5.6 $\partial f/\partial x = 593.6509$, exact $= 593.65264$; $\partial f/\partial y = 148.407$, exact $= 148.4131$.

5.7 | x | y' | Exact |
       |---|---|---|
       | 1 | -2.425543 | -2.4255188 |
       | 2 | -4.77433 | -4.7743992 |
       | 3 | $-2.032018\text{E}{-2}$ | -0.0203195 |
       | 4 | -1.34051 | -1.3405501 |
       | 5 | -11.42771 | -11.4278817 |

5.8   By numerical differentiation, 3.215969E–2; by differentiating series term by term, 3.217137E–2; from tables, 3.233839747E–2.

Chapter 6

6.1   When $t = 10$, exact $= 30.326533$.

Method	$h = 1$	$h = 0.1$	$h = 0.01$
Euler	29.93685	30.28851	30.32271
Modified Euler	30.32338	30.32654	
Runge-Kutta	30.32654		

6.2   When $t = 2$, exact $= 109.1963$, Classical $= 108.9078$, Gill $= 108.9078$, Merson $= 109.0707$, Butcher $= 109.1924$, Ralston $= 108.8753$.

6.3   When $t = 50$, exact $= 4.1042499$, Milne $= 4.10425$, Adams $= 4.104226$, Hamming $= 4.104219$

6.4   When $t = 5$, exact $= -0.0368817$, Milne $= -0.03569949$, Adams $= -0.0359842$, Hamming $= -0.0353289$

6.5   Let $y_1 = y'$ and $y_2 = y$. Thus $xy_1' - y_1 - 8x^3y_2^3 = 0$. Hence

$$y_1' = (y_1 + 8x^3y_2^3)/x$$
$$y_2' = y_1$$

Initial conditions $y_1 = -1/2$ and $y_2 = 1/2$.
When $x = 2$, exact $= 0.2$, Runge-Kutta $= 0.1999831$.

6.6   For 6.1, when $t = 10$, Hermite $= 30.32654$, exact $= 30.326533$.
For 6.2, when $t = 2$, Hermite $= 108.8907$, exact $= 109.1963$.
For 6.3, when $t = 50$, Hermite $= 4.104251$, exact $= 4.1042499$.

6.7   (a) Ans $= 7998.572$, exact $= 8000$.
(b) Ans $= 109.1963$, exact $= 109.1963$.

6.8 $k = 0$. Numerov $= -0.8388709$, exact $= -0.83907153$.
$k = 1$. Numerov $= -0.1004559$, series approximation $= -0.1009196$

6.9

v	x	Finite difference	Exact
0.5	0.25	0.8246248	0.824361
1	0.5	1.359359	1.35914
1.5	0.75	2.240986	2.240844

6.10   It is necessary to calculate the distances between the boundary and adjacent
       nodes. There are 11 (internal) nodes and 2 boundary nodes and $f(x,y) = 0$.
       Then, using program 6.20.1 with an error of 1E–3 the following results should
       be obtained. Note that the solution is independent of R.

Node number	Finite diff sol'n	Exact
1	92.07	92.08
2	84.57	84.40
3	82.17	81.93
4	87.90	88.20
5	70.75	70.48
6	62.14	61.66
7	59.85	59.03
8	59.17	59.03
9	40.91	40.03
10	33.69	33.05
11	31.74	31.19

6.11   Use the same data as problem 6.9 but make $f(x,y) = 2$ and make both boundary
       nodes zero. Using program 6.20.1, with $R = 1$ (i.e. $h = k = 0.25$), gives the
       following nodal values:

Node number	Finite diff sol'n
1	0.0615
3	0.1401
5	0.1344
6	0.1794
7	0.1938
8	0.0710
9	0.1185
10	0.1436
11	0.1515

By varying the value of R it is not difficult to show that these nodal values are
multiplied by R^2 The numerical integral of the stress function over the surface

is approximately $(2.11R^2)(R/4)^2 = 0.132R^4$. Thus $\theta G/TL$ is approximately
$1/(0.264R^4)$. The exact value is $1/(0.296R^4)$. Thus the approximation is about

12% high. Note that if the mesh size is halved then $\theta G/TL$ is approximately
$1/(0.288R^4)$, about 3% high.

Chapter 7

7.1

	(a)	(b)	(c)	(d)	(e)
a_0	0	0	1	0.03125	0
a_1	1.000022	1.004883	1.020833	0.9999999	1.015625
a_2	0	0	0.5	−0.75	0
a_3	−0.1668402	0.1276042	−8.333334E−2	0.3333333	0.2083329
a_4	0	0	−0.125		0
a_5	8.680532E−3	0.1531249			0.4500001
a_6	0	0			0

By Chebyshev, sin 1 = 0.841862.
By using 3 terms in series, sin 1 = 0.8416666 (Exact, sin 1 = 0.841471).

7.2

	(a)	(b)	(c)	(d)
a_0	2.279585	0.7651977	−0.6065036	1.2031
a_1	3.181274	0	1.825829	0.964373
a_2	1.377897	−0.2298069	−0.8231307	0.1875771
a_3	0.4254793	7.450581E−9	0.4851439	−1.721731E−2
a_4	0.101456	4.953086E−3	−0.311612	−1.541419E−2
a_5	0.0196498	−7.450581E−9	0.2121278	−2.96982E−3
a_6	3.197816E−3	−4.214048E−5	−0.122741	1.380779E−4
a_7	4.410539E−4	−9.313226E−9	5.822757E−2	1.672655E−4

By Chebyshev, exp(sin 1) = 2.319754. (Exact value = 2.319777).
By Chebyshev, exp(sin 0.25) = 1.280709. (Exact value = 1.280696).

7.3 If m is small: a_0 = 1.490116E−8, a_1 = 0.8794492, a_2 = 0,
a_3 = −3.925656E−2, a_4 = 0, a_5 = 5.471036E−4, a_6 = 0, a_7 = −6.757677.
sn(0.5, 0.01) = 0.4792514. Exact = 0.47925163.

If $(1 − m)$ is small: a_0 = −1.490116E−8, a_1 = 0.8123885, a_2 = 0,
a_3 = −5.407581E−2, a_4 = 0, a_5 = 4.486062E−3, a_6 = 0, a_7 = −4.103705E−4.
sn(0.5, 0.99) = 0.4623079. Exact = 0.46228938.

Chapter 8

8.1 Using the conditions $x = 2, y = 3, y' = 0$, $x = 5, y = 3.5, x = 7.5, y = 5$,
$y' = 2/3$ leads to
$y = 2.764087 + 0.3773604x − 0.2103899x^2 + 0.045326x^3 − 2.49102E−3x^4$

8.2 erf(0.76) = 0.7175426. (Exact value = 0.7175367)
erf(1.35) = 0.9437620. (Exact value = 0.9437622).

8.3   Calculated root = 1.015325. Exact = 1.015323.

8.4   Coeff   Least squares   Minimax

of	Degree 2	Degree 4	Degree 2	Degree 4
x^0	0.9736814	0.99946	0.9739806	0.9995316
x	4.626E–8	–2.23E–7	–7.33E–4	2.63E–4
x^2	–0.9992598	–1.223033	–0.9992669	–1.222824
x^3		3.28E–7		–7.32E–7
x^4		0.2237729		0.2232925

Note that the cosine function is even and thus we would expect odd powers of x to be absent from the polynomial approximations.

8.5   Coefficients of a b c

	a	b	c
function (i)	3.214539	–9.369182	5.270893
function (ii)	40.3255	–122.7113	86.37751
function (iii)	0.8102945	–0.0964889	6.945692E–2

8.6   If $y = a \exp[b(x-c)^2]$ then $\partial y/\partial a = \exp[b(x-c)^2]$,
        $\partial y/\partial b = a(x-c)^2 \exp[b(x-c)^2]$ and $\partial y/\partial c = -2ab(x-c)\exp[b(x-c)^2]$. Using
        program 8.14.1 iteration $a \approx 3.034$, $b \approx 1.977$ and $c \approx 0.857$. Alternatively,
        taking the logarithm of the equation and letting $\log_e y = Y$ and $\log_e a = A$ we
        have $Y = a_0 + a_1 x + a_2 x^2$ where $a_0 = A + bc^2$, $a_1 = -2bc$ and $a_2 = b$.
        Using 8.13.1 to fit to this linearised form gives
        $a_0 = 2.572989$, $a_1 = -3.47861$ and $a_2 = 2.02273$.
        From these values we obtain $a = 2.937$, $b = 2.02273$ and $c = 0.8599$

8.7   Using program 8.13.1:
        $f(x) = 2248.073 - 579.8266x + 49.61205x^2 - 1.407519x^3$.
        When $x = 10.6$, $y = -5.676571E-2$. Very inaccurate due to ill-condition-
        ing.
        $f(z) = 2.04329 + 3.640234z - 1.025974z^2 - 1.688215z^3$.
        When $x = 10.6$, $z = -1.4$ and $y = -0.4314861$.
        Using program 8.16.1 we obtain the same coefficients for both cases thus:
        $p_0 = 0.3333333$, $p_1 = 2.68$, $p_2 = -2.393939$, $p_3 = 3.545252$,
        $p_4 = 2.0007586E-2$. When $x = 10.6$, $y = -0.4314855$ when we take 4 terms
        in the series, i.e. using a cubic polynomial.
        If we use double precision in program 8.13.1 then
        $f(x) = 2727.856 - 701.04548x + 59.74978x^2 - 1.6882155x^3$
        and when $x = 10.6$, $y = -0.43291$. This is a significant improvement over
        the single precision result but is still not very accurate.

8.8

	Coefficients of $(x + 2)$ Range –2 to 2	Coefficients: of $(x – 2)$ Range 2 to 3	Coefficients: of $(x – 3)$ Range 3 to 5
3rd power	–2.434783	10.08696	–5.217391
2nd power	14.6087	–14.6087	15.65217
1st power	–21.47826	–21.47826	–20.43478
Constant	4	–4	–30

$$y = -5.265\text{E}{-}4 + 31.12294x - 12.79109x^2 - 6.986602x^3$$
$$+ 3.19775x^4 - 0.3235536x^5$$

In the range $x = -2$ to 2 thre are significant differences between the 5th degree polynomial and the cubic spline. The spline seems a better representation in this region.

8.9

k	$A(k)$	$B(k)$
0	1.09375	
1	–7.322929E–2	–1.207107
2	–0.375	–0.5
3	–0.4267767	–0.207107
4	–0.21875	

Note that $A(0)$ and $A(4)$ have been divided by 2.

In program 8.12.1, T(1,K) etc. must be defined. Let us assume the following definitions:

```
T(1,K) = 1, T(2,K) = COS(2*X(K)*PI/L),
T(3,K) = COS(4*X(K)*PI/L), T(4,K) = COS(6*X(K)*PI/L)
T(5,K) = COS(8*X(K)*PI/L), T(6,K) = SIN(2*X(K)*PI/L)
T(7,K) = SIN(4*X(K)*PI/L), T(8,K) = SIN(6*X(K)*PI/L)
```

We will then find that the coefficients for these functions are identical to those determined above, (except for rounding errors). If one or more terms, chosen arbitarily, are removed from the analysis we will find that the coefficients of the remaining terms are unchanged. Note that for programming purposes, when terms are removed he variables T(1,K) etc must be redefined, if necessary, so that the remaining terms are numbered consecutively.

8.10   Using the coefficients obtained in problem 8.9, $y = 1.000006$ when $t = 1$ and $y = 2.250000$ when $t = 1.5$. Using 8 terms in the Fourier series gives $y = 1.009669$ when $t = 1$ and 2.253833 when $t = 1.5$. The small errors in the values obtained from the discrete or finite Fourier series are rounding errors whereas the larger errors in the values obtained from the truncated series are, of course, truncation errors.

Chapter 9

Note that in this chapter only exact answers are quoted since, due to the statistical nature of the methods used in the evaluations, the computer estimations are not repeatable.

9.1   Conventional methods require the value of the function when $t = 1$ whereas there is a high probability that the Monte Carlo method will not evaluate the function at that value.

9.4   When $k = 5$ the function contains a large number of oscillations in the range of integration.

9.5 $J_0(1) = 0.76519, J_0(3) = 0.26005, J_0(5) = -0.17759$.

9.6 $\text{erf}(0.2) = 0.22270, \text{erf}(0.76) = 0.71754, \text{erf}(1.35) = 0.94376$.

9.7 $\text{Si}(0.5) = 0.49310, \text{Si}(1) = 0.94608, \text{Si}(2) = 1.60541$.

9.8   (i) $2 \log_e 2 - 1 = 0.386294$, (ii) 10/3.

Chapter 10

10.1
Accuracy	1E–3	1E–5
By Rayleigh quotient	19.17537	19.17542
By iteration	19.17080	19.17537

10.2 $n = 3$: $\lambda = 9.071106, -5.071068, 2$. Exact $= 9.0710678, -5.0710678, 2$.
(Shift = 3): $n = 5$: $\lambda = -6.660298, 10.66025, -3.000004, 7, 2$.
(Shift = 0): $n = 5$: $\lambda = 10.66028, 6.999966$ (237 iter'ns), $-6.660254, -3, 2$.
Exact $= 10.66025404, 7, 2, -3, -6.660254038$.
Two eigenvalues are close and the shift of origin helps to separate them.

10.3   Eigenvalues are 1.05675, –0.9557598, 0.09901.

10.4   Eigenvalues are 0.7107661, 0.4999998, 0.3517324, 0.5

10.5   The square of the given matrix is

$$\begin{bmatrix} 129 & 80 & 240 \\ -96 & -56 & -189 \\ -32 & -20 & -59 \end{bmatrix}$$

This matrix does have eigenvalues 1, 4, 9.

10.6 $\lambda = 22.40688$, $\mathbf{x}^T = [.4260008, .5239224, .785226, 1, .9639754]$
$\lambda = 7.513724, \ 4.84895, \ 1.327045$.

10.7 Converges to the real solution only, i.e. $\mathbf{x}^T = [0.5, \ 1, \ 0.5]$ $\lambda = 1$.

10.8 Rearrange complex matrix into the form

$$\begin{bmatrix} 2 & 0 & 0 & 0 & -1 & 0 \\ & 2 & 0 & 1 & 0 & 0 \\ & & 3 & 0 & 0 & 0 \\ & \text{Symm} & & 2 & 0 & 0 \\ & & & & 2 & 0 \\ & & & & & 3 \end{bmatrix}$$

This system has eigenvalues 1, 1, 3, 3, 3, 3. Since we expect the eigenvalues of the original system to be duplicate this system must have eigenvalues 1, 3, 3

10.9 After 137 iterations $\lambda = -197 \pm i100$.

Program index

Index

Mathematics and its Applications
Series Editor: G. M. BELL, Professor of Mathematics, King's College London (KQC), University of London

Numerical Analysis, Statistics and Operational Research

Editor: B. W. CONOLLY, Emeritus Professor of Mathematics (Operational Research), Queen Mary College, University of London